HORACE W. DAVENPORT, Ph.D., D.Sc. (Oxon.)

Professor and Chairman, Department of Physiology
The University of Michigan

THE DIGESTIVE TRACT

An

Introductory Text

FOURTH EDITION

YEAR BOOK MEDICAL PUBLISHERS INCORPORATED

35 EAST WACKER DRIVE, CHICAGO

Reprinted, September 1962
Second Edition, 1966
Reprinted, January 1968
Reprinted, May 1969
Third Edition, 1971
Reprinted, August 1971
Reprinted, August 1973
Reprinted, May 1975
Fourth Edition, 1977

Library of Congress Catalog Card Number: 76-15696
Cloth: 0-8151-2327-2   Paper: 0-8151-2328-0

PHYSIOLOGY OF

PHYSIOLOGY
OF THE
DIGESTIVE
TRACT

Preface to the Fourth Edition

THE PURPOSE of this book is to make the basic facts of gastroenterological physiology readily available to medical and graduate students who are beginning their study of the digestive tract. The standards I have used are expressed in the questions: What should a generally well-informed student of physiology know of alimentary functions? What is the body of knowledge a student must command when he prepares himself for advanced study of the physiology of digestion? The standard is subjective and its application equally so. This book is intended to contain material answering the first question and to serve as a point of departure for the student who needs to know more.

Since the first edition of this book was published in 1961 there has been an enormous increase in our knowledge of human gastroenterology, and man, well or ill, has become the chief subject of observation and experiment. As a result, this edition contains much information on the disordered physiology of the diseased digestive tract. Unless otherwise specified, a statement can be assumed to apply to man, and in those instances in which our knowledge of human function must be inferred from animal experiments I have tried to warn the reader by citing the species concerned. I have not explored the wide and fascinating field of comparative physiology of digestion.

The plan of this book is to conduct the reader on three journeys through the alimentary canal, the first two showing him the mechanisms available for propulsion and secretion. Finally, we repeat the trip, accompanying aliments in the process of being digested and absorbed, observing to the best of our ability how and to what extent the mechanisms of motility, secretion and their control are actually used. The usually relentlessly downward gradient of the digestive tract imposes tedious repetitiveness on this method of exposition, but I find the task of adding variety by beginning with an enema and ending with an eructation beyond my powers.

I cannot document with references the thousands of assertions of fact contained in this book. A fragment of the original literature has filtered through my understanding, and the casual student must trust me. The serious scholar can begin with the massive *Handbook of Physiology: The Alimentary Canal,* edited by Charles F. Code and published by The American Physiological Society. This contains references to most of the literature up to 1964–66, and it will remain the starting point for a search of the literature. The references I have added to each chapter include recent reviews whose text and references are useful. The papers I cite, in addition to a few amusing oddities

vii

such as the one in Chapter 5 on the rheology of human feces, are usually the latest ones in an important series. The student can easily attain a view of a large field by working backward through the paper's references.

It is a pleasure to acknowledge help received. Edith Bulbring and Alan Hofmann have diligently provided me with illustrative material. I am grateful to many whose advice is buried anonymously in the text, but I am especially indebted to Morton I. Grossman and Charles F. Code, who together know more about the physiology of the digestive tract than all the rest of us.

H.W.D.

Table of Contents

TABLE OF CONTENTS

PART I

MOTILITY

1

Chewing and Swallowing

FOOD IS TAKEN into the mouth in bites ranging from a few cubic millimeters upward; there it is broken up, mixed with saliva and lubricated. The act of chewing is partly voluntary and partly reflex, but the act of swallowing, once initiated, is entirely reflex. From the mass in the mouth, a bolus of 5–15 cu cm is separated and projected into the pharynx, where it is engulfed by the muscles controlled by the stereotyped swallowing reflex. Pressure generated by muscular contraction moves the bolus through the hypopharynx past the hypopharyngeal sphincter into the esophagus. During swallowing, the passages from the pharynx into the nose and trachea are closed, and respiration is briefly inhibited. In man in the upright position, a liquid or slippery bolus moves rapidly down the esophagus under the influence of gravity; a solid or sticky bolus is slowly propelled by a peristaltic wave. The lower esophageal sphincter opens before the bolus, and the sphincter closes after it, preventing regurgitation.

Chewing

The extent to which a mouthful of food is chewed varies among species; in some, such as the dog and the cat, food is reduced in size only enough to permit swallowing. In man, particles are usually reduced to a few cubic millimeters, but the amount a mouthful is chewed depends on the nature of the food, habit, incidental conversation and early training. The extent of chewing has a negligible effect on the chemical processes of digestion; but the bolting of unchewed food, especially when nervous tension is high, often causes epigastric distress.

The jaw is usually closed, in opposition to gravity, by contraction of the masseter, the medial pterygoid and the temporalis muscles. Pressure of food against the gums, the teeth, the anterior part of the hard palate and the surface of the tongue stimulates receptors and generates afferent impulses which evoke reflex relaxation of the jaw-closing muscles and reflex contraction of the digastric and lateral pterygoid muscles, which open the jaw. This reduces pressure on receptors, and the frequency of afferent impulses falls. Relaxation of jaw-opening muscles and rebound contraction of jaw-closing muscles follow. Although jaw-closing muscles contain spindles that send impulses centrally, the pattern of chewing is centrally determined. During the period of jaw opening, there is a brisk discharge from muscle spindles in the temporalis and masseter muscles, which are being stretched. The discharge stops abruptly during the initial phase of rapid contraction of these muscles, but it begins again and continues during grinding movements of the jaw.

Destruction of the spindle afferents does not change the pattern or relative timing of jaw-closing muscles. As food is broken by the teeth, there is immediate inhibition of the jaw-closing muscles, which prevents the teeth from banging suddenly together. This load-compensating mechanism is not affected by destruction of afferents from jaw-closing muscles. In man, the duration of the chewing cycle is 1 sec.

If in an experimental animal the jaw is divided, the reflex is found to be unilateral, for stimulation on one side causes relaxation, followed by contraction, on that side and not on the other. As a result, the force of chewing is brought to bear asymmetrically on the side of the mouth that is full. However, voluntary chewing and chewing movements caused by stimulation of the cerebral cortex are bilaterally synchronized. Most persons chew a bolus on one side of the mouth at a time, but some divide the bolus to chew it on both sides simultaneously.

The pressure exerted on a lower molar during reflex chewing of cooked meat is 3.9–15.7 newtons per mm², and the vertical load taken by a single tooth is 70–120 newtons. Very much higher pressures are exerted when hard foods are chewed.

Mouth Movements in Swallowing

The digestive tract moves its contents by creating a pressure gradient. At the beginning of a swallow, the tip of the tongue separates a bolus from the rest of the material in the mouth and brings it into the midline between the anterior portion of the tongue and the hard palate. The jaw shuts, and the soft palate is elevated. The forepart of the tongue is pressed firmly against the roof of the mouth and, together with the closed lips, seals off the anterior portion of the mouth. The palate and the contracted palatopharyngeal muscles form a partition between the mouth and the nasal cavity, which prevents pressure generated in the mouth from being dissipated through the nose. When selective paralysis of motor components occurs, as in poliomyelitis, the bolus is regurgitated into the nasopharynx. The bolus, which lies in a groove in the midportion of the tongue, is propelled into the oral pharynx as the tongue rolls backward on the hyoid bone, pressing against the palate progressively more posteriorly. Respiration is now briefly inhibited, the larynx is abruptly raised, and closure of the glottis cuts off the laryngeal airway. As the larynx rises, the epiglottis is thrust into the path of the oncoming bolus. The bolus tilts the epiglottis backward; and in response to further upward movement, the epiglottis is retroverted until it hangs slantwise over the closed glottis. Although the epiglottis acts as a lid over the closed glottis, its presence is not necessary to prevent food from entering the trachea. Firm closure of the glottis seals the airway, and this prevents aspiration of food even if the epiglottis is surgically removed. Driven by a pressure difference of 4–10 mm Hg generated by movement of the tongue, the bolus pours over and around the epiglottis. The hypopharyngeal sphincter, formed by the cricopharyngeal muscles, is opened by upward movement of the structures to which it is attached, and the bolus moves with a wedge-shaped leading edge through the hypopharynx into the upper esophagus. After it passes the level of the clavicle, the larynx descends, the upper respiratory passage and the glottis open, the tongue moves forward and respiration resumes (Fig 1–1). The whole of the pharyngeal part of swallowing occurs within 1 sec.

Mouth Movements in Drinking

When a liquid is drunk by suckling or through a straw, subatmospheric pressure is generated in the mouth by retraction of the tongue without breaking the lingual-palatine seal. This fills the mouth; then, suddenly, the posterior portion of the tongue is depressed, allowing fluid to run into the pharynx. The mandible is raised, closing the jaw firmly, and the swallowing movement begins with a sweeping posterior movement of the tongue.

The champion beer drinker, who steadily

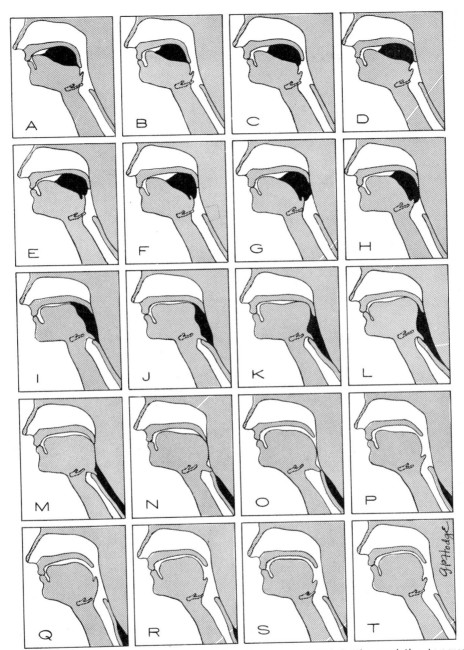

Fig 1–1. — Sequence of events during swallowing. **A** and **B,** the soft palate forms a partition extending to the base of the tongue. **C, D, E** and **F,** the soft palate is elevated to obstruct the nasopharynx as the bolus moves backward over the tongue. **G, H** and **I,** the bolus tilts the epiglottis backward. From **H** through **R** the glottis, not shown, cuts off the laryngeal airway. **J** and **K,** the bolus passes smoothly over the convex epiglottis, and the tongue moves backward as a piston. **K,** the bolus is slightly delayed at the hypopharyngeal sphincter. **O** and **P,** the soft palate relaxes, and the epiglottis ascends. **Q–T,** the bolus moves down the esophagus. The entire sequence occupies one and a third seconds. (Adapted from Rushmer, R. F., and Hendron, J. A.: J. Appl. Physiol. 3:622, 1951.)

pours the contents of a bottle into his mouth, does not, contrary to appearances, down the beer by opening a passage directly into his esophagus. He elevates and extends his jaw to increase the capacity of his mouth and pharynx, and he keeps the level of liquid in his mouth even with his lower lip. Before any beer enters the pharynx, the hyoid bone, larynx and pharynx drop down as in yawning, and the dorsum of the tongue is wedged against the descended soft palate. The glottis is closed, breathing is suspended, and then the dorsum of the tongue moves forward and the soft palate elevates, allowing a flood of beer to flow into the pharynx, where it is held by the closed hypopharyngeal sphincter. The tongue now moves like a cam at the rate of once a second to force fluid through the

Fig 1–2.—Special swallowing technique used by competitive beer drinker. **A,** fluid is being poured into the wide-open mouth. The oropharynx is open, permitting the laryngeal pharynx to fill; the laryngeal vestibule is distended *(black arrow)*. **B,** the dorsum of the tongue bulges backward *(black arrow);* fluid in the mouth is forced down into the already distended laryngeal pharynx. **C,** the *black arrow* points to the nearly completed, cam-like action of the tongue, the hypopharyngeal sphincter is open and fluid has passed into the esophagus. **D,** the final swallow occurs with the mouth closed, the larynx elevated and the epiglottis folded over. (From Ramsey, G. H., et al.: Radiology 64:498, 1955.)

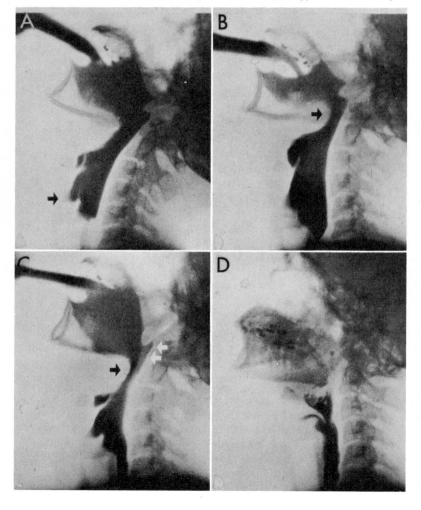

sphincter and into the esophagus. This pumping process continues until the mouth is empty; then an ordinary swallow travels from mouth to stomach to sweep away the foam (Fig 1–2).

Air Swallowing

Air normally present in the pharynx is trapped when a bolus is delivered into the pharynx by the tongue. Most of it passes into the trachea the instant before the glottis closes, and only occasionally a bubble is forced ahead of the bolus into the esophagus. Most swallowed air passes no farther than the esophagus and is soon expelled by belching, but some does pass through the intestines. Air in frothy saliva or in food, particularly meringues and soufflés, is swallowed. In fact, the "dry swallow" consists of air and saliva. During periods of excessive salivation, such as those accompanying nausea, the large volume of air swallowed with saliva may cause distress. As much as 500 cc may be swallowed with a meal. Air may also be voluntarily swallowed by trapping it with the tongue while the lips are closed. Persons lacking a larynx may learn to fill the esophagus with air, which is then expelled slowly as the basis of a form of speech. On the other hand, those with respiratory paralysis can be trained to swallow air with the glottis open, a form of respiration called *glossopharyngeal* or *frog breathing*. The first breath of life may be swallowed, for in 5 of 8 newborn babies observed by successive radiophotographs, the initial inflation of the lungs was found to follow first distention and then compression of the pharyngeal cavity.

The Swallowing Reflex

Swallowing is probably always a reflex act. Voluntary efforts to initiate swallowing are

Fig 1–3.—Outline of afferent and efferent systems involved in swallowing, and the requirement of interaction between swallowing and other synergies. (From Doty, R. W.: Neural organization of deglutition, in Code, C. F. (ed.): *Handbook of Physiology:* Sec. 6. *Alimentary Canal,* Vol. IV [Washington, D. C.: American Physiological Society, 1968].)

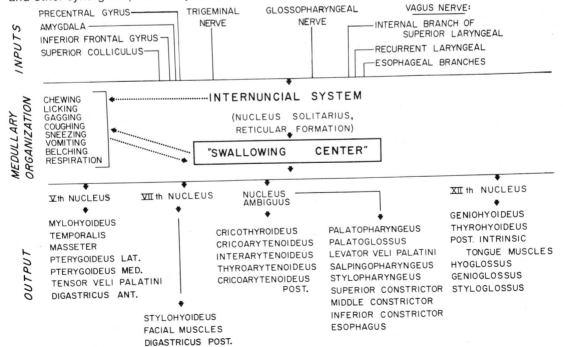

ineffective unless there is something, if only a few milliliters of saliva, to swallow. The reflex is started by stimulation of a large number of receptors in the mouth and pharynx, and afferent impulses travel in the glossopharyngeal nerve and the superior laryngeal branch of the vagus. Afferent impulses go to the swallowing center in the medulla and evoke the complex act of swallowing, in which there is discharge through the six nuclei and the motor neurons listed in Figure 1–3. Afferent impulses in the superior laryngeal nerve travel through relay stations in the pons and thalamus to the frontal cortex, and the swallowing reflex can be facilitated through efferent pathways from the frontal cortex to the medullary swallowing center. Discharges from the medullary center through efferent pathways shown in Figure 1–3 form a pattern of sequential contraction of buccopharyngeal muscles lasting about 0.5 sec. The muscles involved are poorly provided with proprioceptors, and they contain no gamma-efferent system. Their pattern of discharge is substantially unaffected by procainization or transection of the muscles. Feedback from the muscles plays no significant part in determining the sequence or strength of their contraction. The pharyngeal phase of swallowing is followed by the esophageal phase, which lasts several seconds and is responsible for the peristaltic wave moving slowly down the esophagus at the end of a swallow. Messages reaching the swallowing center by way of afferent nerves from the esophagus quantitatively modify the peristaltic wave.

The rigidly ordered pattern of discharge is illustrated in Figure 1–4. Five muscle groups in the pharynx contract concurrently for 250–500 msec. This generates high pressure behind the bolus and propels it toward the hypopharyngeal sphincter. At rest, the esophagus is closed at the pharyngoesophageal junction by passive elasticity of surrounding structures; the cricopharyngeal muscle is not usually contracted. This muscle, however, is easily activated reflexly by manual displacement or by the presence of a

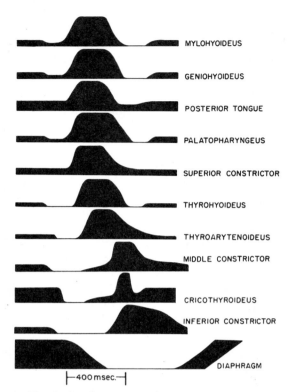

Fig 1–4.—Summary of outflow of the swallowing center of the dog, recorded electromyographically. Height of line for each muscle indicates relative intensity of activity, but contours of rise and fall are schematic. The upper five muscles represented are part of the pharynx; these contract almost simultaneously, creating high pressure above the bolus. At the same time, the lower five muscles shown, which belong to the hypopharynx and hypopharyngeal sphincter, relax. After the bolus has passed, these muscles contract, firmly shutting the sphincter. The diaphragm relaxes while the bolus is passing through the hypopharynx. Pressure changes caused by this muscular activity are shown in Figure 1–7. (From Doty, R. W., and Bosma, J. F.: J. Neurophysiol. 19:44, 1956.)

foreign object in the lumen. If the muscles surrounding the junction are initially contracted, they are reflexly inhibited during the early part of swallowing as the bolus approaches and passes through the sphincter. The sphincter is opened by contraction of the muscles that raise the larynx and cricoid cartilage. After the bolus has passed through the sphincter, the muscles belonging to it contract, firmly shutting the sphincter. Pressure within the sphincter reaches 100 mm Hg in the anterior-posterior direction and 33 mm Hg in the lateral direction. The sphincter remains firmly closed until the peristaltic wave in the esophagus, beginning just below the sphincter, has carried the bolus onward, thus preventing regurgitation.

Structure of the Esophagus

In man, the upper quarter of the muscular wall of the esophagus consists of striated muscle arranged as a layer of longitudinal fiber bundles surrounding a layer of circular fiber bundles. Between these muscle layers and the stratified squamous epithelium forming the lining of the esophagus is a thick submucous elastic and collagenous network. This, together with the thick muscularis mucosae, throws the epithelium into folds whose surfaces appose one another, obliterating the lumen. When swallowing occurs, the folds are smoothed out in the part of the esophagus occupied by the bolus.

The muscular wall of the lower third of the esophagus is smooth muscle arranged in two layers, the inner being roughly circular but containing many helical, elliptical or oblique bundles and the outer being roughly but irregularly longitudinal. Transition between striated muscle above and smooth muscle below generally occurs in the middle third, and it is higher for the inner layer than for the outer. In some animals—dogs, mice and elephants, for example—the whole esophageal muscle is striated. There is no muscular structure that forms an anatomical sphincter separating the esophagus from the stomach; nevertheless, the lower end of the esophagus does behave like a physiological sphincter.

At rest, the upper seven eighths of the esophagus, below the pharyngoesophageal junction, is relaxed. When an esophagoscope is passed downward, the esophagus is found to be closed about 2 cm above the diaphragmatic hiatus and 3–4 cm above the cardia.

There is sometimes a slight thickening of the esophageal muscle at this point, which marks the upper boundary of the physiological lower esophageal sphincter. The lower one eighth of the esophagus, between the constriction and the cardia, is sometimes called the esophageal vestibule.

Motor Innervation of the Esophagus

The esophagus has three distinct parts: the upper third, which is striated muscle; the lower two-thirds, which is smooth muscle; and the lower esophageal sphincter, which is also smooth muscle. These three have different innervation. In addition, hormones may influence the lower esophageal sphincter, but they have no important effect upon the rest of the esophagus.

Those portions of the glossopharyngeal nerve and vagus nerve that innervate the striated muscles of the pharynx and upper esophagus are not autonomic in character; they contain the efferent fibers of motor neurons, and they end in motor end-plates.

The smooth muscle of the lower esophagus, excluding the lower esophageal sphincter, is supplied by the vagus nerve in characteristically autonomic fashion: preganglionic fibers in the vagus nerve end in synapse with ganglion cells in the myenteric plexus. Postganglionic fibers from the plexus innervate the smooth-muscle cells. Both the preganglionic and the postganglionic fibers are cholinergic; parasympathomimetic drugs stimulate contraction, and atropine prevents it. The wave of contraction passing down the esophagus during peristalsis is caused by sequential excitation of esophageal smooth muscle by sequential firing of postganglionic fibers.

Between swallows, the lower esophageal sphincter is partially contracted. During swallowing it relaxes and then contracts more vigorously. Both resting contraction and that following relaxation are partly mediated by acetylcholine liberated by postganglionic fibers from the myenteric plexus. However, excitatory fibers are also contained in the sympathetic innervation of the lower esophageal sphincter. About one third of the resting contraction can be attributed to the action of the mediator norepinephrine upon alpha receptors in the sphincter's smooth-muscle cells.

Relaxation of the sphincter during swallowing is not adrenergically mediated, and all fibers carrying inhibitory impulses to the sphincter are contained in the vagus nerve. It is likely that vagal inhibitory fibers release 5-hydroxytryptamine (5-HT) at their synapses with inhibitory postganglionic cells and that the terminal postganglionic fibers inhibit the smooth muscle of the sphincter by releasing adenosine triphosphate (ATP) or a closely related purine.

Both sympathetic and parasympathetic

stimulation and their mediators, norepinephrine and acetylcholine, cause the muscularis mucosae of the esophagus to contract.

Methods of Measuring Intraluminal Pressure

Pressure in the digestive tract can be measured by passing into it a small tube connected to a pressure-sensing device, such as a strain gauge. The tube, which is usually made of polyethylene or polyvinyl plastic, has an outside diameter of less than 2.5 mm and an inside diameter of about 1.5 mm. The tube has an opening 1.5 mm in diameter on its side just proximal to its tip, and the tip is sealed. The tube is filled with fluid, and fluid is forced through it at a rate of the order of 50 μl per min. Continuous flow of fluid through

Fig 1–5. — Pressures at the pharyngoesophageal junction. Pressures were recorded by means of a transducer whose position at the tip of the tube is shown in the roentgenograms. As the transducer is withdrawn from the esoph-

agus, it records increasingly higher pressures until, at 16.5 cm from the incisors, the upper limit of the superior esophageal sphincter is passed. (From Fyke, F. E., and Code, C. F.: Gastroenterology 29:24, 1955.)

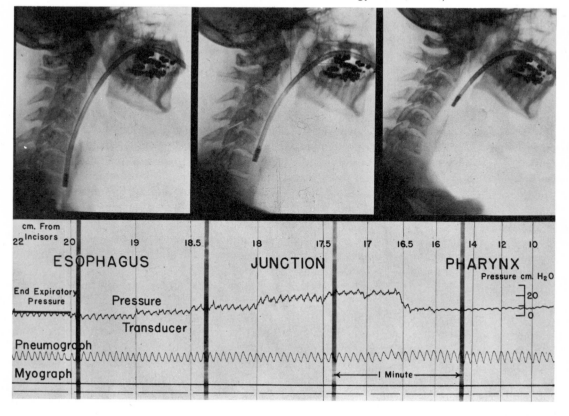

the tube is necessary to keep the distal opening of the tube unobstructed and to permit accurate recording of the pressure in the sphincter. Often several tubes are bound into a bundle with their tips known distances apart. Four such tubes with openings separated by 5 cm were used to obtain the record shown in Figure 1–7.

In some instances, such as the arrangement described in Figure 1–5, the miniature transducer is itself placed within the lumen of the digestive tract.

To identify and quantitate the action of sphincters, a tube with an opening near its tip is passed through the sphincter, its position being determined by fluoroscopy. The tip is slowly withdrawn as pressure is recorded. This method was used to obtain the data shown in Figure 1–6. In this type of study, fluid must be forced continuously through the tube; otherwise, the mucosa of the sphincter seals the opening in the tube, and while the opening is in the sphincter the pressure obtaining immediately before the opening was sealed continues to be recorded.

The opening of the tube may be placed under fluoroscopic control within the sphincter to be studied. Fluid is slowly forced through the tube into the sphincter. The pressure recorded rises steadily until at some value it abruptly levels off. This value is the *yield pressure,* and it is a measure of sphincter competence. The higher the yield pressure, the more firmly is the sphincter closed.

The process of measurement, particularly within sphincters, may itself disturb the physiological function to be measured. For example, it is possible that presence of the tube and transducer within the hypopharynx shown in Figure 1–5 stimulated reflex contraction of muscles surrounding the sphincter, with the result that the sphincter was not truly at rest.

Resting Pressure in the Esophagus

Pressure in the resting midesophagus is equal to intrathoracic pressure because the esophageal muscle is flaccid. In fact, respiratory physiologists use esophageal pressure as a measure of intrathoracic pressure. During quiet breathing the pressure is 5–10 mm Hg below ambient pressure. It falls on inspiration and rises on expiration.

Pressure at each end of the esophagus is higher than that within the esophagus. Pres-

Fig 1–6.—Resting pressures at the gastroesophageal junction in an adult man, recorded by passing an open-tip catheter connected with a transducer into the stomach and slowly withdrawing it. The position of the tip was observed fluoroscopically, and the position of the diaphragm was determined by inversion of respiratory excursions on the record. End-expiratory pressures are on the left; the esophageal pressure is above atmospheric. End-inspiratory pressures are on the right; esophageal pressures are subatmospheric. At both stages of respiration, there is a zone of pressure higher than that in the body, or fundus, of the stomach, and the zone moves upward in expiration. (Adapted from Vantrappen, G., et al.: Gastroenterology 35:592, 1958.)

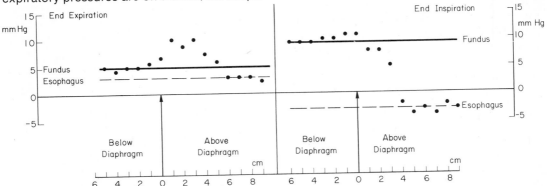

sure in the mouth and pharynx, when these are at rest, is approximately atmospheric; and were not the upper end of the esophagus closed by the hypopharyngeal sphincter, air would flow from the mouth into the esophagus until esophageal pressure became equal to atmospheric pressure.

Intragastric pressure prevails beyond the lower end of the esophagus. This is usually 5 – 10 mm Hg above ambient pressure; it rises in inspiration and falls on expiration. Because intragastric pressure is higher than intraesophageal pressure, gastric contents would flow from stomach to esophagus unless their junction were closed by the lower esophageal sphincter.

The Lower Esophageal Sphincter at Rest

If the open tip of the pressure-recording tube is passed all the way through the esophagus into the stomach, it registers the basic intragastric pressure of 5 – 10 mm Hg above ambient pressure (see Fig 1 – 6). Superimposed on this basic pressure are variations caused by respiratory movements: quiet inspiration causes an increase in intragastric pressure of about 5 mm Hg, and quiet expiration causes a fall toward ambient pressure.

As the tip is withdrawn upward from the stomach into the subdiaphragmatic esophagus, which is 2 cm long in a normal man, the base-line pressure rises until, at a point close to the hiatus of the diaphragm through which the esophagus passes into the thorax, it is higher than intragastric pressure. Above this point it falls until it is below ambient (and therefore below intragastric) pressure. When the subject is in end expiration, the elevation in pressure is first detected 3 cm below the diaphragm and ends 3 cm above it. When the subject is in end inspiration, the zone of increased pressure begins 3 cm below the diaphragm and ends 1 cm above it. Thus, a zone of elevated pressure 4 – 6 cm long separates the stomach and esophagus and acts as a barrier between the two. This is the physiological lower esophageal sphincter, and it is not

identical with any anatomical structure at the cardia or with the hiatus.

In normal persons, the resting pressure in the lower esophageal sphincter is between 15 and 35 mm Hg, and it remains within that range in repeated observations on the same subject over a 6-month period. In persons with esophageal reflux, pressure in the sphincter is between 1 and 10 mm Hg.

Bolus injection of the hormone gastrin causes pressure in the sphincter to rise about 20 mm Hg, but the question whether gastrin is a normal determinant of sphincter pressure is unsettled. Sphincter pressure rises 4 – 8 mm Hg during digestion of a high-protein meal, but if the meal is acidified to pH 1.2 before ingestion, a process that inhibits release of gastrin, rise in pressure does not occur. Patients with gastrinomas who have very high plasma gastrin concentrations likewise have higher than normal resting sphincter pressure. However, there is very little correlation between plasma gastrin concentrations in normal persons and sphincter pressure.

The hormones secretin, glucagon and cholecystokinin inhibit the sphincter when injected, as do prostaglandins of the E_2 series and dibutyryl cyclic adenosine monophosphate.

When the tip of a pressure-sensing catheter is in the stomach, changes in pressure coincident with respiration are superimposed upon the base-line pressure. These are positive upon inspiration and negative upon expiration. As the tube is withdrawn into the esophagus, the respiratory excursions invert, becoming negative upon inspiration and positive upon expiration. The respiratory-pressure inversion point occurs at variable points along the lower esophageal sphincter, and it is not a reliable landmark of the hiatus of the diaphragm.

Pharyngeal Pressures During Swallowing

Pressure in the pharynx rises abruptly during the act of swallowing to a maximum of

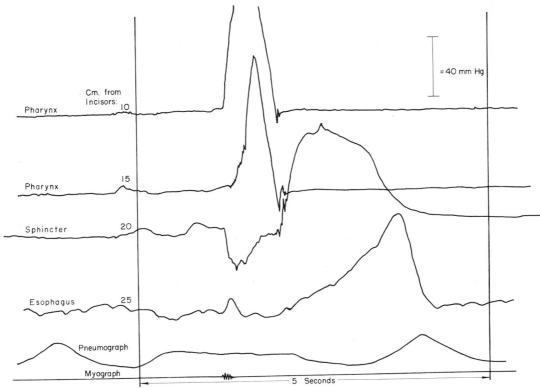

= 40 mm Hg

Cm. from
Incisors: 10

Pharynx

Pharynx   15

Sphincter   20

Esophagus   25

Pneumograph

Myograph

5 Seconds

Fig 1–7.—Pressures in the normal human pharynx and upper esophagus during a swallow. Four transducers are in place at distances from incisors given on the left (see Fig 1–5). The beginning of the swallow is signaled by the myograph, which records action potentials in the jaw muscles. Respiration is interrupted. Pressure in the pharynx (10-cm trace) rises to a high level, and this increased pressure travels quickly down the pharynx (15-cm trace). The third transducer, in the pharyngoesophageal junction (20-cm trace), shows that the sphincter relaxes at the time the pharyngeal pressure is high. As soon as the pressure in the pharynx has fallen to base-line level, pressure in the sphincter rises, and remains high for more than a second. While the sphincter is tightly closed, the peristaltic wave (25-cm trace) begins to move slowly down the esophagus. (Courtesy of C. F. Code.)

about 100 mm Hg as the bolus passes through it (Fig 1–7). The pressure elevation lasts only about 0.5 sec. A fraction of a second before the wave of high pressure reaches the pharyngoesophageal junction, the sphincter is opened by contraction of muscles raising the cricoid cartilage. If muscles of the sphincter are contracted, they are reflexly inhibited as the pressure wave approaches the sphincter (see Fig 1–4). The bolus passes through the open sphincter into the esophagus. The sphincter remains open only during the half-second that high pressure exists in the pharynx above it. Then the sphincter closes, and within another half-second its pressure rises to 90–100 mm Hg, or about twice its resting pressure. While pressure in the sphincter is highest, a peristaltic wave begins at the upper end of the esophagus, generating a pressure of about 30 mm Hg. The higher pressure in the sphincter prevents reflux from esophagus to pharynx. After the peristaltic wave has passed farther down the esophagus and pressure in the upper esophagus is again low, the sphincter's pressure subsides, over 2 or more sec, to its resting level. Thus, the sphincter is closed before swallowing begins, relaxes and opens as the bolus is propelled toward it, remains open as the bolus passes, snaps shut again and maintains a high pressure, preventing reflux until esophageal peristalsis has cleared the upper esophagus.

Esophageal Peristalsis

There are three kinds of esophageal peristalsis. *Primary* peristalsis is the wave of contraction moving down the esophagus as a continuation of oral and pharyngeal swallowing movements. *Secondary* peristalsis is a similar wave originating just below the hypopharyngeal sphincter without any antecedent mouth or pharyngeal movements. *Tertiary* peristalsis is the wave occurring only in the smooth-muscle part of an esophagus that has been vagotomized.

The peristaltic wave beginning just below the pharyngoesophageal sphincter pushes a solid bolus ahead of it. The pressure generated ranges from 30–120 mm Hg. It rises to

a peak in 1 sec, lasts at the peak 0.5 sec and subsides in about 1 sec (Fig 1–8). The whole course of rise and fall may occupy one point in the esophagus for 3–7 sec. The length of esophagus contracting at any one instant is 10–30 cm. Because the longitudinal as well as the circular muscle contracts, the esophagus shortens. The peak of the wave of contraction moves down the esophagus at 2–4 cm per sec, reaching the lower end of the esophagus of an adult man about 9 sec after swallowing has begun. Consecutive swallows produce similar waves; but when the pharyngeal phase of swallowing is rapidly repeated, the esophagus remains relaxed, and an esophageal wave occurs only after the last swallowing movement.

The pressure wave in the esophagus is caused by sequential activation of its muscles by efferent nerves firing in a pattern determined by higher centers, which may not be identical with the swallowing center that governs the pharyngeal stage of swallowing. Continuity of the esophagus is not necessary, for if its muscle, but not the nerve sup-

Fig 1–8.— Pressures in the lower part of the normal human esophagus during a swallow. Three transducers are in place at distances from the incisors given on the left. The beginning of the swallow is signaled by the myograph, which records action potentials in the jaw muscles. Almost immediately, the pressure in the lower esophageal sphincter (39-cm trace) falls, and it remains low for about 8 sec.

In the meantime, the peristaltic wave is passing down the lower esophagus (29- and 34-cm traces). After the pressure in the lower esophagus has fallen to resting level, the pressure in the sphincter rises, and it remains elevated for about 10 sec. (Adapted from Fyke, F. E., Jr., Code, C. F., and Schlegel, J. F.: Gastroenterologia 86:135, 1956.)

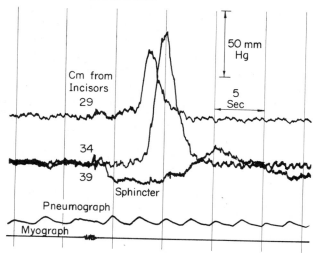

ply, is cut across, the wave begins at the distal end of the esophagus below the cut as it dies out at the proximal end above the cut. If the esophagus is chronically denervated by vagotomy, the denervated parts contract simultaneously, and repetitive responses are frequent. Afferent impulses from receptors within the esophageal wall are not essential for progress of the coordinated wave of peristalsis. Nevertheless, impulses carried to higher centers by afferent nerves from the esophagus modify the reflex, and esophageal peristalsis is affected by bolus volume, consistency and temperature. When a person swallows 2–20 ml, the velocity of his peristaltic wave is slower and its duration and force greater than when he swallows nothing.

Secondary peristalsis is the result of afferent impulses from the esophagus. If the normal esophagus is distended at any point, as by a piece of peanut butter sandwich left behind from previous swallows, afferent nerves are stimulated, and contraction begins with forcible closure of the hypopharyngeal sphincter and sweeps downward. Secondary peristalsis occurs without any movements of the mouth or pharynx and without the subject's awareness of it.

In man, the propulsive force of esophageal peristalsis is feeble. If a man attempts to swallow a bolus attached by a string to a counterweight, the maximal weight he can overcome is 5–10 gm. A dog's esophagus can successfully oppose 50–500 gm.

In old age, the pressure generated by esophageal peristalsis is diminished, but its velocity and duration are the same as in a young person.

Lower Esophageal Sphincter During Swallowing

When a normal man swallows, the lower esophageal sphincter relaxes. Relaxation begins at the upper edge of the sphincter 1.3 sec after the start of swallowing; at this time, esophageal peristalsis is just beginning at the top of the esophagus. A wave of relaxation moves down the sphincter and continues into the upper part of the stomach. The pressure barrier between stomach and esophagus is not completely abolished by relaxation of the sphincter, for the lowest pressure occurring at the hiatus during swallowing is slightly above intragastric pressure. This residual barrier prevents reflux during inspiration, when the pressure difference between stomach and esophagus is greatest. Then, about 6 sec after swallowing has begun, a wave of contraction starts at the top of the sphincter and reaches its maximal force 2–3 sec later, as esophageal peristalsis dies out in the muscle just above it. Contraction moves downward through the sphincter at the rate of 0.6 cm per sec, and its force diminishes as it moves. Pressure generated by contraction is greatest above the hiatus, where 20–30 mm Hg above resting pressure may be reached. This high pressure dies out over 5–10 sec. The 39-cm trace in Figure 1–8 was recorded from a part of the sphincter showing this behavior: early and prolonged relaxation followed by late and equally prolonged contraction. A record made from the sphincter below the hiatus would show only relaxation followed by return to resting pressure. Thus, the barrier between stomach and esophagus is reduced but not obliterated very early in swallowing, remains low until the esophagus has completed its propulsive wave, is raised high after the wave has died out and slowly falls to its resting level.

During rapidly repeated swallowing, the lower esophageal sphincter relaxes at the beginning of swallowing movements, and it remains relaxed until the single peristaltic wave that follows the last swallow has reached it.

Gastrin, the hormone from the pyloric glandular mucosa that stimulates gastric secretion of acid and pepsinogen, also causes the pressure within the lower esophageal sphincter to rise.

Bolus Movement During Swallowing

The actual movement of the bolus effected by swallowing depends on its consistency and the position of the subject. When a man in the upright position swallows water, it is shot rapidly into the esophagus by bucco-pharyngeal movements, and in the esopha-gus the head of the column reaches the lower esophageal sphincter within 1 sec (Fig 1 – 9). Descent of the bolus is aided by gravity. It is sometimes held up briefly if the sphincter has not yet completely relaxed. If a stethoscope is placed on the abdomen over the lower esophageal sphincter of an erect subject, a gurgle is heard 5-8 sec after he swallows water; water waiting at the bottom of the esophagus is pushed through the sphincter by the slow peristaltic wave. Water also moves along the esophagus of a man in the

Fig 1 – 9. – Lateral oblique view of the normal human esophagus during swallowing of a thin mixture of barium sulfate and water. The sub-ject is upright; and the mixture, falling by grav-ity, fills the whole esophagus well ahead of the peristaltic wave. (Courtesy of F. J. Hodges and J. N. Correa.)

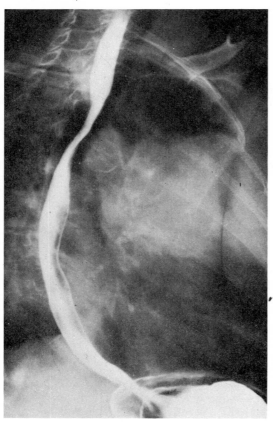

horizontal position much faster than the peri-staltic wave, apparently as the result of momentum derived from the force of pharyn-geal contraction. When a man in the head-down position swallows water, the esopha-gus fills, the column of water being supported by the closed pharyngoesophageal sphincter. The weak propulsive force of esophageal peristalsis is incapable of raising the whole column of fluid into the stomach. On repeat-ed swallowing in this position, more water is forced into the esophagus by pressure de-veloped in the pharynx, while an equal vol-ume at the top of the column is delivered into the stomach.

With the subject in any position, a pasty mass moves more slowly through the phar-ynx and esophagus, taking, on the average, about 5 sec to reach the stomach. Solid masses move still more slowly, and small particles may be left behind in the esophagus to be gathered up by a succeeding peristaltic wave.

Drinking water at 0.5 – 3 C slows or abol-ishes esophageal peristalsis. Nevertheless, the lower esophageal sphincter relaxes and contracts with each swallow although no peristalsis occurs above it. Sphincteric re-laxation is delayed, and water often pools above the sphincter. Relaxation, when it oc-curs, is prolonged and is followed by exag-gerated contraction. In contrast, drinking water at 58 – 61 C accelerates movements of the esophagus as compared with those oc-curring when water at 23 – 26 C is drunk. Rate of propagation of peristalsis is in-creased, the waves have a shorter duration, relaxation of the lower esophageal sphincter is briefer and the subsequent contraction is feebler.

Reflux into the Esophagus

During quiet respiration, the intragastric pressure is about 10 mm Hg higher than in-traesophageal pressure, but reflux does not occur. This is because pressure in the lower esophageal sphincter is higher than intragas-tric pressure and because the sphincter resists distention. The yield pressure of the lower esophageal sphincter was measured in two groups of subjects: 9 normal persons and 6 complaining of frequent reflux of gastric contents into the esophagus. The resting pressures in the lower esophageal sphincter were the same in both groups. Among the normal subjects, the yield pressures ranged from 10–38 mm Hg, with a mean of 25 mm Hg. In those with reflux, the range was 5–15 mm Hg, and the mean was only 10 mm Hg.

During eating, intragastric pressure rises as the stomach fills. At the same time, resting pressure in the lower esophageal sphincter rises. Factors responsible may include an increase in plasma gastrin concentration, reflex effect of acid in contact with the mucosa just distal to the sphincter and reflex contraction of the sphincter as the result of the increase in intra-abdominal pressure.

External abdominal compression raises intragastric pressure, but it does not alter the gradient between the subhiatal portion of the sphincter and the stomach. The reason is that intra-abdominal pressure is transmitted equally to the subdiaphragmatic portion of the gastroesophageal sphincter and to the stomach. Although the absolute pressure in the two areas rises, the pressure in the sphincter remains higher than that in the stomach. The same mechanism prevents reflux when the pressure difference between stomach and esophagus is raised by forced inspiration on a closed glottis. In deep inspiration there is also a steep rise in pressure in a narrow subhiatal zone of the sphincter, probably caused by pinch-cock action of the contracted diaphragm. Contraction of the diaphragm may also increase pressure in the sphincter during quiet inspiration.

When intra-abdominal pressure rises, there is a reflex increase in pressure the whole length of the lower esophageal sphincter. In normal subjects, the increment in sphincter pressure is greater than the increment in intra-abdominal pressure, and reflux is thereby prevented. This mechanism is effective if the sphincter is entirely in the thorax, as it is in hiatus hernia. Consequent-

ly, patients with hiatus hernia who have a normal response to an increase in intra-abdominal pressure do not have reflux when intra-abdominal pressure rises. On the other hand, in some persons the increment in lower esophageal pressure is less than the increment in intra-abdominal pressure; in them, reflux occurs whether the sphincter is in its normal position or is in the thorax.

Prevention of reflux has also been attributed to valve-like mucosal folds at the cardia, to the acute angle the stomach sometimes makes with the esophagus and to contraction of oblique sling-like fibers of smooth muscle at the cardia. None of these is important in man.

The stomach can accept a large volume of food without an appreciable rise in pressure; but a heavy meal, large amounts of swallowed gas or carbon dioxide generated by the taking of sodium bicarbonate may raise intragastric pressure and bring the lower edge of the gas bubble normally present in the gastric fundus below the level of the cardia. Then, when the sphincter relaxes during swallowing, eructation into the esophagus follows, and esophageal pressure rises abruptly. Carminatives, the food seasoners and ingredients of after-dinner liqueurs that produce the sensation of warmth and facilitate eructation reduce intrasphincteric pressure and allow reflux. If the subject is in the supine position, gastric contents quickly run into the esophagus. Afferent nerves are stimulated, with the result that a secondary peristaltic wave begins at the pharyngoesophageal sphincter and sweeps down the esophagus to empty it again.

To study belching, large volumes of air were run through a tube into the stomach of normal human subjects. Intragastric pressure rose 4–7 mm Hg during entry of the first 200–600 cc, but it did not rise further until the total volume was about 1,600 cc. Sooner or later (5–157 sec) after the pressure plateau was reached, there was sudden reflux of air into the esophagus, and intragastric and esophageal pressures became equal. Then, if the subject did not belch, secondary peristalsis stripped the esophagus and returned the air to the stomach. Belching occurred in 18 subjects during reflux of air into the esophagus. At the moment of belching, pressure in the esophagus, stomach and abdomen rose abruptly, apparently following contraction of somatic muscles, and gas was expelled from the common gastroesophageal cavity. Pressure in the esophagus remained equal to intragastric pressure for 6–10 sec after the belch, until secondary peristalsis emptied the esophagus.

During reflux of gastric contents, the pH within the esophagus may be as low as 2, and in some men and nonpregnant women the burning sensation known as heartburn is felt if the pH is below 4. It is not known whether the sensation is aroused by acid stimulation of receptors within the mucosa or by increased tension in the esophageal muscle. Perfusion with 0.1 N HCl of the lower esophagus of patients subject to heartburn reproduces the symptoms; these are invariably accompanied by motor abnormalities, including nonpropulsive synchronous contractions of the esophagus, prolonged peristalsis and raised intraluminal pressure. No pain or motor abnormalities occur in normal subjects similarly perfused. Heartburn attributable to acid reflux occurs in about 50% of pregnant women in the last five months of pregnancy. Secondary peristalsis has been found in 45% of pregnant women examined who complained of heartburn but never in those free from it. The major cause of heartburn in pregnancy is diminished resistance at the lower esophageal sphincter. In addition, no intra-abdominal esophageal segment can be identified. Whenever abdominal pressure increases during bending, stooping or lying, pressure is transmitted to the stomach; because there is no subdiaphragmatic portion of the esophagus, the pressure cannot at the same time reinforce the sphincter. Heartburn subsides in the last weeks of pregnancy as the uterus descends, and it ceases immediately after delivery.

If acid is present in the lower esophagus of a normal person, it is quickly cleared by 4–12 swallows or secondary peristaltic waves. A person with symptomatic reflux may require 28 or more swallows to clear the same amount of acid from his esophagus. A normal person has 15 or so episodes of reflux in 15 hours, and in each the refluxed gastric contents are returned to the stomach in 5–15 min. In a patient with abnormal reflux, the pH of the lower esophagus may be less than 3.0 for 30 min to an hour after each episode of reflux, because he has difficulty clearing his esophagus.

Mechanisms controlling adult esophageal function are incomplete at birth. The infant's superior laryngeal sphincter is not strongly closed; peristalsis does not occur in his lower esophagus; there is little or no abdominal esophagus; and the lower esophageal sphincter does not close between swallows. On the other hand, the crural muscle fibers of the diaphragm are well developed, and they act as an accessory closing mechanism during expiration as well as during inspiration. Reflux into the esophagus is infrequent, but when it does occur, absence of secondary esophageal peristalsis and weakness of the superior laryngeal sphincter allow gastric contents to bubble out the mouth in the manner of a fumarole overflowing.

Reverse Peristalsis and Rumination

Reverse peristalsis does not occur in the esophagus of man or other nonruminating animals. A ruminator, such as the cow, aspirates rumenal contents into its esophagus by relaxing its lower esophageal sphincter and inspiring against a closed glottis. This steepens the gradient between rumen and esophagus by dropping the intraesophageal pressure 30–40 mm Hg. Neither rumen nor abdominal muscle contracts, and intra-abdominal pressure does not rise. The cud is then carried to the mouth by reverse peristalsis of the esophagus, traveling 1 m a second. Other animals are less hasty. The llama "chews, placidly swallows and a lemon-sized ball moves measuredly down the neck; after a short interval the ball reappears at the base of the neck and, with equally measured pace, rises to the mouth."*

Fermentation in the rumen generates a large volume of gas, which must be expelled. A thousand-pound cow produces 1–2 liters per min, chiefly carbon dioxide and methane. Eructation begins with movements of the rumenoreticulum, which force gas into the area of the cardia. Relaxation of the lower esophageal sphincter allows gas to fill the esophagus. Then, after the sphincter has closed again, a powerful antiperistaltic wave, sweeping upward at 1.6 m per sec, forces gas into the pharynx. Because the glottis is open and the nasopharyngeal airway is closed, more than half the gas is forced into the lungs, where carbon dioxide and methane are absorbed.

Achalasia of the Esophagus

In achalasia, or failure of the lower esophageal sphincter to relax, the swallowed bolus may have difficulty passing into the stomach (Fig 1–10). The pharyngeal swallowing reflex and the action of the pharyngoesophageal sphincter are normal, but in the esophagus itself, peristalsis does not occur. If the esophagus does contract, the contractions, lasting 2–8 sec, engage the whole muscle at once. Pressure generated by such massive contractions is 10–20 mm Hg, and four or five of them may occur repetitively after each swallow. Resting pressure in the lower esophageal sphincter is twice normal. When swallowing occurs, the sphincter fails to relax; in fact, it may contract instead, reaching a pressure about 10 mm Hg above its resting level. This contraction is similar to that in normal swallowing, but it is not preceded by relaxation, and its onset is early rather than late. Therefore, the bolus fails to pass into the stomach, and the esophagus above the sphincter is enlarged. A whole meal may lodge above the sphincter and pass slowly into the stomach over a period of hours. Mechanical dilatation or surgi-

*Ingelfinger, F. J.: Physiol. Rev. 38:533, 1958.

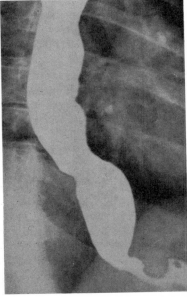

Fig 1–10. — Frontal view of an abnormal human esophagus; achalasia of the lower esophageal sphincter, with dilation of the esophagus. (Courtesy of F. J. Hodges and J. N. Correa.)

cal weakening of the sphincter may be necessary.

The pathology underlying this disorder seems to be absence of the myenteric plexus. The smooth muscle of the esophagus is therefore denervated. This makes it more sensitive to acetylcholine. The uncoordinated action of the esophagus is probably the response of sensitized muscle fibers to acetylcholine liberated by intact preganglionic fibers. In a normal person, injection of 5–10 mg of the parasympathomimetic drug methacholine has no effect upon the esophagus; in persons with achalasia, that dose causes prolonged, vigorous and painful contractions of the esophagus.

Other Causes of Delayed Emptying

In some patients who complain of inability to swallow or of pain on swallowing, the esophagus may appear anatomically and physiologically normal during x-ray examination while a suspension of barium sulfate is being swallowed. Some patients have been considered neurotic on the basis of such negative findings. Conscious rejection is frequently expressed in the phrase "I can't swallow that," and states of rejection below the level of consciousness can be expressed by inability to swallow. Therefore, esophageal dysfunction may be a neurotic symptom, and in patients with true achalasia, exacerbation and remission of symptoms are often related to emotional stress and the return of relative security. However, in some persons who can swallow water without difficulty, swallowing a bite of meat or a peanut butter sandwich may evoke abnormal contraction of the distal esophagus. The solid bolus, instead of falling through the esophagus as does a liquid, is moved by the peristaltic wave, and in these individuals, afferent impulses aroused by the bolus reflexly arouse premature and exaggerated contraction of the lower esophageal sphincter.

Esophageal emptying is also delayed by pain. The esophagus of a normal subject that requires 12 sec or less to empty may take 25–102 sec when the subject's hand is plunged in ice water.

Sensations from the Esophagus

Afferent fibers mediating esophageal sensations travel centrally with the sympathetic nerves; the only afferents from the esophagus traveling centrally with the vagus nerve are those concerned with secondary peristalsis. Distention of the esophagus at any level causes pain in the midsternum. Pain is aroused by strong and continued contractions. In some persons, normal peristalsis of the lower half of the esophagus is replaced by simultaneous contraction of the whole esophageal muscle, which is prolonged and often repetitive. This is called "diffuse spasm of the esophagus." These contractions develop pressures of 150–200 mm Hg, and they are accompanied by severe, aching substernal pain, which may extend between the tips of the clavicles and upward into the throat and jaws. The pain subsides as esoph-

ageal pressure falls. A sense of fullness or substernal constriction is caused by dilatation of the upper and middle two thirds of the esophagus. Pain caused by abnormal esophageal contractions may be precipitated by exercise or emotional upsets; and because it closely resembles cardiac pain with identical distribution, it is often confused, particularly by laymen, with pain arising in the heart. Differential diagnosis is made difficult by the fact that in many patients such pain arises both from the esophagus and from coexistent ischemic heart disease.

REFERENCES

Anderson, D. J.: Mastication, in Code, C. F. (ed.): *Handbook of Physiology:* Sec. 6. *Alimentary Canal,* Vol. IV (Washington, D.C.: American Physiological Society, 1968), pp. 1811–1820.
Bennett, J. R., and Atkinson, M.: The differentiation between oesophageal and cardiac pain, Lancet 2:1123, 1966.
Car, A., Jean, A., and Roman, C.: A pontine primary relay for ascending projections of the superior laryngeal nerve, Exp. Brain Res. 22: 197, 1975.
Code, C. F., and Schlegel, J. F.: Motor action of the esophagus and its sphincters, in Code, C. F. (ed.): *Handbook of Physiology:* Sec. 6. *Alimentary Canal,* Vol. IV (Washington, D.C.: American Physiological Society, 1968), pp. 1821–1840.
Cohen, B. R., and Wolf, B. S.: Cineradiographic and intraluminal pressure correlations in the pharynx and esophagus, in Code, C. F. (ed.): *Handbook of Physiology:* Sec. 6. *Alimentary Canal,* Vol. IV (Washington, D. C.: American Physiological Society, 1968), pp. 1841–1860.
Cohen, S., and Harris, L. D.: The lower esophageal sphincter, Gastroenterology 63:1066, 1972.
DiDion, L. J. A., and Anderson, M. C.: *The "Sphincters" of the Digestive System* (Baltimore: Williams & Wilkins Co., 1968).
Doty, R. W.: Neural organization of deglutition, in Code, C. F. (ed.): *Handbook of Physiology:* Sec. 6. *Alimentary Canal,* Vol. IV (Washington, D.C.: American Physiological Society, 1968), pp. 1861–1902.
Dougherty, R. W.: Physiology of eructation in ruminants, in Code, C. F. (ed.): *Handbook of Physiology:* Sec. 6. *Alimentary Canal,* Vol. V (Washington, D.C.: American Physiological Society, 1968), pp. 2673–2694.
Edwards, D. A. W.: The oesophagus, Gut 12: 948, 1971.

Goodwin, G. M., and Luschei, E. S.: Discharge of spindle afferents from jaw-closing muscles during chewing in alert monkeys, J. Neurophysiol. 38:560, 1975.

Lamarre, Y., and Lund, J. P.: Load compensation in human masseter muscles, J. Physiol. 253:21, 1975.

Stevens, C. E., and Sellers, A. F.: Rumination, in Code, C. F. (ed.): *Handbook of Physiology: Sec. 6. Alimentary Canal,* Vol. V (Washington, D.C.: American Physiological Society, 1968), pp. 2699–2704.

Sturdevant, R. A. L.: Is gastrin the major regulator of lower esophageal sphincter pressure?, Gastroenterology 67:551, 1974.

2

The Neuromuscular Apparatus
of the Lower Digestive Tract

EXCEPT AT ITS pharyngeal and anal ends, the motor function of the digestive tract is performed by smooth muscle. In order to understand the propulsion of food through the gut, it is necessary to know the structure, innervation and physiological properties of intestinal muscle. Gastrointestinal smooth-muscle structures are extremely diverse in form and function, and only the general properties of the neuromuscular apparatus will be considered here. Later, coordinated activity as expressed in peristalsis, spasm and other movements will be discussed, together with special nervous and humoral factors influencing them.

Structure of Smooth-Muscle Cells

Throughout life, intestinal smooth-muscle cells maintain some degree of tension, and they have no characteristic resting length. After death, intestinal smooth-muscle cells greatly elongate, with the result that the length of the dead intestine is far greater than that of the living one.

An intestinal smooth-muscle cell at its greatest extension is probably less than 250 μ long and 6 μ thick at the center. Its nucleus is eccentrically placed at the thickest part. The cell membrane is approximately 0.015 μ

thick; and immediately external to it, an irregular array of collagen fibrils forms an external reticulum attached to the membrane. The cell membrane contains a large number of pinocytotic vesicles; many of these are open to the extracellular space, and they enormously increase the surface area of the cells. Mitochondria are less abundant in smooth muscle than in cardiac muscle, and they are grouped near the ends of the nucleus. Their structure and function are the same as those in other mammalian cells. Both thin filaments of actin and thick filaments of myosin are present. In some cross-sections, about 15 thin filaments are seen clustered around a thick filament, but there is no regular organization corresponding to the myofibrils of striated muscle. Longitudinally, there is no regular transverse alignment or crossbanding. An important physiological consequence is that the filaments may slide to an unlimited extent with respect to each other; this accounts for the great range of length over which tension can be exerted. The myofilaments do not converge into bundles at the cell tips; instead, they tend to maintain generally parallel organization, with individual filaments terminating as they approach the cell wall obliquely at the tapering tips of the cell. Their contractions cause lateral indentations on the cell border, so that the cell seems to fold on itself like an accordion.

Smooth muscle cells contain a poorly developed system of T tubules, which release and sequester calcium during the cycle of excitation and relaxation. The very large surface-to-volume ratio of the cells allows calcium to activate actomyosin simply by diffusing into the cell at the time of the action potential.

Connections of Smooth-Muscle Cells

Individual smooth-muscle cells are surrounded by a sheath of reticular fibers attached to neighboring connective tissue. The force of contraction is transmitted through the reticulum to the connective tissue. Groups of muscle cells are organized as bundles. In the circular muscle of the cat intestine, these bundles are 500 μ thick and contain about 7,000 cells in cross-section. In other species, bundles are from 100–400 μ thick. Cells may be connected with each other through gap junctions or nexuses, and gap junctions are particularly abundant in the circular layer. Gap junctions are very labile, readily breaking and re-forming, but on the average they occupy 6% of the cell surface. The electric resistance between cells through gap junctions is less than the resistance of the cell membrane, and therefore gap junctions permit electric coupling between cells in a bundle. In some bundles, membranes of adjacent cells are closely joined without fusion of their outer membranes. Bundles, not individual cells, are the functional effector unit. Bundles are attached to adjacent bundles by connective tissue. Larger units, made up of many bundles, form the circular layer on the mucosal side and the surrounding longitudinal layer on the serosal side. In some tissues, thin muscle strands connect the circular and longitudinal layers and are responsible for interaction between the layers.

The details of organization of bundles into layers are disputed. Some anatomists believe the circular layer is wound in a helix, whose direction is counterclockwise (viewed from the oral end), and that the longitudinal layer consists of bundles twisting in a more open, elongated helix. Others think that the circular layer is formed by closed rings and the longitudinal layer by bands, whose axes are parallel with that of the gut. The muscularis mucosae, lying beneath the glandular mucosa, consists of both circular and longitudinal layers of muscle fibers. The layers vary greatly in thickness in different species and parts of the tract, and they are especially thick in the pig's esophagus and in the human stomach. Smooth-muscle cells pass from the circular layer of the muscularis mucosae into the mucosa.

The submucosa is composed of interlacing collagenous fibers, which, when the intestine is relaxed, form helical coils at an approximately 45-degree angle with the axis, half clockwise and half counterclockwise. In the dog, this braided sheet is 18–24 layers thick.

Sympathetic Innervation

The autonomic innervation of intestinal smooth muscle is summarized in Table 2–1 and Figure 2–1. Efferent sympathetic fibers go to the stomach from the celiac plexus, to the small intestine from the celiac and superior mesenteric plexuses and to the cecum, appendix, ascending colon and transverse colon from the superior mesenteric plexus. The remainder of the colon receives sympathetic fibers from the superior and inferior hypogastric plexuses. Most sympathetic fibers to the intestine are postganglionic, and their cell bodies are in the ganglia named. Some sympathetic fibers entering the gut may be preganglionic ones, and their postganglionic fibers would be those of the cell bodies of the intramural plexuses upon which the preganglionic fibers terminate.

Sympathetic fibers innervating the intestine end in one of four places: (1) Some sympathetic fibers enter glandular tissue where they appear to innervate some secretory cells. (2) Some sympathetic fibers innervate smooth-muscle cells of blood vessels, causing vasoconstriction, and smooth-muscle

TABLE 2-1. FUNCTIONAL CLASSIFICATION OF NERVES AFFECTING THE LOWER DIGESTIVE TRACT

Systems Mediating Extrinsic Sympathetic Innervation (Fig 2-1, A):

Preganglionic cholinergic fibers pass
  from cell bodies in lateral columns of thoracolumbar region of cord, via white rami communicantes, to
    (chiefly) the prevertebral ganglia, where they synapse with
Postganglionic adrenergic (and cholinergic?) fibers, which pass
  from cell bodies in prevertebral ganglia, via thoracic splanchnic nerves,
  to endings on the following effectors:
    smooth-muscle cells of blood vessels, constrictor (and dilator?)
    smooth-muscle cells of muscularis mucosae, excitatory (and inhibitory?)
  cell bodies of neurons within the myenteric plexuses whose activity they modulate, inhibitory
    gland cells of salivary glands, excitatory
    gland cells of pancreas, small intestine and colon (inhibitory and excitatory?)
    smooth muscle cells of some circular layers, inhibitory
    smooth muscle cells of some sphincters, excitatory

Systems Mediating Extrinsic Parasympathetic Innervation (Fig 2-1, B):

Preganglionic cholinergic fibers pass
  from cell bodies in cranial division of neuraxis, via vagus nerves
  and from sacral region of the cord, via pelvic and splanchnic nerves
  to the digestive tract, where they synapse with
Postganglionic cholinergic fibers, which pass
  from cell bodies in ganglia of nerve plexuses
  to synapse with
    other ganglion cells in the same ganglion or in the same plexus or other plexuses
    and to the following effectors:
      smooth-muscle cells in all muscle layers, chiefly excitatory
      gland cells in gastric antrum (and elsewhere?) that liberate hormones of the digestive tract
      gland cells that liberate external secretions (but not directly to blood vessels?)
Postganglionic fibers in the lower esophagus and stomach, inhibitory

Systems Mediating Reflexes of Peripheral Location (Fig 2-1, C):

Afferent limb, consisting of
  nerve terminals from which impulses originate
    in the mucosal epithelium
    in the muscle layers and plexuses
  cell bodies in the submucous (and myenteric?) plexus
    potentially effective stimuli of which are:
      stretch or distention
      pH of contents of viscus
      specific chemical constituents, such as amino acids, peptides, fats
  axons synapsing with
Efferent limb, consisting of
  cell bodies in myenteric and submucous plexuses
  second-order effector neurons and
  other (identical?) effector neurons of myenteric and submucous plexuses, which pass to
  endings that innervate the following effectors:
    smooth-muscle cells of the digestive tract
    gland cells of internal secretion in gastric antrum (and elsewhere?) that liberate hormones of the digestive
      tract
    gland cells that liberate external secretions
    (smooth-muscle cells of blood vessels?)

Systems Mediating Reflexes through Celiac Plexus (Fig 2-1, D):

Afferent limb, consisting of
  nerve terminals from which impulses originate in epithelium of duodenal and other mucosa
  cell bodies in intrinsic plexuses
    potentially effective stimuli of which are:
      chemical composition of gastrointestinal contents, pH, osmotic pressure, digestion products of protein and
      fat
  axons synapsing in celiac plexus with

TABLE 2–1—*Continued*

Efferent limb, consisting of
  cell bodies in celiac plexus
  postganglionic sympathetic fibers, which pass to endings on neuronal cell bodies within the intrinsic plexuses
    of stomach, inhibitory

Systems Mediating Reflexes through Central Neuraxis (Fig 2–1, E):

Afferent limb, consisting of
  nerve terminals from which impulses originate
    in epithelium
    in plexuses
    in smooth muscle, potentially effective stimuli of which are distention or
      chemical composition of contents of viscus
  fibers that pass centrally,
    via thoracic, lumbar or pelvic splanchnic nerves and dorsal roots, to cord, and having cell bodies in dorsal
      root ganglia or
    via vagus nerves to brain stem,
  mediating the following:
    vasodilatation via axon reflexes
    pain and other visceral sensations
    vomiting reflex
    defecation reflex
    gallbladder reflex
    response to obstruction, and
  initiating specific action on
    digestive tract reflexes, facilitating and inhibiting
    somatic reflexes
    cardiovascular reflexes and
    centripetal fibers from cell bodies within intrinsic plexuses

cells of the muscularis mucosae, causing contraction. (3) Most sympathetic fibers terminate in contact with neuronal cell bodies of the intramural plexuses and with presynaptic fibers surrounding the cell bodies. They inhibit ganglionic activity, possibly by presynaptic inhibition. (4) There is some sympathetic innervation of the circular muscle layers of the small and large intestine, where its function is chiefly inhibitory. In some organs such as the sphincters sympathetic innervation is excitatory.

Sphincters of the gastrointestinal tract are adrenergically innervated, and the density of innervation is often greater than in adjacent muscle. Although there is considerable species difference, in most species examined the effect of sympathetic innervation is to excite the lower esophageal sphincter, the choledochoduodenal sphincter and the internal anal sphincter. Excitation is mediated by the action of norepinephrine upon alpha receptors of the smooth muscle cells.

Adrenergic nerve terminals avidly take up norepinephrine, thereby reducing the effectiveness of circulating norepinephrine.

Parasympathetic Innervation

Efferent parasympathetic innervation to the stomach, small intestine, cecum, appendix, ascending colon and transverse colon is by way of the vagus, whose fibers follow blood vessels to end in the myenteric plexus. The rest of the colon receives parasympathetic innervation from the pelvic nerves via the hypogastric plexuses, and these, too, end in the myenteric plexus. The fibers are all preganglionic, and most are cholinergic and excitatory. However, vagal and pelvic nerves contain nonadrenergic, noncholinergic inhibitory fibers.

Plexuses

Five nerve plexuses are distinguished: subserous, myenteric, deep muscular, internal and submucous. The deep muscular and

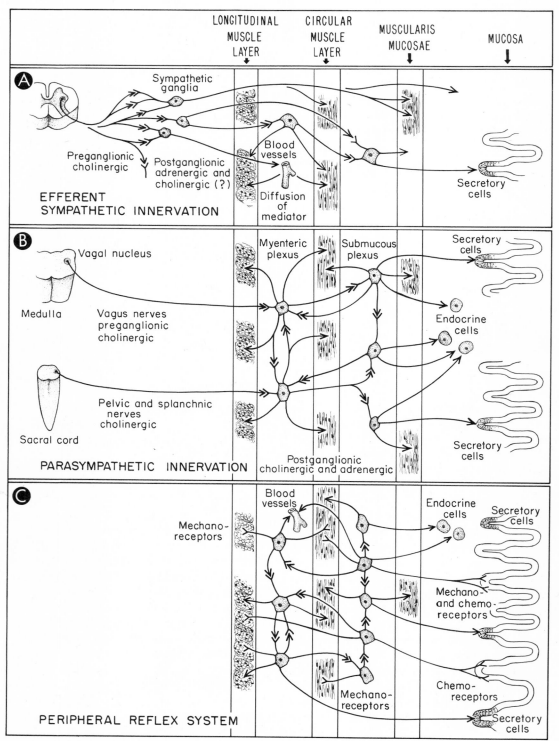

Fig 2–1.—Efferent and afferent innervation of the digestive tract. **A,** efferent sympathetic innervation. **B,** parasympathetic innervation. **C,** peripheral reflex system. *(Continued.)*

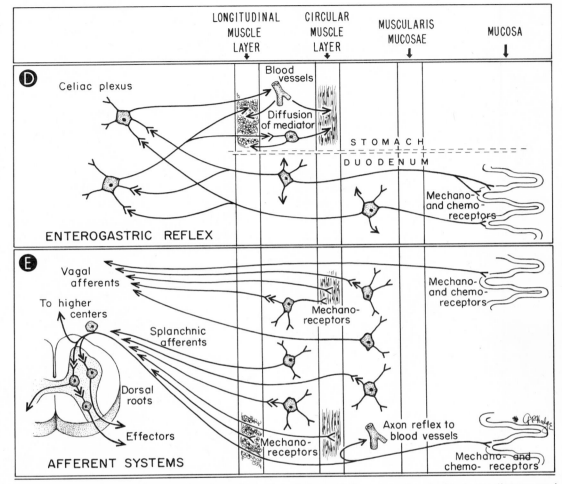

Fig 2–1 (cont).—D, enterogastric reflex. **E,** afferent systems. See Table 2–1 for detailed listing of the nerve tracts, their mediators and their effects.

the internal plexuses are probably formed by terminal fibers of the myenteric and submucous plexuses. Postganglionic sympathetic and preganglionic parasympathetic fibers make up a large fraction of the plexuses and disappear after extrinsic denervation. The myenteric plexus in the interspaces between longitudinal and circular muscle layers consists of a mesh of fibers, with ganglion cells at the nodal points, all ensheathed by the cytoplasm and membranes of a connective tissue syncytium comparable with the neurilemma of myelinated fibers. The plexuses themselves are composed of individually distinct cells, not of a syncytial network. Nerve fibers extend within the plexus for several centimeters, and from the plexus arises the reticulum of fine fibers passing to muscle cells of the two layers.

Branches of the preganglionic parasympathetic fibers synapse with the ganglion cells, some with the first cell encountered after they enter the plexus and others with more remote cells. Still other preganglionic fibers pass through the myenteric plexus to the submucous plexus. In turn, fibers go from the reticulum of ganglion cells and from fibers forming the submucous plexus to the muscularis mucosae, to the gland cells of the mucosa, to unidentified endocrine cells and to muscular tissue within the mucosa. In addition, axonal processes from one ganglion cell synapse with cells in the same node or in more distant ganglia; the result is that when extrinsic fibers have degenerated after denervation, ganglion cells are still functionally connected and are capable of mediating local reflexes. Afferent limbs of the reflex arcs are provided by cells in the submucous plexus whose dendrites are receptor organs in the mucosa and in the muscle layers and whose axons project to synapses formed with cells of either the submucous or the myenteric plexus.

Function of Sympathetic Innervation

Sympathetic fibers to mucosal glands control to some extent the secretion of mucus, and those to the muscularis mucosae govern its contraction. Otherwise, the chief function of sympathetic innervation is to inhibit contraction of intestinal smooth muscle. This is effected in three ways. Inhibition of intestinal motility following general sympathetic stimulation is caused by epinephrine and norepinephrine liberated from the adrenal medulla. Stimulation of sympathetic fibers to blood vessels causes vasoconstriction within the intestine; and after a long latent period, of 20–40 sec, circular and longitudinal muscle is inhibited. This inhibition is attributed to effects of vasoconstriction and diffusion of adrenergic mediator from sympathetic nerve endings on blood vessels. The action of the numerous sympathetic fibers that end in contact with ganglion cells within the intramural plexuses is to modulate the activity of those cells. Sympathetic nerves do not inhibit ganglion cells by direct action upon their membrane; rather, the norepinephrine liberated from sympathetic nerve terminals acts on presynaptic, excitatory nerve terminals to block the release of acetylcholine. Because acetylcholine is the excitatory transmitter at ganglion cells, blockade of its release reduces the excitatory drive.

Consequently, an effect of sympathetic stimulation is seen when there is ongoing activity within the plexuses. If, for example, gastric motility is enhanced by stimulation of vagal excitatory nerves to the stomach, stimulation of the hypothalamic defense area or of the surrounding pressor area produces prompt and often complete inhibition of gastric motility. When these hypothalamic areas are stimulated in the absence of simultaneous vagal excitation of the intramural plexuses, stimulation rarely has any effect on gastric motility. Reflex or direct stimulation of adrenergic fibers to the gut does not inhibit the effects of acetylcholine or its congeners upon gastric or intestinal motility, for the reason that the cholinergic mediator acts directly upon intestinal smooth muscle.

Afferent Fibers

Nerves of the digestive tract contain many visceral afferent fibers, which can be divided into two classes: those having their cell bodies in centrally located ganglia and those having their cell bodies within the myenteric plexuses. Those of the first class have sensory endings in the mucosal epithelium, in the plexuses and within the muscle layers. Their fibers pass centrally with the vagus and with the sympathetic rami to the dorsal roots. Cell bodies of afferent vagal fibers are in the nodose ganglion, and those of nerves traveling centrally with the sympathetics are in the dorsal root ganglia. Among the latter are the dorsal root dilators, whose action on intestinal blood vessels is the same as that of cutaneous dorsal root dilators on the vessels of the skin. Of the 30,000 fibers contained in the vagus nerves of the cat, at least 80% are afferent, and so are 50% of the 30,000 fibers in sympathetic nerves to the gut.

The sympathetic nerves contain afferent fibers whose cell bodies lie in the intramural plexuses. Some of their axons may pass through the celiac and other sympathetic ganglia and enter the central neuraxis before synapsing. Others appear to synapse in the celiac ganglion with postganglionic sympathetic fibers, and they are probably the afferent limb of reflex arcs that regulate gastrointestinal motility.

Nerve Endings

Each postganglionic fiber in the plexuses branches many times and ramifies extensively. Branching is the anatomical basis of divergence of nervous effects, and stimulation of only a few parasympathetic preganglionic fibers or sympathetic postganglionic fibers influences many effector cells. Bundles of terminal branches enter the muscular coats accompanying capillaries. Fibers leave the neurilemma and pass as naked filaments between muscle fibers. Terminal fibers containing vesicles do not penetrate the plasma membrane but intrude into cell pockets, where their tips and the cell membrane meet. The vesicles probably contain stores of chemical mediators that are liberated when the nerve is excited. Large numbers of naked nerve fibers pass close to all muscle cells, and these may also liberate transmitters at intervals along their length. Nerve fibers approaching one cell are derived from many cell bodies, and influences having widespread origins converge on a single cell.

Membrane Potentials

Contraction of a smooth-muscle cell follows electric changes at its membrane; and electric changes, in turn, depend on electrolyte distribution between the cell and the extracellular fluid. The major determinant of the resting membrane potential is the potassium ion. Other ions, chiefly sodium and chloride, also play a part. The ion that diffuses most rapidly through the membrane dominates its potential; at rest, potassium diffuses through the membrane much more rapidly than the other ions, and the resting membrane potential approaches the potassium equilibrium potential.

Potassium is at higher concentration in intracellular water than in extracellular water, and therefore it tends to diffuse across the cell membrane from its higher to its lower concentration. Within the bulk of the cell, K^+ ions are electrostatically balanced by an equal number of anions; some of the anions are high molecular weight organic compounds, and some are Cl^- ions. At the cell membrane, a fringe of K^+ ions diffuses outward, carrying positive charges across the membrane. This outward diffusion of K^+ ions establishes an electric gradient across the membrane, positive outward and negative inward, which exists because positive K^+ ions are physically separated across the membrane from anions remaining within the cell. The electric gradient itself opposes further diffusion of K^+ ions, and an equilibrium is quickly reached in which the outward diffusion force, created by the concentration difference, is balanced by an electrostatic force in the opposite direction. Thus, the potential difference across the cell membrane is maintained by a tendency for K^+ to diffuse, but only a very small separation of charges actually occurs. The general equation expressing the potential difference (E_K) if only K^+ ions diffuse is

$$E_K = \frac{RT}{F} \ln \frac{P_K[K^+]_o}{P_K[K^+]_i} \qquad (2.1)$$

where R is the gas constant (8.3 joule · degree^{-1} · mole^{-1}), T is the absolute temperature (311 K) and F is the Faraday constant of electric charge carried by 1 mole of univalent ions (96,500 coulombs· equivalent^{-1}). The logarithmic term is to the base e, which is 2.3 times logarithms to the base 10. The potential E_K, expressed in millivolts, is called the *potassium equilibrium potential*. When the appropriate numerical values are substituted, the equation becomes

$$E_K = 61 \log_{10} \frac{P_K[K^+]_o}{P_K[K^+]_i} \qquad (2.2)$$

The values in brackets are K^+_i and K^+_o, the concentrations (more accurately, the activities) of K^+ ions inside and outside the cell. Each concentration is multiplied by the constant P_K, which expresses the permeability of the membrane to K^+ ions. This is an essential factor, for if the membrane were not at all permeable to K^+ ions, no fringe of K^+ ions could diffuse through it, and there could be no potential difference across the membrane, despite the existence of a concentration difference. When the contribution of K^+ ions alone is being considered, values of P_K greater than zero cancel out. In cat circular intestinal muscle, the electrolyte concentration of extracellular fluid is equal to that of an ultrafiltrate of plasma, and the K^+_o concentration is 4 mEg/kg H_2O. Intracellular concentrations are difficult to measure, but a good estimate of K^+_i is 164 mEq/kg of cell water. When these values are substituted in equation (2.2), a potassium equilibrium potential of -101 mV is obtained. This is the maximum potential difference K^+ could contribute.

The measured resting membrane potential of intestinal smooth muscle is always less than the potassium equilibrium potential, and consequently there must be additional determinants of the membrane potential.

Sodium and Chloride Potentials

The two other ions whose diffusion affects the resting membrane potential are sodium and chloride.

The concentration of Na^+ ions in extracellular fluid is about 153 mEq/kg H_2O. The best estimate of the concentration of Na^+ ions within intestinal smooth-muscle cells is 19 mEq/kg H_2O, which is greater than the concentration in nerve or striated muscle. Because the diffusion gradient is inward, diffusion of Na^+ ions causes a potential difference across the cell membrane opposite to that of K^+ ions. The equation for the Na^+ equilibrium potential is

$$E_{Na} = 61 \log_{10} \frac{P_{Na}[Na^+]_o}{P_{Na}[Na^+]_i} \qquad (2.3)$$

Here, the factor P_{Na} is the permeability of the membrane to Na^+ ions. Using the values given for Na^+_o and Na^+_i, the maximal contribution of Na^+ to the membrane potential is calculated to be 55 mV.

Chloride concentration outside the cell is 132 mEq/kg H_2O, and its concentration inside intestinal smooth-muscle cells is estimated to be 55 mEq/kg H_2O. The diffusion gradient, like that of sodium, is inward; but, because Cl^- ions carry a negative charge, the

diffusion potential is, like that of K^+ ions, positive outward. The equation for the Cl^- equilibrium potential is

$$E_{Cl} = 61 \log_{10} \frac{P_{Cl}[Cl^-]_i}{P_{Cl}[Cl^-]_o} \qquad (2.4)$$

Here P_{Cl} is the permeability to Cl^- ions, and the difference in charge is allowed for by placing the inside concentration in the numerator. The calculated contribution of Cl^- ions to the membrane potential is -23 mV.

Because the actual resting membrane potential is more negative than the chloride equilibrium potential, chloride cannot be passively distributed between extracellular and intracellular fluids. It is probable that chloride ions are actively accumulated within the cells.

Resting Membrane Potential

The general expression for the membrane potential produced by the tendency of the three ions to diffuse is

$$E = 61 \log_{10} \frac{P_K[K^+]_o + P_{Na}[Na^+]_o + P_{Cl}[Cl^-]_i}{P_K[K^+]_i + P_{Na}[Na^+]_i + P_{Cl}[Cl^-]_o} \qquad (2.5)$$

This equation is clearly not simply the sum of the three equations for the individual ions; it is, rather, the equation relating the three ions to the membrane potential derived on the assumption that the electric field within the membrane is constant with respect to the distance through the membrane. It is called the *constant field equation*. The assumption is equivalent to saying that the charge density within the membrane has negligible effects on movement of ions through it. If the equation is rewritten

$$E_m = 61 \log_{10} \frac{[K^+]_o + \dfrac{P_{Na}}{P_K}[Na^+]_o + \dfrac{P_{Cl}}{P_K}[Cl^-]_i}{[K^+]_i + \dfrac{P_{Na}}{P_K}[Na^+]_i + \dfrac{P_{Cl}}{P_K}[Cl^-]_o} \qquad (2.6)$$

the role of the permeability constants is obvious: the actual membrane potential depends on the ratios of the permeability of sodium and chloride to that of potassium.

The potassium permeability constant for guinea pig taenia coli has been calculated to be 11×10^{-8} cm $\cdot$ sec^{-1}; $P_{Na} = 1.8 \times 10^{-8}$ and $P_{Cl} = 6.7 \times 10^{-8}$ in the same units. The ratio P_{Na}/P_K is 0.16, and the ratio P_{Cl}/P_K is 0.61. Substitution of these values in equation (2.6) gives a calculated membrane potential of -37 mV. The calculated value is lower than that determined by inserting a microelectrode into smooth-muscle cells. Potentials up to -80 mV have been measured in cat intestinal smooth muscle, and the average of a large number of measurements is -56 mV. In guinea pig taenia coli, the resting membrane potential is -60 mV.

There are probably two reasons for the discrepancy between calculated and observed membrane potentials. The first is that the other ions, particularly calcium, may contribute to the diffusion potential across the membrane. The other is that there is probably an electrogenic pump in the membrane that transfers sodium ions from the inside of the cell to the outside. If positively charged sodium ions are pumped out of the cell unaccompanied by anions, the act of pumping establishes a potential difference across the membrane, and this potential adds to that produced by the tendency of ions to diffuse through the membrane. Consequently, the membrane potential is higher than the diffusion potential.

The membrane potential is reduced by application of acetylcholine and by stretching of the muscle. The usual effect of epinephrine is to stabilize the membrane so that reduction of potential by other agents is made more difficult. Epinephrine may do this by making more energy available for the extrusion of sodium by the electrogenic pump. When epinephrine is applied to guinea pig taenia coli that has previously been loaded with radioactive ^{24}Na, the rate of loss of the isotope is doubled and the rate of sodium uptake is reduced. On the other hand, when the supply of high-energy phosphate is reduced, as by exhaustion of carbohydrate

stores or poisoning with iodoacetate, the effect of epinephrine is to depolarize the membrane.

Calcium

Calcium has at least three functions in intestinal smooth muscle: it controls the permeability of the membrane to sodium, it carries current inward during the action potential and it couples electric and mechanical activity.

The concentration of ionized calcium in extracellular fluid is $2-3$ mEq per liter, and its concentration is controlled by the same factors controlling its concentration in plasma. Calcium is bound to the membrane of smooth-muscle cells to the extent of 10 mEq per kg of cells. Although the total concentration of calcium within the cells is $2-3$ mEq per kg of cell water, calcium is almost entirely bound within the cells, and the concentration of ionized calcium is 10^{-3} or 10^{-4} mM.

If calcium is entirely removed from smooth-muscle cells, the membrane is depolarized, and electric and mechanical activities are abolished.

Potential Changes: The Action Potential

The membrane potential recorded by means of a microelectrode thrust into the cell may fall abruptly and then quickly return to the resting level. This variation is known as the action potential; and because it appears sharp in records, it is often called a spike. In mammalian nerve and striated muscle, the spike is caused by a sudden increase of about 200-fold in the permeability of the membrane to sodium. The membrane potential, as expressed by equation (2.6), thereupon approaches the sodium equilibrium potential; and because the sodium equilibrium potential is positive inward, the membrane potential actually reverses, or overshoots.

The most thoroughly studied intestinal smooth muscle is the taenia coli of the guinea pig. In this tissue, sodium is not responsible for the action potential. If the taenia coli is bathed in a solution containing only 10 mN Na, its intracellular sodium concentration falls to about 24 mEq per kg H_2O. At these sodium concentrations the sodium equilibrium potential reverses to become 22 mV, outside positive. Nevertheless, if the muscle is electrically stimulated, an action potential with an overshoot of 20 mV, outside negative, occurs. The action potential and its overshoot depend upon the presence of calcium in the external medium and bound to the membrane. When the action potential is initiated, calcium is released from the membrane, and the permeability of the membrane to calcium increases. Calcium ions cross the membrane into the cells, and as ions they carry current into the cells. The entry of calcium is responsible for the rising phase of the action potential. As calcium is released from the membrane, the permeability of the membrane to other ions increases, and the falling phase of the action potential is probably the result of a large increase in the membrane's potassium conductance.

The amplitude of the action potential of intestinal smooth muscle is variable. In some instances, depolarization may be incomplete, so that at the height of the spike, there may be residual polarization of the membrane, positive outward. Although some of the vari-

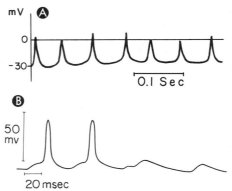

Fig 2–2.—Membrane potential and spontaneous discharges recorded by means of a microelectrode impaling a single smooth-muscle cell of cat intestine. **A,** spontaneous spike discharges. Membrane potential is about −30 mV; spikes vary in their height, and some overshoot. **B,** action potentials and slow waves. Slow waves seem to initiate the two spikes but are unable to do so for the third and fourth times. (From Barr, L.: Am. J. Physiol. 200:1251, 1961, and J. Theoret. Biol. 4:73, 1963.)

ability in records of spikes of smooth-muscle cells results from artifacts created by technical difficulties of recording from such small cells, some of it reflects the inconstant behavior of the cells. Figure 2–2 shows spikes of several sizes produced by a single cell.

Spike discharges may be started by elec-

Fig 2–3.—Membrane potential and tension recorded from strips of guinea pig taenia coli. **Top,** the effect of adding epinephrine. **Bottom,** the effect of adding acetylcholine. (Adapted from Bulbring, E., and Kuriyama, H.: J. Physiol. 166:59, 1963.)

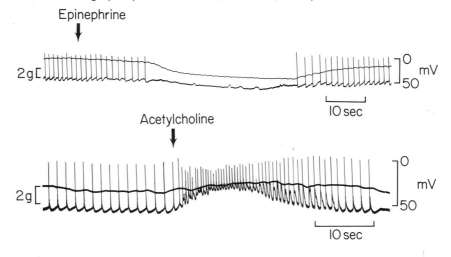

tric shocks. Excitation occurs at the cathode as it does in other excitable tissues. This means that the effect of an electric current is to reduce the membrane potential to some critical level, and then depolarization continues spontaneously. Once a part of the membrane is depolarized, current flows through that part from adjacent areas, depolarizing them. Thus, depolarization spreads over the whole cell, and the rate of spread is sufficiently rapid so that the membrane of one cell is totally depolarized before repolarization occurs on any part of it.

Spike discharges also occur spontaneously at frequencies ranging from 1–10 per sec. Discharges are started by acetylcholine and, when spontaneously occurring, may be stopped by epinephrine (Fig 2–3).

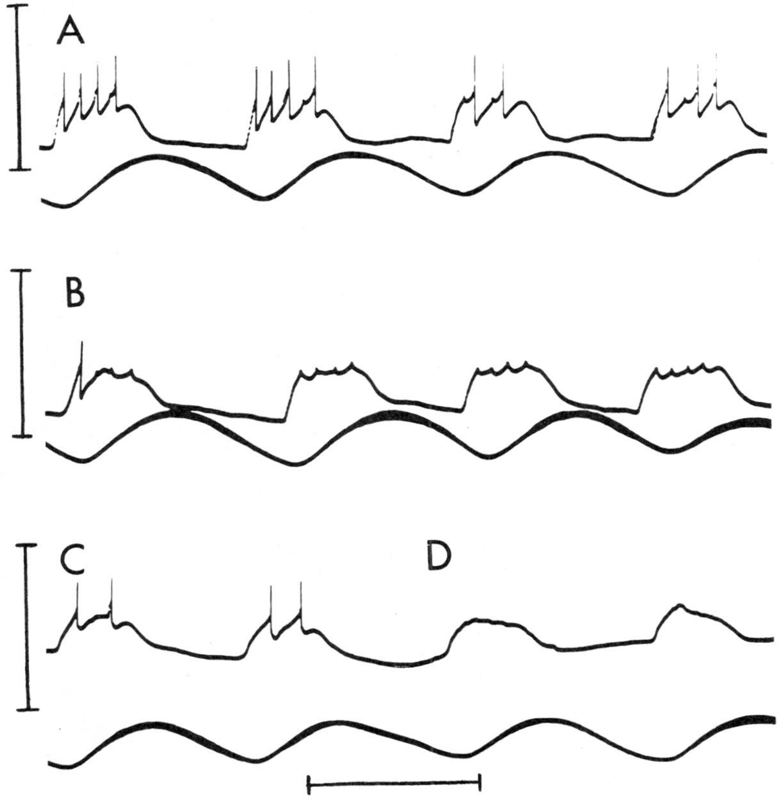

Fig 2–4.—Spontaneous electric and mechanical activity recorded from sheets of longitudinal muscle of rabbit jejunum. The electric record is the transmembrane potential of a single cell, but the tension record is that of a sheet of cells. Consequently, tension may increase as the result of contraction following action potentials in many cells other than the one from which the electric record is made. **A** and **B** are continuous records; **C** and **D** are taken from two other preparations. Electric waves are above and mechanical response is below in each frame. Each slow electric wave consists of prolonged depolarization, which usually gives rise to several generator-type potentials. Generator potentials can give rise to spikes, **A,** or decay, as in the last three slow waves of **B**. In some cases, where a single generator potential fails to elicit a spike, a second generator potential seems to sum with the first, and a spike follows as in the third slow wave of **A**. In some cases, miniature generator potentials occur at the top of slow waves, none being large enough to elicit a spike, as in **D**. All spikes are followed by undershoots. Vertical calibrations: 0 to −40 mV; horizontal calibrations: 5 sec. Mechanical records are uncalibrated, but upward deflection indicates increase in tension. (From Bortoff, A.: Am. J. Physiol. 201:203, 1961.)

Potential Changes: The Prepotential and Junction Potentials

The membrane of a smooth-muscle cell often decreases spontaneously and rhythmically by a few millivolts. Such changes may or may not be followed by spike discharges. In Figures 2–3, 2–4, 2–5 and 2–6 each spike is preceded by a small, slow depolarization, and in Figure 2–4 some of the small depolarizations are followed by spikes, whereas others are not. These prepotentials resemble those occurring in pacemaker tissue of the heart.

A single electric shock to a parasympathetic nerve twig going to intestinal smooth muscle evokes, after a latent period of 400 msec, a slow depolarization, which may be followed by a spike. This slow potential, which resembles the end-plate potential of striated muscle, decays with a half-life of 150–300 msec. It is probably caused by acetylcholine liberated by the nerve endings.

Fig 2–5. — Tension and electric activity of a strip of taenia coli. Tension increases as spike frequency increases and falls as spike frequency decreases. (From Bulbring, E.: *Lectures on the Scientific Basis of Medicine*, Vol. 7 [London: Athlone Press, 1959].)

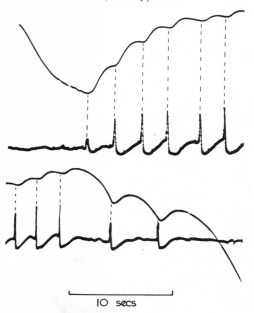

10 secs

When repetitive shocks are given, the slow potentials sum. On the other hand, stimulation of sympathetic nerves to intestinal smooth muscle hyperpolarizes the membrane, suppresses spontaneous prepotentials and abolishes action potentials. This effect is mediated by the action of norepinephrine on alpha receptors of smooth-muscle cells and is mimicked by circulating epinephrine.

Conduction of Excitation

Rhythmic changes in length and tension move along intestinal smooth muscle. Ability to propagate excitation is a property of the muscle itself, for propagation occurs in sheets or tubes of completely denervated muscle and in nerve-free embryonic tissue. Mechanical stretch of a resting cell by active ones is not essential for propagation; excitation passes over immobilized strips of muscle, and it jumps the cut between two separate but closely apposed rings of muscle that are not mechanically linked.

Conduction is the coordinated activity of many cells. It is fast in tissue like pig esophageal muscularis mucosae, which has closely packed cells, and slow in cat longitudinal muscle, whose extracellular space is relatively large. Velocity of conduction is about 10 times as fast in the longitudinal direction of muscle fibers as it is in the direction perpendicular to the fiber axis. The moving wave of activity is approximately 1 mm long in the longitudinal direction, the length of 5 or so muscle cells. The minimal diameter of a fiber bundle that can be excited to conduct is 0.1 mm, which includes 200–300 cells in cross-section. Thus the wave passing through the tissue involves at least 1,000 cells at a time.

Conduction of excitation within the bundle is by electrotonic spread of current. There is a graded potential along each active cell at the leading edge of spreading activity (Fig 2–7). The depolarized part of the membrane is a current sink. Current flows from a neighboring resting cell which, because it is still polarized, has a positively charged mem-

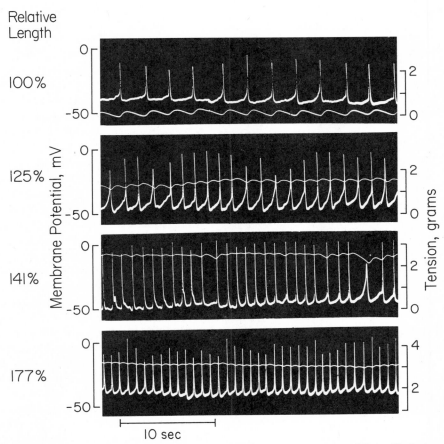

Fig 2–6. — Tension and action potentials of a strip of taenia coli as it is stretched from its resting length. (Adapted from Bulbring, E., and Kuriyama, H.: J. Physiol. 169:198, 1963.)

brane. The flow partially depolarizes the resting cell; and when the degree of depolarization reaches a critical value over a sufficient area of the cell membrane, abrupt depolarization of the resting cell occurs. The cell now becomes an active one, and it is a current sink for adjacent resting cells.

Because smooth-muscle cells are very small, there is a very large surface-to-volume ratio in the tissue. Current flowing across a depolarized area of one cell membrane spreads over a wide area of adjacent cell membranes. This means that the density of current flowing across one or only a few cells is not great enough to depolarize the critical area of the membrane of a resting cell. Converging electric fields from many cells are needed to depolarize a sufficient area of the cell membrane so that excitation and conduction can occur. This is the reason that bundles of cells are the conducting unit in intestinal smooth muscle.

If current flows from a polarized membrane to a current sink in an active cell, there must be a return flow of current to complete the electric circuit. Such flow would be greatly facilitated were there low-resistance pathways between adjacent cells. Such a pathway could be provided by gap junctions, or nexuses, between cells. These provide low-resistance pathways between cells.

Gap junctions between intestinal smooth-muscle cells are very labile; they are broken by stretch and they can re-form when condi-

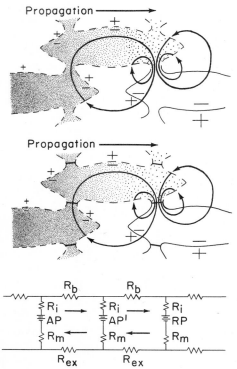

Fig 2–7.—Two models of electric relations between smooth-muscle cells. **Top,** cells connected by low-resistance gap junctions. **Middle,** the situation when cell membranes are closely apposed but are not fused to form low-resistance gap junctions. There is high resistance to current flow between cells. **Bottom,** the equivalent circuit. (Adapted from Barr, L.: J. Theoret. Biol. 4:73, 1963.)

tions return to normal. The process of preparing smooth muscle for electronmicroscopy often breaks gap junctions, and none may be seen in a section of tissue that had abundant gap junctions in life. Nevertheless, gap junctions may be scarce or entirely absent in some smooth-muscle tissues in which excitation is propagated from cell to cell. In that case, current must flow from the interior of one cell to the interior of its neighbor through high-resistance pathways.

Potential Changes: Slow Waves, or Electric Control Activity

An electrode placed in a cell of the longitudinal muscle of the stomach or small intestine records a conducted wave of depolarization distinct from the prepotentials or action potentials. The magnitude of the wave is 10–15 mV; its duration is 1–4 msec; and its frequency, which is characteristic of the tissue, is 3 per min in the human stomach, 5 per min in the dog stomach and 17–19 per min in the dog duodenum. Figure 2–4 shows these waves recorded from the rabbit duodenum; they are seen to occur with and without prepotentials and action potentials superimposed upon them. Bursts of action potentials, when they occur, are local and are followed within 0.5 sec by muscular contraction. Although the waves occur only in longitudinal muscle in the stomach and small intestine, contraction of circular muscle follows contraction of longitudinal muscle. Longitudinal and circular muscles are connected by two means: small bundles of muscle fibers pass from one layer to the other, and current spreads electrotonically from longitudinal to circular layer. In the latter case, partial depolarization of cells of the longitudinal layer forms a current sink, and the consequent flow of current from circular muscle cells partially depolarizes them. Because action potentials normally occur only during a period of depolarization, the frequency of the conducted wave of depolarization determines the frequency of contraction. The velocity of the conducted wave determines the velocity at which the wave of contraction travels over the tissue.

There is no general agreement about the name of the conducted wave of depolarization. It is frequently called the "slow wave," but other waves are slow as well. Because it determines the rhythm of contraction it has been called the Basic Electric Rhythm (BER) or Pacesetter Potential (PSP). Because it controls contraction, it is called Electric Control Activity (ECA) to distinguish it from action potentials, which are sometimes called Electric Response Activity. Until gastroenterologists agree, the student must be prepared to recognize all the names.

Slow waves may be caused by variations in the electrogenic sodium pump. They disappear when sodium is removed from the tissue or when ouabain, which inhibits sodium pumping, is added. In the interval between waves the membrane is partially permeable to sodium, and sodium leaks into the cells. At the same time, the sodium pump extrudes sodium at a basal rate. Net extrusion of sodium is greater than is inward leakage of sodium, and the action of the pump tends to hyperpolarize the cells. Thus, a fraction of the resting membrane potential is generated by active extrusion of sodium. For some unknown reason, the rate of sodium extrusion falls. This leaves the inward flux of sodium unbalanced; the membrane is partially depolarized, and the rising phase of the slow wave is created. Now the rate at which sodium leaks into the cell is greater than the rate at which it is extruded, and sodium ions carry current into the cells. Eventually, the pump returns to its basal level, and the membrane voltage, which lags behind the current flow on account of membrane capacitance, reaches a peak of depolarization before returning to its resting value.

Another explanation of the slow wave is that it begins with a calcium-dependent increase in sodium conductance. This partially depolarizes the membrane. The wave concludes with an increase in chloride conductance, which repolarizes the membrane. Other explanations are possible, and completely conclusive experiments have not yet been done.

When the slow waves are recorded using extracellular electrodes, they are the first derivative of the transmembrane potential changes. Slow waves recorded with extracellular electrodes are shown in Figure 2–8, and the figure also shows the relation between electric control activity, spiking and pressure development.

Local factors determine whether the positive deflection of the electric control activity is followed by action potentials and contraction. The most important of these is the activity of the local intramural plexuses. When many impulses carried in the plexuses liberate acetylcholine, the excitability of smooth-muscle cells is increased, and the cells respond with action potentials and contraction. Stimulation of the vagus nerve or administra-

Fig 2–8.—Electric activity and pressure recorded from duodenum of an unanesthetized dog. At an earlier operation, the duodenum was brought out through the abdominal wall and covered by a skin flap so that it formed a permanent, handle-like loop. Electric recording was done by leading from two pairs of needle electrodes; pressure was recorded from a transducer inserted by mouth to the site of the lower electrodes. **Left,** slow waves without spikes sweep over the duodenum at a frequency of 17 per min and with conduction velocity of 15 cm per sec. There are no spikes (the small deflections on the upper record are electrocardiogram), and intraluminal pressure is atmospheric. **Right,** spontaneous smooth-muscle spikes follow the positive phase of each slow wave; contraction and pressure increases at lower electrode coincide with spiking and are proportional to spike frequency. (Courtesy of P. Bass and C. F. Code.)

DOG DUODENUM, ELECTRODES 68 mm APART

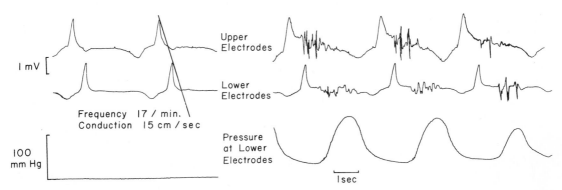

tion of cholinergic drugs also causes bursts of action potentials and contraction in phase with the electric control activity. Sympathetic stimulation, by depressing plexus reflexes, reduces spiking and contraction without affecting the electric control activity.

Groups of longitudinal muscle fibers are capable of generating the slow waves. The frequency with which they do so is characteristic of the group's anatomical position. The role of the electric control activity in governing gastrointestinal motility is discussed in subsequent chapters.

Cell Potentials and Tension; Mechanical Properties

Records in Figure 2–5 show each spike followed by an increase in tension. Over the short term, there is a direct relation between frequency of spiking and tension. When a muscle is stimulated electrically, chemically by acetylcholine or mechanically by stretch, spike discharges and increased tension follow. When the muscle is inhibited by epinephrine, spike frequency and tension both diminish.

The chemical reactions generating tension can be stimulated independently of electric changes at the membrane. If a smooth muscle is placed in isotonic potassium chloride solution, its membrane is depolarized, and it shows no conducted electric changes. Nevertheless such preparations contract when placed in electric fields of 5–25 volts per cm in the longitudinal direction of the fibers. The fields necessary are about 10 times those established by the action potential in a normally polarized muscle. Depolarized intestinal smooth muscle responds to stimulating drugs such as acetylcholine, histamine, 5-hydroxytryptamine and oxytocin, and to inhibiting drugs as well. On the other hand, if glucose is removed, smooth muscle cannot develop tension although electric activity at its membrane is normal. Over the long term, measured in minutes, the sustained tension of intestinal smooth muscle is more closely related to the membrane potential than to

spike frequency. When intestinal smooth muscle is stimulated through its parasympathetic nerve supply at frequencies greater than 10 per sec, the prepotentials fuse into steady depolarization of the membrane; no spikes occur, yet tension is maximal.

The link between membrane activity and tension development in smooth muscle may be similar in outline but different in details from that in striated muscle. During the smooth-muscle action potential, ionized calcium enters the cell, and the intracellular concentration of ionized calcium probably rises. The ionized calcium activates the actin-myosin-ATP system so that tension develops. In striated muscle, relaxation occurs when ionized calcium is sequestered by the membrane of the T-system or is extruded from the cell. The slowness with which calcium is sequestered may be responsible for the slow decay of tension in smooth muscle.

Mechanical properties of smooth muscle are qualitatively similar to those of striated muscle but quantitatively different. Resting intestinal smooth muscle is greatly elongated by stretch, but a large fraction of the elongation results from reorientation of cells and fasciculi to a direction parallel with the direction of stretch, and from stretch of connective tissue components. A smaller part results from elongation of muscle fibers themselves. The tension-length diagram of smooth muscle is similar in shape to that of striated muscle; when tension developed is plotted against initial length, a curve, concave downward, is obtained with maximal tension at resting length. Initial velocity of shortening with a 2-gm load is of the order of 0.4 cm per sec, whereas that of frog sartorius is about 4 cm per sec with the same load.

Response to Stretch and Release of Stretch

Four properties of smooth muscle permit the viscera to accommodate to changes in volume with minimal changes in transmural pressure: (1) elastic behavior of the muscle, (2) stress relaxation, (3) active but intrinsic changes in muscle tension and (4) reflexly mediated contraction and relaxation.

Pressure in a hollow organ is directly proportional to tension in its wall. For a sphere, which can serve as a model for the stomach, the relation is $P = 2T/r$, where T is the tension or total force in dynes per cm along a length of wall and r is the radius. For a cylinder such as the intestine, the relation between transmural pressure and tension in the circumferential direction is $P = T/r$. The relation among volume, pressure and tension is much more complex. As volume increases, the wall becomes thinner, and the number of tension-producing elements per unit of cross-sectional area of the wall diminishes. Consequently the number of elements exerting tension along the unit length of the wall also diminishes. Although tension along the length of the wall increases as the wall is stretched, the rise in tension, and therefore in transmural pressure, for an increment in volume is not so steep as would be expected if wall thickness remained constant. These relations among transmural pressure, tension in the wall and contained volume hold for any distensible structure, living or dead. In life, the tension in the wall of the digestive tract is maintained by active contraction of its muscle, and adjustments in

active tension are the means by which pressure and volume are regulated.

When smooth muscle is quickly stretched, its length and tension increase, and it obeys the law that increase in length is directly proportional to tension. However, if the muscle is held at constant length, its tension falls; or if it is held at constant tension, its length slowly increases (Fig 2–9). This slow response is stress relaxation; it is the result either of sliding of myofibrils past one another or of inhibition of active contraction. After an increase in volume has raised tension in the wall of the stomach or intestine, stress relaxation reduces tension, and transmural pressure falls toward the level prevailing before volume increased.

Reflexly regulated receptive relaxation is described in Chapter 3.

If tension is being maintained by active and regular spike discharge, stretch may inhibit spiking for about 30 sec, during which the muscle relaxes to accommodate itself to the new length. On release of tension, smooth muscle frequently contracts, so that tension is restored at shorter length (see Fig 2–6). On the other hand, stretch of intestinal muscle may be followed not by relaxation but by a burst of spikes and increased tension (see Fig 2–6).

Fig 2–9. — Response of smooth muscle to stretch and release of stretch.

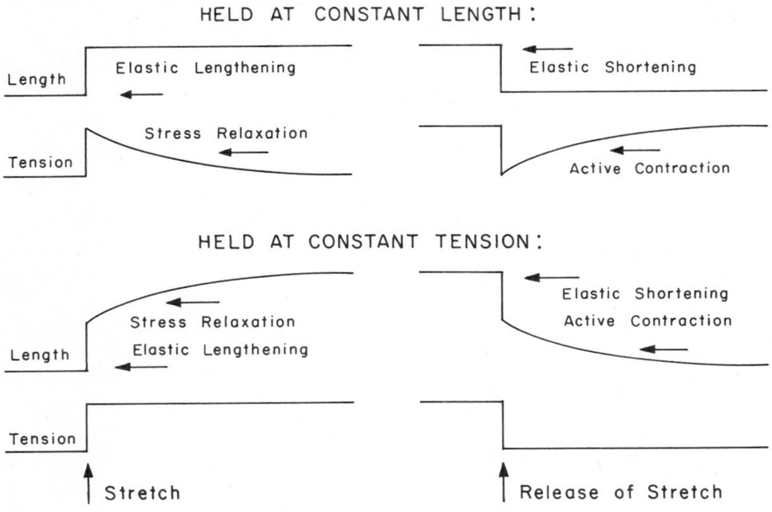

REFERENCES

Bennett, M. R., and Burnstock, G.: Electrophysiology of the innervation of intestinal smooth muscle, in Code, C. F. (ed.): *Handbook of Physiology:* Sec. 6. *Alimentary Canal,* Vol. IV (Washington, D.C.: American Physiological Society, 1968), pp. 1709–1732.

Bennett, A., and Stockley, H. L.: The intrinsic innervation of the human alimentary tract and its relation to function, Gut 16:443, 1975.

Bortoff, A.: Digestion: Motility, Ann. Rev. Physiol. 34:261, 1972.

Bulbring, E., Brading, A. F., Jones, A. W., and Tomita, T. (eds.): *Smooth Muscle* (Baltimore: Williams & Wilkins Co., 1970).

Burnstock, G.: Evolution of the autonomic innervation of visceral and cardiovascular systems in vertebrates, Pharmacol. Rev. 21:247, 1969.

Burnstock, G., and Costa, M.: Inhibitory innervation of the gut, Gastroenterology 64:141, 1973.

Connor, J. A., Prosser, C. L., and Weems, W. A.: A study of pace-maker activity in intestinal smooth muscle, J. Physiol. 240:671, 1974.

Furness, J. B., and Costa, M.: The adrenergic innervation of the gastrointestinal tract, Ergeb. Physiol. 69:1, 1974.

Gabella, G.: Fine structure of the myenteric plexus in the guinea-pig ileum, J. Anat. 11:69, 1972.

Goodford, P. J.: Distribution and exchanges of electrolytes in intestinal smooth muscle, in Code, C. F. (ed.): *Handbook of Physiology:* Sec. 6. *Alimentary Canal,* Vol. IV (Washington, D.C.: American Physiological Society, 1968), pp. 1743–1766.

Hillarp, N.-A.: Peripheral autonomic mechanisms, in Field, J. (ed.): *Handbook of Neurophysiology,* Vol. II (Washington, D.C.: American Physiological Society, 1960), pp. 979–1006.

Holman, M. E.: An introduction to electrophysiology of visceral smooth muscle, in Code, C. F. (ed.): *Handbook of Physiology:* Sec. 6. *Alimentary Canal,* Vol. IV (Washington, D.C.: American Physiological Society, 1968), pp. 1665–1708.

Kuriyama, H.: Ionic basis of smooth muscle action potentials in Code, C. F. (ed.): *Handbook of Physiology:* Sec. 6. *Alimentary Canal,* Vol. IV (Washington, D.C.: American Physiological Society, 1968), pp. 1767–1792.

Prosser, C. L.: Smooth muscle, Ann. Rev. 36:503, 1974.

Ruegg, J. C.: Smooth muscle tone, Physiol. Rev. 51:201, 1971.

Sarna, S. K.: Gastrointestinal electrical activity: Terminology, Gastroenterology 68:1631, 1975.

Schofield, G. C.: Anatomy of muscular and neural tissue in the alimentary canal, in Code, C. F. (ed.): *Handbook of Physiology:* Sec. 6. *Alimentary Canal,* Vol. IV (Washington, D.C.: American Physiological Society, 1968), pp. 1579–1628.

Smith, B.: Disorders of the myenteric plexus, Gut 11:271, 1970.

Wood, J. D.: Neurophysiology of Auerbach's plexus and control of intestinal motility, Physiol. Rev. 55:307, 1975.

3

Gastric Motility

THE EMPTY STOMACH is small and relaxed; feeble contractions passing over its muscle usually have negligible effects on intragastric pressure. When a man is hungry, vigorous contractions, whose frequency and rate of progress are governed by a basic electric rhythm originating in a pacemaker high on the greater curvature, sweep over the antrum at the frequency of 3 per min. Occasionally, pressure in the empty stomach rises briefly. As the stomach fills, its fundus and body relax to accommodate its contents, and pressure increase in the stomach is small. Food swallowed tends to form layers in the body of the stomach, and gastric contents are well mixed with gastric secretions only in the antrum. Immediately after the stomach is filled, its peristaltic waves diminish in force; they gradually become stronger, the more quickly the less fat in the meal. As they engage the strong muscle of the antrum, they may deepen sufficiently to mix the stomach's contents and to propel small portions into the duodenum. Emptying is effected by these waves in the antrum and pylorus, and the rate of emptying is chiefly regulated by factors controlling the vigor of antral contraction. In addition to adventitious influences such as pain, nausea and emotional disturbances, the major ones affecting gastric motility are the composition and volume of gastric contents delivered to the duodenum. These, operating through nervous and humoral pathways, regulate the rate of emptying so that delivery of chyme is adjusted to duodenal motility and emptying.

Structure and Innervation of the Stomach

The stomach acts as a unit because the longitudinal and circular muscles of its wall are continuous sheets. The excitatory wave, which can be recorded as the basic electric rhythm of the stomach, begins high on the greater curvature and moves over the longitudinal layer to the pylorus, setting the rhythm of gastric contraction. Its velocity, which is low in the body and high in the terminal antral segment, determines the rate at which gastric peristaltic waves move. The strength of contraction is governed by both anatomical and physiological characteristics of the muscle. Contractions are usually weak in the body, whose muscle is thin, and strong in the antrum, whose muscle is thick. The myenteric and submucous plexuses are continuous throughout the stomach, and in some unknown way they regulate the response of the muscle to excitation. The extrinsic innervation, sympathetic via the celiac plexus and parasympathetic via the vagus, is distributed to the whole of the stomach. In man, the effect of sympathetic stimulation is to depress all movement of the stomach. Stimulation of some fibers in the vagus nerves enhances gastric motility, and stimulation of

41

other fibers inhibits it. Both sets of nerves have small effects upon the frequency and rate of progress of the basic electric rhythm, and both, acting through the intrinsic plexuses, reduce or enhance contraction of gastric muscle.

There is no functional division between antrum and pylorus; the pylorus is simply the last part of the antral muscle. In particular, the pylorus and the rest of the stomach are not affected in opposite fashions: nervous influences, mediated through either branch of the autonomic nervous system, have the same effects on the pylorus as on the rest of the stomach muscle.

There is a partial discontinuity between stomach and duodenum. At the pylorus, the circular muscle of the antrum ends in a thick ring, which is anatomically and physiologically identified as the pyloric sphincter. Beyond the muscle, there is a thick ring of connective tissue, the hypomuscular segment, which almost completely separates the stomach and the duodenum. The longitudinal muscle layer dips down into this barrier, and a few bundles pass through it to the duodenum. The myenteric plexus of the stomach is continuous with that of the duodenum. The basic electric rhythm of the stomach can be recorded 2–3 mm beyond the pylorus, but the rhythm of the more distal duodenum is independent of it. Duodenal contractions are influenced by gastric peristalsis, for about 85% of duodenal contractions occur when the gastric terminal antral segment is contracting. This rough coordination is mediated by the few longitudinal fibers passing the barrier, by the plexuses and by the mechanical effect of chyme entering the duodenum. The connective tissue ring at the pylorus extends into the submucosa, and there is no connection between the submucous lymphatic systems of stomach and duodenum.

Types of Movements of the Stomach

The volume of the chamber of the stomach of a fasting man is 50 ml or less. There is little or no tension in its wall, and intraluminal pressure is equal to intra-abdominal pressure. No movement of the muscle of the body can be seen by x-ray examination. When either a small balloon or an open-tip manometer is placed in the antrum, pressure changes reflecting contraction of the antral muscle at the rate of 3 per min are recorded (Fig. 3–1). Although the sequence may appear to be nonrhythmic, waves are separated by some multiple of 20–22 sec.

Pressure waves in the digestive tract are empirically classified by size:

Fig 3–1.—Sequence of large and small waves passing over the antrum of a normal fasting human subject. The record was made using a small balloon. Vertical lines mark minute intervals. The waves occur at intervals which are multiples of 21 sec. (Courtesy of C. F. Code.)

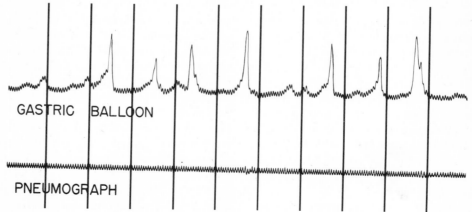

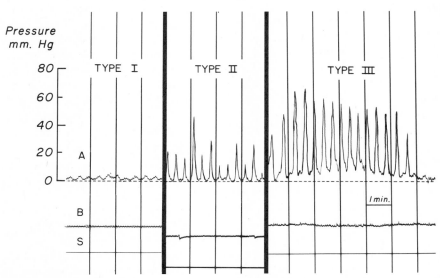

Fig 3–2.—Empiric classification of three types of pressure waves recorded from the digestive tract. Trace *A*, pressures in mm Hg; trace *B*, pneumograph; trace *S*, base line. Vertical lines are at minute intervals. (From Code, C. F., *et al.*: Am. J. Med. 13:328, 1952.)

Type I waves (Fig 3–2) are simple monophasic waves of low amplitude, 3–10 mm Hg, lasting 5–20 sec. Their rhythm is characteristic of that part of the bowel from which they are recorded. The waves may or may not be caused by progressive movement in the wall; independent evidence, such as x-ray observations or recording from two adjacent points, is necessary to establish this. In the stomach, the contraction that produces type I waves is always progressive; it may or may not be progressive in the small intestine.

Type II waves are simple ones of greater amplitude than type I, 8–40 mm Hg, lasting 12–60 sec. In the stomach their rhythm is the same as type I, but in the colon they are slower, about 2 per min.

Type III are complex waves whose most important characteristic is a rise in base-line pressure, which lasts for a few seconds to some minutes. The rise is less than 15 mm Hg in the stomach and about 20–30 mm Hg in the small intestine. Type III waves occur 10–30% of the time in the antrum of a normal man after an overnight fast, and each wave lasts about 1 min.

Waves are recorded because contraction of the viscus displaces fluid into the open-tip manometer or deforms the balloon. If the open-tip or the small balloon is pulled back from the antrum into the body of the stomach, the only pressure variations it transmits are those of the abdomen coincident with respiration; although strong waves are passing over the antrum, they have little effect on general intragastric pressure.

Movements of the Empty Stomach

As the stomach empties, its contractions die out; and in the interval before hunger appears, shallow type I waves pass over the antrum about 25% of the time. Type II waves occur about 15% of the time, and only rarely (less than 1%) does the antral muscle contract enough to raise the base-line pressure above intra-abdominal pressure for a minute or more. The remaining 60% of the time, the stomach is quiet. As the fast proceeds, transitory increases in activity appear. The feeble type I waves grow into strong type IIs, and these become more frequent, so that 50% of the time they are sweeping the antrum. There may be incomplete relaxation between waves, and base-line pressure rises over 1–2 min. The stomach relaxes, and then another cycle of strong peristaltic contractions begins. A person often experiences the pangs of hunger during one of these strong contractions.

The nature of the sensations of hunger, appetite and satiety is a problem of physiologic psychology too esoteric to be discussed here.* Many persons, when hungry, feel intermittent pangs or sharp gnawing sensations referred to the lower midchest or epigastrium. Afferent impulses from the stomach signal tension in its wall; impulses in these fibers could be the input to the central nervous system arousing the sensation. However, the correlation between contractions of the stomach and presence or intensity of the hunger sensation is very poor, and we may reject the naive equation of gastric contractions with hunger pangs.

Increased gastric motility during fasting is aroused by impulses carried over the vagus nerves, which make the gastric muscle more responsive to its own basic electric rhythm. The signal for the nervous response is probably a fall in the rate of glucose utilization. Contractions appear in fasting men or animals when the arteriovenous (A-V) difference in glucose concentration is small. Increase of the A-V difference, as by injection of glucagon, stops contractions. This effect is mediated through the satiety center in the medial hypothalamus. When glucagon is given to cats having electrodes implanted in this center, electric activity is found to increase with increasing A-V difference. If the center is destroyed, either by electrolytic lesions or by administration of gold thioglucose, glucagon no longer abolishes gastric motility. Animals with the center destroyed are hyperphagic, and if allowed unlimited access to food, they become enormously obese. The absolute level of blood glucose is not

*The subject is treated authoritatively and at length in Code, C. F. (ed.): *Handbook of Physiology: Sec. 6. Alimentary Canal,* Vol. I (Washington, D.C.: American Physiological Society, 1967). See also Tepperman, J.: *Metabolic and Endocrine Physiology* (3d ed.; Chicago: Year Book Medical Publishers, Inc., 1973), pp. 199–224.

important in regulating gastric contractions. Diabetic subjects with a high blood glucose level but low A-V difference have typical contractions. In normal subjects, elevation of blood glucose content by intravenous glucose injection, unless it is also accompanied by increased A-V difference, does not inhibit gastric contractions. Injection of insulin into dogs and rats, but not in man, augments gastric contractions.

Pressure in the Full Stomach

The body of the stomach relaxes with each swallow, and food is accommodated with little rise in intragastric pressure. The efferent pathway is the vagus nerve, and the fibers responsible are nonadrenergic, noncholinergic.

Figure 3–3 shows the pressure recorded in a vagally innervated pouch of the fundus of a dog's stomach, the fundus being defined as that portion of the stomach lying orad to the entrance of the esophagus. In this instance, when the dog swallowed, the fundus relaxed along with the lower esophageal sphincter, but there was no change in antral motility. Such receptive relaxation, which was observed in 26 of 40 trials, provides some additional volume for swallowed food; on eating, the volume of the body increases as well. The drinking of 225 ml of water raises pressure in the human stomach less than 2 mm Hg. When air is rapidly injected into the human stomach by way of a tube, the pressure rises by 4–7 mm Hg for the first 200–600 cc; after this, there is no further rise in pressure until a volume of 1,600 cc is reached. Intra-abdominal pressure rises at the same time, so that gastric transmural pressure remains approximately constant. Tension in the stomach wall must rise, for in a hollow organ at constant transmural pressure, tension in the wall is directly proportional to the radius.

Afferent impulses from the mouth and esophagus are relayed through the hypothalamus to the vagal nucleus. Sham feeding and distention of the esophagus cause relaxation of the body of the stomach. Afferent im-

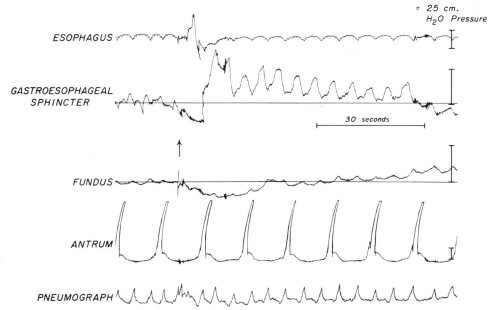

ESOPHAGUS

GASTROESOPHAGEAL
SPHINCTER

= 25 cm.
H₂O Pressure

30 seconds

FUNDUS

ANTRUM

PNEUMOGRAPH

Fig 3–3.—Simultaneous records made from a dog's lower esophagus, lower esophageal sphincter, a vagally innervated pouch of the fundus of the stomach and the gastric antrum. At the arrow, the dog swallowed; 5 sec later the peristaltic wave reached the lower esophagus. The lower esophageal sphincter relaxed before the peristaltic wave reached it, and it subsequently closed at high pressure for a long period. The fundus of the stomach relaxed as soon as the swallow began and remained relaxed for 25 sec. Antral contractions did not change. (From Lind, J. F., et al.: Am. J. Physiol. 201:197, 1961.)

pulses from the heart and chemoreceptor trigger zone (see Chapter 6) also result in nonadrenergic, noncholinergic impulses in vagal fibers to cause the relaxation of the stomach that precedes vomiting. Distention of the stomach itself, either body or antrum, stimulates slowly adapting tension receptors, and afferent impulses from these receptors traveling centrally in the vagus nerve are reflected in efferent vagal impulses, which cause gastric relaxation. However, the stomach may also respond to distention with contractions. In an experiment on a fistulous human subject whose stomach was distended by a balloon at a pressure of 8 mm Hg, a small increase in volume sufficient to double intragastric pressure was followed 2 of 10 times by strong contraction.

The abdominal contents are relatively mobile, and their specific gravity is approximately that of water. The physical result is that the viscera apply hydrostatic pressure to the serosal surface of the stomach, proportional to the vertical height of the viscera. The contents of the stomach, with the exception of the gas bubble, have nearly the same specific gravity as the viscera. Although the gastric contents are acted upon by gravity pulling them downward, this force is balanced by an equal and opposite force of the abdominal viscera applied to the serosal surface of the stomach, pushing them up. The stomach behaves like a water-filled bag suspended in water. Gravity can affect only contents whose specific gravity differs from that of the surrounding viscera.* The gas bubble rises, and dense materials sink. Because the density of most food is close to that of water, the contents of the stomach resemble a thick

---

*The same physical principle applies to the esophagus, but the specific gravity of the thoracic contents averages about half that of water. Consequently, the force of gravity does help to move material through the esophagus.

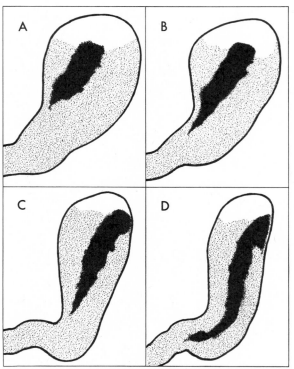

Fig 3–4.—Tracings of x-ray photographs of the stomach of a young man. Immediately before **A,** he ate 200 gm of fried meat balls, then 50 gm of meat balls mixed with powdered barium sulfate and finally another 100 gm of meat balls without contrast medium. **B,** 10 min later. **C,** 35 min later. **D,** 64 min later. The layers remain stratified in the body of the stomach, but mixing occurs in the antrum. (Adapted from Nielsen, N. A., and Christiansen, H.: Acta Radiol. 13:678, 1932.)

Irish stew with small lumps of meat and vegetables submerged in gravy. However, the density of the common x-ray contrast medium, barium sulfate, is 4.50, and it rapidly sinks to the lowest part of the stomach unless it is finely dispersed as a stable suspension in water. Viscous food forms layers in the stomach, stratified in the order in which it was swallowed. The strata are sometimes vertical and sometimes oblique. Figure 3–4 shows that the relative position of two layers of fried meat balls remained constant in the stomach of a Swedish medical student for 65 min, although considerable emptying occurred. Solid lumps stay in the body of the stomach; a relatively large lump entering the antrum is squirted back by the retropulsive action of antral peristalsis, and small lumps are ground by the terminal antral contrac-

tion. Fluid moves more quickly into the antrum and, when swallowed on a stomach containing solids, seems to pass around the solid mass (Fig 3–5). Acid and pepsin secreted by the mucosa penetrate the mass slowly; but, judging from the low pH of antral contents, secretions readily reach the antrum.

Movements of the Full Stomach

Three kinds of movements occur in the full stomach: peristaltic waves, systolic contraction of the terminal antrum and diminution of size of the fundus and body. A peristaltic wave follows the basic electric rhythm of the stomach, which begins at a pacemaker in the longitudinal muscle high on the greater curvature and moves over the longitudinal mus-

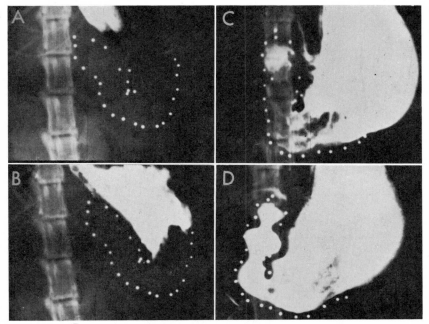

Fig 3–5.—X-ray photographs of the stomach of a cat. The pyloric portion and antrum are outlined by lead shot embedded in the gastric wall. The cat was given 5 gm of meat, then 30 ml of barium sulfate suspension. **A,** the fluid enters the stomach. **B,** the fluid fills the body. **C,** the fluid passes around the meat. **D,** the fluid completely surrounds the meat. Observe four peristaltic waves simultaneously passing over the antrum in sequence. The duodenal cap is outlined by a wisp of barium, giving an arched, horizontal shadow, but the pyloric canal is empty. (From Gianturco, C.: Am. J. Roentgenol. 31:735, 1931.)

cle of the body, antrum and pylorus. In man, the frequency of gastric peristalsis is 3 per min, and the frequency and rate of progress are affected by nervous and humoral factors, to be described below. The electric wave consists of a moving band of partial depolarization of the smooth-muscle cells of the longitudinal layer (see Fig 2–4). When recorded by means of extracellular electrodes, it appears as a diphasic or triphasic complex that is the derivative of the transmembrane potential variations (see Fig 2–6). The wave moves over the muscle of the body of the stomach at less than 1 cm per sec, more rapidly over the greater than over the lesser curvature. It lasts about 1.5 sec at any one point and consequently occupies about 1–2 cm of muscle. The wave accelerates in the antrum to 3–4 cm per sec.

Excitation spreads from the longitudinal to the circular muscle, and the latter contracts with variable intensity. Consequently, a circular band of contraction, about 2 cm wide, moves as a peristaltic wave over the body and the antrum, following the electric wave. On account of the relation between the size of the stomach and the frequency and rate of peristalsis, two or three waves are usually present at any one time (see Fig 3–5). The waves of contraction are no more than shallow ripples as they pass over the body, and during the first hour of digestion they usually remain shallow as they pass over the antrum. Later in digestion, a wave deepens as it nears the angle, and it begins to push gastric contents ahead of it. As the wave moves over the antrum, the lumen is constricted but not obliterated. The wave of electric excitation accelerates as it passes over the terminal antral segment (a length of

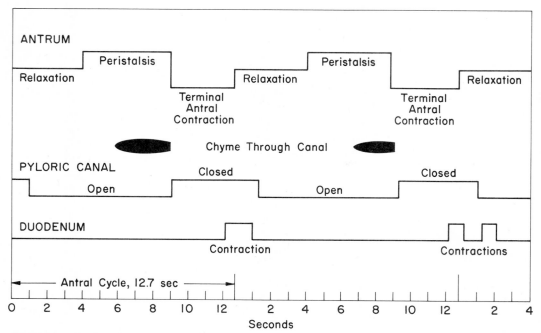

Fig 3–6.—Cycle of events occurring in the antrum, pyloric canal and duodenum of a dog during digestion of a meal. The representative cycles were derived from study of many individual ones, such as that shown in Figure 3–7. The designation of "open" or "closed" applied to the pyloric sphincter refers not to force of contraction, which was not measured, but to whether the canal could be seen to be patent, i.e., to contain barium-impregnated chyme. (From Carlson, H. C.: Ph.D. Thesis, University of Minnesota, 1962.)

about 5 cm in man), so that the terminal antral segment and the pylorus, which is the last part of the terminal antrum, contract almost simultaneously. This is the systolic contraction of the terminal antral segment. Because the pylorus is narrow, it closes early (Fig 3–6).

Hypoglycemia induced by insulin administration speeds the rate of propagation of the wave of excitation over the antrum, and it deepens the peristaltic wave by increasing the frequency of action potentials. Vagotomy completely abolishes this effect of insulin hypoglycemia. Distention of the stomach by filling it with water slows the rate of propagation but increases the frequency of action potentials. Vagotomy does not change the consequences of distention. Administration of cottonseed oil not only slows the rate of propagation, but it reduces the strength of peristaltic contraction by reducing the incidence of action potentials.

Gastric contents are viscous; and as they are pushed into the terminal antrum by the peristaltic wave, pressure in the antrum rises. About 60% of the time, the rise in pressure is enough to overcome the pressure barrier at the pylorus, and a small fraction of antral contents passes into the duodenum. Then as the pylorus contracts, passage of chyme into the duodenum is abruptly stopped. Contraction of the terminal antral segment continues, and pressure within it rises by about 10–25 mm Hg. The contents of the segment, prevented from passing through the pylorus, are forcibly squirted by the terminal antral contraction back through the patent lumen into the proximal antrum. This retropulsion of chyme is highly effective in mixing food and digestive juices. Then the terminal antral segment and pylorus relax, and both remain relaxed until the next peristaltic wave comes down the antrum.

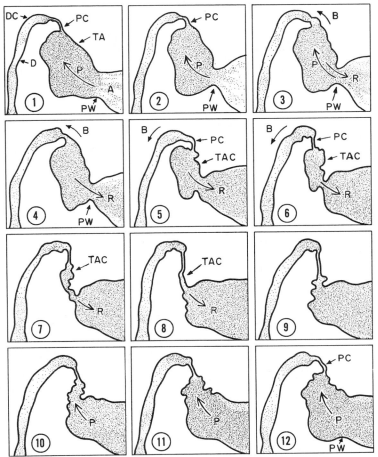

Fig 3–7.—Tracings made from a cineradiographic study of the motions of a dog's antrum during digestion of a meal made opaque with barium sulfate. Pictures were taken at 15 frames per sec, and every 15th frame was traced; consequently, each picture represents the shape of the antrum 1 sec later than the preceding one. **1,** a peristaltic wave of type II intensity *(PW)* indents the antrum. The barium mixture has been propelled *(P)* into the terminal antrum *(TA),* leaving only a small amount of barium along the folds of the antral mucosa *(A).* The pyloric canal *(PC)* is empty. Some barium is in the duodenal cap *(DC)* and duodenum *(D).* **2,** the pyloric canal *(PC)* is open but nearly empty, and the peristaltic wave *(PW)* has moved farther along the terminal antrum. **3,** the peristaltic wave *(PW)* is still farther along. The pyloric canal is open, and a small bolus of chyme *(B)* is in it. The largest part of the barium has begun to move backward *(R)* into the antrum. **4,** the bolus *(B)* continues to move through the open pyloric canal, and more of the content of the terminal antrum returns to the antrum *(R).* **5,** the pyloric canal *(PC)* begins to close and is nearly empty. The terminal antrum begins a systolic contraction *(TAC),* which increases retropulsion of its contents. **6,** contraction of the terminal antrum continues *(TAC)* with more retropulsion *(R),* and some chyme moves down from the duodenal cap *(B).* **7,** and **8,** the terminal antral contraction is almost complete, and most of the pyloric canal is closed. **9,** terminal antral contraction is complete. **10,** the terminal antrum begins to fill again as the result of the propulsive force of another peristaltic wave in the antrum. **11,** filling of the terminal antrum continues. **12,** the terminal antrum is full, and the pyloric canal *(PC)* begins to open. The whole cycle has taken 12 sec. (From Carlson, H. C.: Ph.D. Thesis, University of Minnesota, 1962.)

The fraction of gastric contents passing through the pylorus at each cycle is small. The rate of emptying is a function of the volume of gastric contents, and the maximal rate for a 750-ml liquid meal is 20 ml per min. With three type II waves per min, this is 7 ml per wave, or 1% of gastric contents. When the stomach contains a smaller volume, about 1–2 ml is emptied per wave.

Sounds above 500 cps are produced by passage of chyme through the pylorus. They do not occur during the 3–4 sec preceding pyloric contraction, but they begin and continue during the 4–5 sec the pylorus is contracting. They are loudest when contraction is most vigorous.

As the stomach empties, its volume becomes smaller. Intragastric pressure does not rise, and this means that tension in the wall of the stomach must diminish with volume and radius of curvature.

Control of Gastric Motility in Man

The foregoing description of gastric motility is based on studies of the dog in which cineradiography can be combined with electric recording. Observations on man are necessarily more limited. Electric recording from the mucosal surface of the stomach is made by means of electrodes carried at the end of a nasogastric tube. Short-term electric records are obtained from the serosal surface of the stomach by means of electrodes placed during peritoneoscopy or laparotomy. In some instances, electrodes have been sewn to the surface of the antrum and duodenum during cholecystectomy; they have been left in place for 6 days, to be removed at the same time that the T tube draining the bile duct is taken out. Observations made by these means show that there are minor quantitative differences between gastric motility in dog and in man.

In man, the frequency of the wave of excitation passing over the antrum ranges from 2.1–3.7 per min with a mean of about 3.1. In any one person at rest, the frequency varies by no more than 10%. Conduction velocity over the antrum is about 0.5 cm/sec with a range of 0.3–1.4. This means that the wave of excitation is about 10 cm long. In man as in the dog, the velocity of propagation increases by about 4 times at the terminal antrum, and that segment contracts as a unit. The frequency found in the duodenum 10–12 cm from the pylorus is 12 per min, but in the more proximal duodenum the gastric rhythm of 3 per min is superimposed on an occasional rhythm of 12 per min.

Intravenous infusion of gastrin increases the frequency to 3.8 per min. Infusion of secretin may abolish electric activity in the longitudinal muscle and reduce motor activity to zero. These are probably pharmacologic effects.

In a group of patients scheduled for vagotomy, regular antral electric activity was found 96% of the time. Immediately after truncal vagotomy it was present only 30% of the time, and after selective vagotomy 38% of the time. After highly selective vagotomy, the regular antral frequency was present in 74% of the records. When such subjects were restudied several weeks later, normal rhythm was found 98% of the time. When antral electric activity is abolished or disorganized by vagotomy, gastric retention may be complete. Consequently, retention is more frequent in patients with truncal vagotomy, and this is the reason surgeons combine pyloroplasty with that operation. In patients with all types of vagotomy, normal rhythm is usually restored as time passes.

The Gastroduodenal Junction

The pylorus divides the gastroduodenal junction into a proximal part composed of the terminal antrum and a distal part composed of the duodenal bulb and the first part of the duodenum proper. The terminal antrum and the pylorus are part of the gastric muscle, and they form the peristaltic pump and funnel through which gastric contents are emptied into the duodenum. The duodenal cap, the first 2–5 cm of the duodenum, is a nearly spherical or ovoid chamber having a capacity of 5–10 ml, which receives the chyme as it is driven through the pylorus. Beyond the duodenal cap, the postbulbar part of the duodenum is a relatively narrow tube whose mucosa is thrown into interdigitating circular folds.

When an infused catheter is withdrawn from duodenum to stomach, a zone of high pressure is found between the two organs. This zone, which averages 1.5 cm in length, has a pressure 5 mm Hg above intra-abdominal pressure. It is a physiological sphincter dividing antrum and duodenum. A brief decrease in pressure in the sphincter occurs when antral peristalsis reaches the terminal antral segment, and at that time contents of the antrum flow through the gastroduodenal junction into the duodenal bulb. Then pressure rises briefly by about 40 mm Hg as the terminal antral segment contracts vigorously. When the duodenum is acidified with 0.1 N HCl, sphincter pressure rises to 25 mm Hg, and there is no regurgitation into the stomach. Olive oil, amino acid solutions and hypertonic glucose solutions in the duodenum also cause an increase in sphincter pressure.

In normal persons, there is little or no regurgitation through the gastroduodenal junction. The resting pressure in the junction in patients with gastric ulcers is normal, but the pressure fails to rise when the duodenum is acidified or irrigated with olive oil or amino acid solutions. In these patients, infusion of cholecystokinin or secretin does not increase sphincter pressure, as it does in normal persons. Regurgitation of duodenal contents is frequent in patients with gastric ulcers, and the high and sustained concentration of bile acids and lysolecithin in the stomach may contribute to their mucosal damage.

In man, the antral peristaltic pump empties chyme into the duodenum in spurts at the gastric frequency of 3 per min. Contractions of the postbulbar duodenum are governed by the duodenal basic electric rhythm. This has a frequency of about 11 per min in man. The duodenal cap lies anatomically and physiologically between two structures, which contract at different frequencies. It is filled and

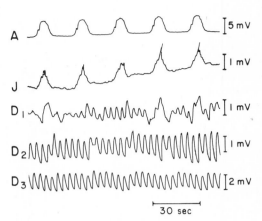

Fig 3–8.—Monophasic potentials recorded simultaneously from points on the antrum, junction and proximal duodenum of a rhesus monkey in situ. The antral electrode, *A,* was 14 mm orad to the junction; *J* was on the junction; D_1, D_2 and D_3 were 4, 10 and 17 mm aboral to the junction, respectively. The antral augmentation of duodenal slow waves is evident as far as 17 mm from the junction. (From Bortoff, A., and Davis, R. S.: Am. J. Physiol. 215:889, 1968.)

distended following antral contraction, and chyme spills from it into the postbulbar part of the duodenum. The cap itself contracts in an apparently irregular, eccentric manner, thereby displacing its contents into the postbulbar duodenum. The cap seldom empties completely.

Two mechanisms are responsible for the fact that 85% of proximal duodenal contractions occur just after the gastric peristaltic wave has delivered chyme to the cap. The first is simply the response of the duodenum to distention. The second is the conduction of the gastric basic electric rhythm through the pylorus into the duodenum (Fig 3–8). Gastric longitudinal muscle bands, particularly those on the lesser curvature, are continuous with those of the duodenal cap. The potential changes occurring at the gastric frequency mix with those occurring at the duodenal frequency for many millimeters beyond the pylorus, augmenting depolarization of duodenal muscle and therefore increasing the likelihood of excitation of duodenal circular muscle.

Control of Gastric Emptying

The rate at which the stomach empties is governed chiefly by the influence of chemical and physical properties of chyme in the duodenum. If chyme is drained from the duodenum through a fistula close to the pylorus, the stomach empties far more rapidly than if chyme is allowed to remain in the duodenum. Reduction in gastric motility mediated through the duodenum ranges from slight depression to complete cessation of all contraction. The terminal antrum, as an integral part of gastric muscle, is affected the

Fig 3–9.—Effect of fat in the duodenum on antral and duodenal motility of a normal dog. The top record shows type I and type II waves in the antrum as measured by a small balloon; there are vigorous contractions in the duodenum similarly recorded. Five minutes after olive oil has been placed in the duodenum, the only pressure changes recorded by the antral balloon are those corresponding to respiratory movements; duodenal contractions are less vigorous. Twenty minutes after the olive oil was placed in the duodenum, both antrum and duodenum are quiet. (Courtesy of C. F. Code.)

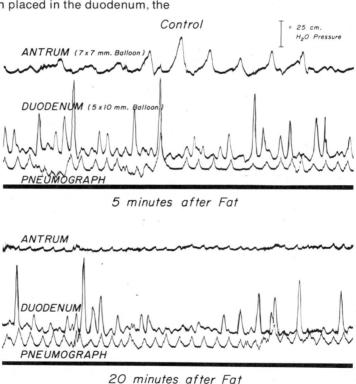

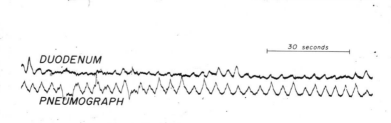

capacity of 5–10 ml, which receives the chyme as it is driven through the pylorus. Beyond the duodenal cap, the postbulbar part of the duodenum is a relatively narrow tube whose mucosa is thrown into interdigitating circular folds.

When an infused catheter is withdrawn from duodenum to stomach, a zone of high pressure is found between the two organs. This zone, which averages 1.5 cm in length, has a pressure 5 mm Hg above intra-abdominal pressure. It is a physiological sphincter dividing antrum and duodenum. A brief decrease in pressure in the sphincter occurs when antral peristalsis reaches the terminal antral segment, and at that time contents of the antrum flow through the gastroduodenal junction into the duodenal bulb. Then pressure rises briefly by about 40 mm Hg as the terminal antral segment contracts vigorously. When the duodenum is acidified with 0.1 N HCl, sphincter pressure rises to 25 mm Hg, and there is no regurgitation into the stomach. Olive oil, amino acid solutions and hypertonic glucose solutions in the duodenum also cause an increase in sphincter pressure.

In normal persons, there is little or no regurgitation through the gastroduodenal junction. The resting pressure in the junction in patients with gastric ulcers is normal, but the pressure fails to rise when the duodenum is acidified or irrigated with olive oil or amino acid solutions. In these patients, infusion of cholecystokinin or secretin does not increase sphincter pressure, as it does in normal persons. Regurgitation of duodenal contents is frequent in patients with gastric ulcers, and the high and sustained concentration of bile acids and lysolecithin in the stomach may contribute to their mucosal damage.

In man, the antral peristaltic pump empties chyme into the duodenum in spurts at the gastric frequency of 3 per min. Contractions of the postbulbar duodenum are governed by the duodenal basic electric rhythm. This has a frequency of about 11 per min in man. The duodenal cap lies anatomically and physiologically between two structures, which contract at different frequencies. It is filled and

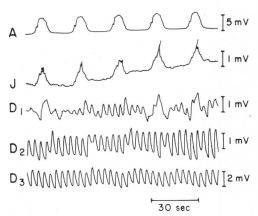

Fig 3–8.—Monophasic potentials recorded simultaneously from points on the antrum, junction and proximal duodenum of a rhesus monkey in situ. The antral electrode, *A*, was 14 mm orad to the junction; *J* was on the junction; D_1, D_2 and D_3 were 4, 10 and 17 mm aboral to the junction, respectively. The antral augmentation of duodenal slow waves is evident as far as 17 mm from the junction. (From Bortoff, A., and Davis, R. S.: Am. J. Physiol. 215:889, 1968.)

distended following antral contraction, and chyme spills from it into the postbulbar part of the duodenum. The cap itself contracts in an apparently irregular, eccentric manner, thereby displacing its contents into the postbulbar duodenum. The cap seldom empties completely.

Two mechanisms are responsible for the fact that 85% of proximal duodenal contractions occur just after the gastric peristaltic wave has delivered chyme to the cap. The first is simply the response of the duodenum to distention. The second is the conduction of the gastric basic electric rhythm through the pylorus into the duodenum (Fig 3–8). Gastric longitudinal muscle bands, particularly those on the lesser curvature, are continuous with those of the duodenal cap. The potential changes occurring at the gastric frequency mix with those occurring at the duodenal frequency for many millimeters beyond the pylorus, augmenting depolarization of duodenal muscle and therefore increasing the likelihood of excitation of duodenal circular muscle.

Control of Gastric Emptying

The rate at which the stomach empties is governed chiefly by the influence of chemical and physical properties of chyme in the duodenum. If chyme is drained from the duodenum through a fistula close to the pylorus, the stomach empties far more rapidly than if chyme is allowed to remain in the duodenum. Reduction in gastric motility mediated through the duodenum ranges from slight depression to complete cessation of all contraction. The terminal antrum, as an integral part of gastric muscle, is affected the

Fig 3–9.—Effect of fat in the duodenum on antral and duodenal motility of a normal dog. The top record shows type I and type II waves in the antrum as measured by a small balloon; there are vigorous contractions in the duodenum similarly recorded. Five minutes after olive oil has been placed in the duodenum, the only pressure changes recorded by the antral balloon are those corresponding to respiratory movements; duodenal contractions are less vigorous. Twenty minutes after the olive oil was placed in the duodenum, both antrum and duodenum are quiet. (Courtesy of C. F. Code.)

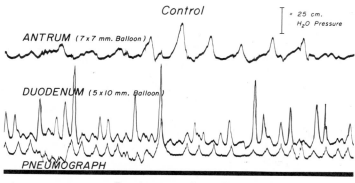

Control

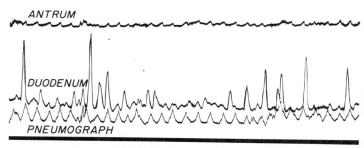

5 minutes after Fat

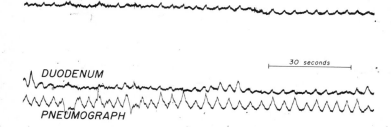

20 minutes after Fat

same way; when it is relaxed, there is no pressure gradient to drive gastric contents into the duodenum. Therefore, emptying is minimal despite relaxation of the pyloric sphincter. The only exception to this occurs when a very strong stimulus, such as 100 mN HCl, is placed in the duodenum; then the pylorus may contract from 1 to several min. After this brief spasm, which has no lasting influence on gastric emptying, the pylorus relaxes and remains flaccid. The duodenum, however, contracts vigorously at its basic rhythm until the acid is removed. In infants, hypertrophy and hyperplasia of the circular muscle of the pylorus sufficient to cause obstruction is sometimes called pylorospasm. Since it is not spasm in the sense that the muscle is abnormally contracted, pyloric stenosis is a better name.

Properties of chyme affecting the rate of emptying include its physical state, the fineness of division of its particles, its osmotic pressure, its acidity and its volume. In general, solutions or suspensions of small particles leave faster than lumps; lumps are reduced by digestion in the antrum and by terminal antral contractions before they leave the stomach. The larger the volume, the faster the initial rate of emptying. Solutions empty fastest if their osmotic pressure is 200 mOsm. Although solutions of lower osmotic pressure sometimes empty more slowly, the difference between water and 200 mOsm is small. More concentrated solutions empty more slowly, the degree of retardation being roughly proportional to the osmotic pressure. The osmotic pressure of solutions emptied into the duodenum appears to affect osmoreceptors there, which in turn affect gastric motility by an unknown path. Solutions having a pH of 3.5 or less greatly reduce gastric motility, and irrigation of the duodenum with 150 mN HCl stops gastric emptying altogether. Solutions between pH 3.5 and 5 may have a small effect. In man, isotonic $NaHCO_3$ has no effect on gastric emptying, and the reported inhibition produced by 400 mM $NaHCO_3$ may be attributed to its osmotic pressure rather than to its alkalinity.

Fat in any digestible form (triglycerides or phospholipids) in the presence of bile and pancreatic juice is the most powerful of the chemical agents that slow gastric motility (Fig 3–9). Fatty acids are more effective than their corresponding glycerides. In the first 15 min after a man has eaten a meal containing 100 gm of fat, strong antral contractions (type II waves) are entirely absent, and thereafter their frequency and strength are about half those of the waves after an ordinary meal. However, during digestion of a fatty meal, small ripples (type I waves) continue to pass over the stomach. Therefore the effect of fat is not to abolish gastric peristalsis but to reduce the amplitude of the waves. If a fatty meal is removed from the stomach, its inhibitory influence disappears in 3–5 min.

The products of protein digestion (amino acids and polypeptides) and the products of carbohydrate digestion (dextrins and oligosaccharides) have an inhibitory effect less powerful than that of fat.

Mechanism of Regulation of Gastric Motility*

In the regulation of gastric motility, three pathways are involved: (1) reflex mechanisms whose afferent and efferent fibers are in sympathetic and parasympathetic nerves, (2) reflexes not involving higher centers but

*This and the next sections summarize the obviously very elementary knowledge of the subject contained in Western literature. Russian physiologists have extensively studied interoception, and they have described many autonomic and somatic reflexes whose afferent limbs arise in the digestive tract. For example, if water is placed in the stomach of a dog through a fistula, diuresis follows as the water is absorbed from the intestine. But diuresis also follows sham administration of water after a conditioned reflex to distention of the stomach is established. A glimpse of this literature, which I am unable to evaluate, can be seen through Bykov, K. M.: *The Cerebral Cortex and the Internal Organs* (trans. W. H. Gantt) (New York: Chemical Publishing Co., Inc., 1957). More recent literature can be read in the English translations of the *Sechenov Physiological Journal of the USSR* and the *Bulletin of Experimental Biology and Medicine, USSR.*

operating through intrinsic plexuses or the celiac plexus and (3) blood-borne hormones liberated from the intestinal mucosa in response to chemical properties of the chyme.

Ninety percent of the 30,000 fibers in the vagus nerves are afferent, and an unknown proportion of these have their receptors in the stomach. Records of activity of the afferent fibers are obtained by dissecting the cervical vagus so that only a few fibers are in contact with recording electrodes. Then the stomach is stimulated in various ways and places until a mode and locus of effective stimulation of the receptor organs connected with the fibers are found. Figure 3–10 shows results of distending the stomach of a goat. Increased pressure in the stomach causes increased rate of firing of the afferent fiber. This slowly adapting fiber and similar ones found in the muscular wall of the cat stomach respond to the increased tension evoked either by passive stretch or by active contraction. Other slowly adapting fibers

respond to distention of the stomach not accompanied by a rise in intragastric pressure, thus signaling the size of the stomach. Still other fibers fire when the pH of gastric contents is 3.0 or lower; others when it is above 8.0. In addition, numerous afferent fibers in the splanchnic nerves enter the cord and ascend in the same region as homologous somatic afferents. Impulses carried by the afferent fibers affect feeding behavior. Distention of the stomach increases electric activity in the satiety center of the hypothalamus and decreases firing in the feeding center. The vagal fibers carry activity that is projected to the orbital cortex and to the amygdaloid region; the activity carried by the splanchnic afferents is projected to the somatosensory cortex and, by way of connections with the reticular formation, to the central excitatory system of the brain. These fibers are the afferent limbs of diverse reflexes affecting blood pressure and contraction of somatic musculature whose central

Fig 3–10.—Intragastric pressure and frequency of discharge in goat afferent fiber. **Top,** reflex isometric contraction of the stomach of a lightly anesthetized goat elicited by 1,200-ml inflation. **Bottom,** frequency of discharge in a slowly adapting fiber from the

stomach, traveling centrally in the cervical vagus nerve. The rate of firing increases from about 15 per sec before the contraction to 30 per sec and reaches peak value before contraction is maximal. (From Iggo, A.: J. Physiol. 128:593, 1955.)

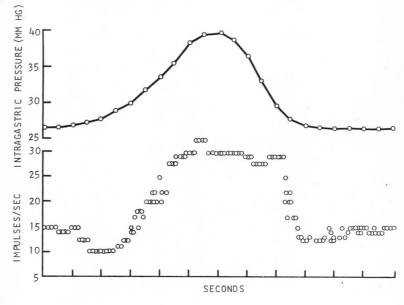

pathways and significance are poorly understood. Activity in the vagal fibers appears to affect the vagal nucleus, which in turn affects gastric motility. In sheep and goats, coordinated reticuloruminal movements are entirely dependent on such reflexes. Afferent stimulation of the vagus elicits rhythmic contractions at 10-sec intervals, whereas similar efferent stimulation merely produces uncoordinated movements. In the cat, stimulation of an afferent branch of the abdominal vagus nerve near the lower part of the esophagus is usually followed by relaxation of gastric muscle, increased secretion of gastric juice and secretion of enzymes by the pancreas.

Afferent nerves mediating pain have their receptors in the gastric muscle wall and perhaps in the mucosa. The major cause of deep gastric pain is tension in the wall of the stomach. Afferents mediating this sensation travel with sympathetic nerves, for the sensation is completely abolished by bilateral preganglionic sympathectomy. No pain or other sensation is caused by acid in the normal stomach. Insensitivity of the mucosa to stimuli causing pain is relative, not absolute. The mucosa may be cut or actually eroded with-

out pain, but crushing of the mucosa (or the muscle) is painful. The pain of gastric or duodenal ulcer, an intense burning sensation referred to the epigastrium, appears when the ulcerated surface is exposed to acid, but its exact cause is unknown. It has been attributed to action of acid on exposed nerves, to spasm of gastric muscle and to inflammation. Although there are abundant nerve endings at the edge of the ulcer crater, they are deep in the tissue and no more exposed to acid than other tissue components. There are few or no nerve endings in the crater itself, and those present appear to be insulated by layers of tissue. Typical pain can occur when there is no acid in the stomach. The pangs of ulcer pain are associated with contractions of the stomach (Fig 3–11), but there is no evidence that the contractions are different from those in the normally painless stomach. Tissue around the ulcer is inflamed and indurated; pressure on such areas is painful. Ulcer pangs may be caused or exacerbated by movements that distort the area. Contraction of the gastric muscle momentarily occludes venous outflow, and increased venous pressure within the ulcer area may also contrib-

Fig 3–11.—Relation between contractions of the stomach and pain in a patient with peptic ulcer. Pressure was recorded by means of a small balloon within the stomach, and the patient signaled when he felt a pang of ulcer pain. At the arrow he was given the anticholinergic drug, propantheline bromide. (Courtesy of C. F. Code.)

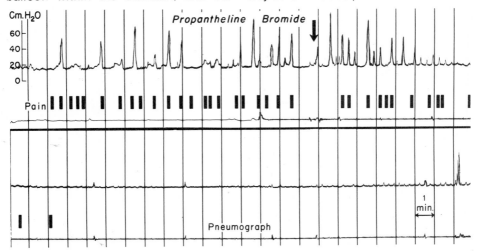

ute. Anticholinergic drugs reduce pain, probably more by diminishing motility than by reducing acid secretion.

Both sympathetic and parasympathetic efferent fibers influence the stomach. In general, sympathetic stimulation reduces the frequency and rate of conduction of the basic electric rhythm of the stomach, and it decreases the strength of contraction of the circular muscle. The major effect of sympathetic stimulation is probably upon the activity of the neurons of the intramural plexuses. Central stimulation of the hypothalamic defense area or of the surrounding pressor area causes, in addition to widespread vasoconstriction, prompt inhibition of vagally induced gastric motility. In the absence of ongoing vagal stimulation, sympathetic discharge produces only small and sluggish inhibition of gastric muscle, perhaps as the result of overflow of adrenergic mediator released near blood vessels. Distention of the duodenum or jejunum also causes slowing of the antral rhythm and weakening of antral contractions. The efferent pathway is in part by way of adrenergic sympathetic fibers. Under usual circumstances, sympathetic nerves have little or no influence on gastric motility, for most of the numerous persons who have surgical or chemical sympathectomies for hypertension have normal gastric function.

Parasympathetic nerves have both inhibitory and excitatory influences on the stomach. Receptive relaxation of the fundus, which is part of the swallowing reflex, has been described above. Vagal inhibitory fibers also go to the body and antrum. Slight distention of the esophagus causes rapid and deep relaxation of the stomach. Afferent impulses activating vagally mediated reflex relaxation of the stomach also arise from nerve endings in many abdominal structures; among them are receptors in the duodenum sensitive to the chemical composition of the chyme. Instillation of cottonseed oil into the duodenum of a dog decreases the frequency of the basic electric rhythm (normally 5.4 cycles per min) by about 1 cycle per min; and

it reduces the number of action potentials, and therefore the strength of contraction, of antral circular muscle. This inhibitory effect is abolished by transthoracic vagotomy. The mediator is neither acetylcholine nor catecholamines.

Nervous and Hormonal Control of Gastric Motility

Both gastric secretion and gastric motility are reduced by acid, fat and protein digestion products and solutions of high osmotic pressure in the duodenum or jejunum. The latent period for inhibition of motility is as short as 20–40 sec. At least part of the inhibitory effect is mediated through nervous connections between small intestine and stomach, and consequently it is called the enterogastric reflex. The inhibition produced by acid in the first 5 cm of duodenum is in part effected through the intrinsic plexuses connecting duodenum and stomach. There is also a reflex arc, described in Figure 2–1, by which stimuli such as 100 mN HCl in the duodenum or jejunum inhibits gastric motility; acid stimulates receptors of afferent fibers, which travel to the celiac plexus and there synapse with efferent sympathetic fibers.

Many forms of central nervous activity having emotional concomitants (e.g., pain and anxiety, sadness, withdrawal, hostility) also have expression in changes in gastric motility and secretion, but the particular effect is unpredictable. Gastric emptying may be completely suppressed for 24 hours following a painful injury; but when an experimental subject experiences severe pain by plunging one hand in ice water or contracting the forearm muscles under ischemic conditions, gastric motility may not be reduced.

Hormonal control of gastric motility is described in Chapter 12.

REFERENCES

Abrahamsson, H.: Studies on the inhibitory nervous control of gastric motility, Acta Physiol. Scand. [Suppl. 88] 390:1, 1973.

Bykov, K. M.: *The Cerebral Cortex and the Internal Organs* (trans. W. H. Gantt) (New York: Chemical Publishing Co., Inc., 1957).

Code, C. F., and Carlson, H. C.: Motor activity of the stomach, in Code, C. F. (ed.): *Handbook of Physiology:* Sec. 6. *Alimentary Canal,* Vol. IV (Washington, D.C.: American Physiological Society, 1968), pp. 1903–1915.

Code, C. F., Carlson, H. C., Szurszewski, J. H., Kelly, K. A., and Smith, I B.: A concept of control of gastrointestinal motility, in Code, C. F. (ed.): *Handbook of Physiology:* Sec. 6. *Alimentary Canal,* Vol. V (Washington, D.C.: American Physiological Society, 1968), pp. 2881–2896.

De Groot, J.: Organization of hypothalamic feeding mechanisms, in Code, C. F. (ed.): *Handbook of Physiology:* Sec. 6. *Alimentary Canal,* Vol. I (Washington, D.C.: American Physiological Society, 1967), pp. 239–248.

Edwards, D. A. W., and Rowlands, E. N.: Physiology of the gastroduodenal junction, in Code, C. F. (ed.): *Handbook of Physiology:* Sec. 6. *Alimentary Canal,* Vol. IV (Washington, D.C.: American Physiological Society, 1968), pp. 1985–2000.

Ingram, W. R.: Central autonomic mechanisms in

Field, J. (ed.): *Handbook of Neurophysiology,* Vol. II (Washington, D.C.: American Physiological Society, 1960), pp. 951–978.

Jansson, G.: Extrinsic control of gastric motility, Acta Physiol. Scand. [Suppl.] 326:1, 1969.

Kelly, K. A., Code, C. F., and Elveback, L. R.: Pattern of canine gastric electrical activity, Am. J. Physiol. 217: 461, 1969.

Sharma, K. N.: Receptor mechanisms in the alimentary tract: Their excitation and functions, in Code, C. F. (ed.): *Handbook of Physiology:* Sec. 6. *Alimentary Canal,* Vol. I (Washington, D.C.: American Physiological Society, 1967), pp. 225–238.

Stoddard, C. J., Smallwood, R., Brown, B. H., and Duthie, H. L.: The immediate and delayed effects of different types of vagotomy on human gastric myoelectric activity, Gut 16:165, 1975.

Thomas, J. E., and Baldwain, M. V.: Pathways and mechanisms of regulation of gastric motility, in Code, C. F. (ed.): *Handbook of Physiology:* Sec. 6. *Alimentary Canal,* Vol. IV (Washington, D.C.: American Physiological Society, 1968), pp. 1937–1968.

Wolf, S.: *The Stomach* (New York: Oxford University Press, 1965).

4

Motility of the Small Intestine

THE MAJOR PART of digestion and absorption occurs in the duodenum and jejunum; undigested and unabsorbed residues are delivered from the ileum to the colon. The duodenum is short, about 20 cm on the average, and the jejunum and ileum together are 7–8 meters long. Chyme moves slowly through this tube, at a rate such that the residue of one meal leaves the ileum as another enters the stomach. As befits its digestive and absorptive functions, the most important motion of the small intestine is segmentation, controlled by a basic electric rhythm originating in the longitudinal muscle near the entrance of the bile duct. This mixes chyme with digestive juices and repeatedly exposes the mixture to the absorptive surface of the mucosa. Although segmentation accomplishes some downward propulsion, chyme is also moved by small peristaltic contractions, which, during normal digestion, move only a few centimeters before dying out. Peristaltic rushes rapidly traversing the whole length of the small intestine are abnormal, as are spasm, reverse peristalsis and complete absence of movement. During the interdigestive period a band of intense segmentation followed by peristalsis sweeps from the stomach through the duodenum to the terminal ileum, and another, almost identical, band of segmentation follows in about an hour. Smooth muscle and intrinsic nerves of the small intestine are alone sufficient to perform segmentation and peristalsis, but extrinsic innervation affects the occurrence and modifies the strength of these movements and mediates reflexes by which stimuli acting in one part of the intestine govern activity in the rest of the bowel.

Fig 4–1.—X-ray photograph of the small intestine of a normal man. The shadow thrown by the barium sulfate shows characteristic patterns of the mucosa. (Courtesy of F. J. Hodges and J. N. Correa.)

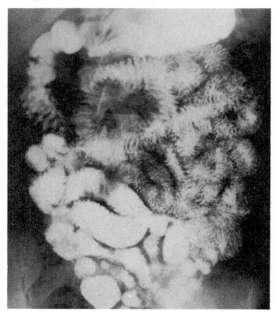

Segmentation

When electrodes are placed in longitudinal array in the duodenal muscle of an unanesthetized dog, the electric control activity, or basal electric rhythm, is found to originate near the duodenal bulb and to move down the duodenum at the rate of 19–20 cm/sec (Fig 4–2). The record obtained when the electrodes are arranged radially shows that all points in cross section are in phase. The electric cycle is 3.4 sec long, and the rhythm is 17–18 per min. Spike potentials signaling muscular contraction occur intermittently at a definite phase of the control rhythm, and the interval between contractions is some multiple of 3.4 sec. Spike potentials appear sequentially in the longitudinal direction, and the wave of contraction moves in the aboral direction. However, 80% of the contractions travel less than 3 cm.

Fig 4–2.—Electric activity and pressure recorded from the duodenum of an unanesthetized dog. At an earlier operation, the duodenum was brought out through the abdominal wall and covered by a skin flap so that it formed a permanent handle-like loop. Electric recording was made from needle electrodes inserted into the muscle through the skin; pressure was recorded by means of an open-tip catheter inserted through a thin-walled

The following is the original description of segmentation seen by fluoroscopic examination in the cat (Fig 4–3):

Rhythmic segmentation is by far the most common . . . mechanical process to be seen in the small bowel. . . . A small mass of food is seen lying quietly in one of the intestinal loops. Suddenly an undefined activity appears in the mass, and a moment later constrictions at regular intervals along its length cut it into little ovoid pieces. . . . A moment later each of these segments is divided into two particles, and immediately after the division neighboring particles rush together, often with the rapidity of flying shuttles, and merge to form new segments. The next moment these new segments are divided, and neighboring particles unite to make a third series, and so on.*

*Cannon, W. B.: *The Mechanical Factors of Digestion* (New York: Longmans Green & Co., 1911), pp. 131–133. Cannon was a first-year medical student at Harvard in 1896 when he made the basic discoveries of modern gastroenterologic physiology.

hypodermic needle. The dog was given morphine by slow intravenous injection. The left half of the record shows the basic electric rhythm with some spiking. Pressure increases following spiking. At the middle of the record, spiking becomes prolonged, and there is a corresponding increase in intraluminal pressure. (From Bass, P., *et al*.: Am. J. Physiol. 201:287, 1961.)

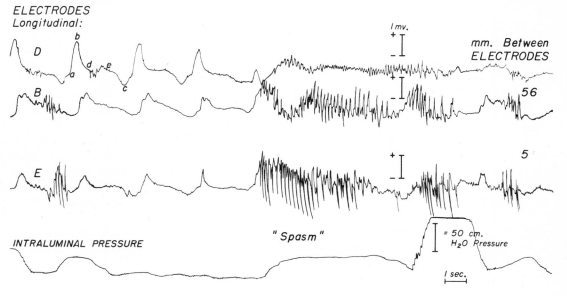

ELECTRODES
Longitudinal:

D

B

mm. Between ELECTRODES

56

E

5

INTRALUMINAL PRESSURE

" Spasm "

= 50 cm. H₂O Pressure

1 sec.

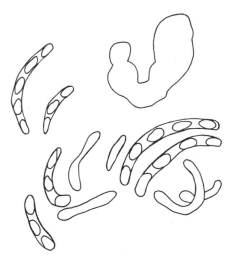

Fig 4–3.—Tracing of the fluoroscopic shadow of a cat's intestine during digestion of a meal of starch mixed with contrast medium. The parts marked with ovals were undergoing segmentation. (From original tracing of W. B. Cannon.)

The process described by Cannon has also been observed in the dog, the rat and man. In the human duodenum, segmental contractions are of two types. The first are eccentric contractions, confined to a segment less than 2 cm long; these do not empty the segment completely and do not cause any change in intraluminal pressure. The second are concentric contractions, which usually succeed in emptying a segment greater than 2 cm in length and which generate luminal pressure as high as 20 mm Hg. The empty segment refills as the annular contraction disappears.

In 12 normal human subjects the frequency of segmentation was found to be 11.80 ± S.D. 0.32 per min at the end of the duodenum and 9.39 ± S.D. 0.18 in the ileum. In some subjects the descent of frequency was stepwise; in others it was linear. The rate of segmentation is little, if at all, affected by extrinsic nerves, by feeding or fasting or by castor oil. However, vigor and amplitude of contractions vary greatly. Segmenting movements may be absent or barely visible in fasting, and they become strong immediately after feeding. Activity may wax and wane in cycles; and influences such as fright, which stimulate the thoracolumbar nervous system, depress segmenting movements. Vagal stimulation or administration of the anticholinesterase drug physostigmine or of quinidine augments segmentation.

Although segmenting movements are only slightly progressive, they move chyme in the aboral direction (Fig 4–4). A single segmental contraction squirts chyme lying beneath it in both directions. Because the frequency of contraction is greater in a higher part of the intestine than in a lower part, on the average more chyme is propelled in a downward than in an upward direction. Flow of chyme is governed by resistance as well as by pressure generated by contraction. A segmental contraction offers resistance; and because segmental contractions are less frequent aborally than orally, the average resistance is less in a lower part of the small intestine than in a higher part. Hence, chyme flows more readily downward than upward.

Nature of the Intestinal Gradient

The gradient of the rate of segmental contraction is determined by interaction of the basic electric rhythms spontaneously generated by cells of the longitudinal muscle layer. The following description of the gradient is based on observations made on the intestine of the anesthetized cat. Essentially similar results have been obtained on dog and monkey.

Recordings from closely spaced electrodes placed on the serosal surface of the intestine, beginning high in the duodenum, show that the frequency of the basic electric rhythm is constant for some centimeters. In the example given in Figure 4–5 the frequency from mid-duodenum to upper jejunum is 18 per min. Below this, a length of jejunum has a frequency of 17 per min. Between the two frequency plateaus is a short length of intestine whose frequency waxes and wanes between the two limits of 18 and 17 per min.

Dog's jejunum segmenting

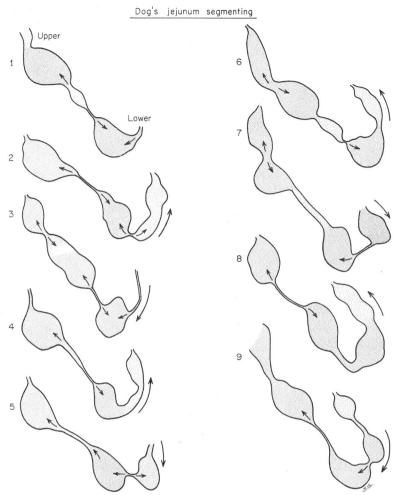

Fig 4–4. – Movement of chyme by segmentation. A dog had been fed a meal mixed with the x-ray contrast medium, barium sulfate, and continuous cinefluorographic pictures were taken as the contrast medium entered the upper jejunum. The tracings in this figure were made from the film to show the locus of the medium at approximately 1-sec intervals. The *arrows* show the direction in which the jejunal contents were moving at the instant represented by the tracing. Progress of the contents into the two distal segments shows that the intestinal contents can be moved downward from above by segmentation without peristalsis. (From Davenport, H. W.: *A Digest of Digestion* [Chicago: Year Book Medical Publishers, Inc., 1975]. Adapted from a film by H. C. Carlson.)

The entire length of intestine exhibits frequency plateaus, each of a lower frequency than the one above it and each connected to its neighbors by a short segment whose frequency waxes and wanes. At the terminal ileum the last plateau has a frequency of only 12.5 per min.

The longitudinal muscle of each segment of the small intestine spontaneously generates a basic electric rhythm whose frequency is characteristic of that segment. If the intestine is cut into small bits, each piece exhibits its own intrinsic rhythm. The frequency of the intrinsic rhythm diminishes in a linear,

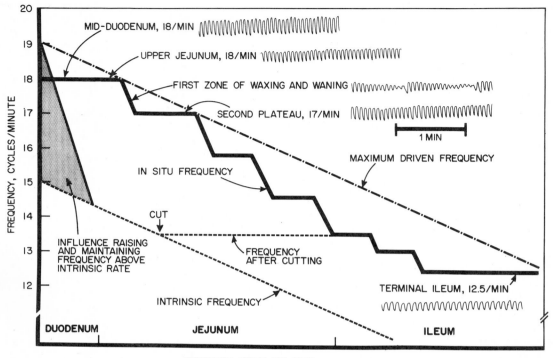

Fig 4–5.—Frequency of the basic electric rhythm of the small intestine of the anesthetized cat. The stepwise *heavy line* shows the observed frequency, 18 per min in the duodenum and upper jejunum and descending to 12.5 per min in the terminal ileum. Segments of the electric record from which the frequency was measured are shown. The *lower slanting* *dotted line* shows the intrinsic frequency displayed by short, isolated segments of the intestine. If the intestine is transected at the point labeled *CUT,* the frequency distal to the cut falls to that of the intrinsic frequency at the point of transection. (Adapted from Diamant, N. E., and Bortoff, A.: Am. J. Physiol. 216:734, 1969. Electric records supplied by A. Bortoff.)

not a stepwise, manner from duodenum to terminal ileum. Each segment of longitudinal muscle is, in effect, an oscillator having a characteristic frequency.

In the intact intestine, an orally located oscillator having a higher intrinsic frequency is coupled with its aborally located neighbor, which has a lower intrinsic frequency. If the coupling is strong enough, the oscillator having the higher frequency drives that having the lower frequency, and the second in turn drives a third. Consequently, a series of points along the intestine has a basic electric rhythm of uniform frequency, that set by the frequency of the most proximal segment. This accounts for the frequency plateaus. At

some distal point along the plateau the driving frequency is so much higher than the intrinsic frequency of the driven cells that the cells having the lower intrinsic frequency are incapable of following the driving frequency. At this point the frequency of the basic electric rhythm falls and another plateau is established. Between the two plateaus is a short segment in which the frequency shifts between that of the higher and that of the lower segment. This is the place at which coupling between the two segments breaks down.

If the intestine is transected, the frequency of the basic electric rhythm and of segmental contraction of the intestine below the cut falls, for coupling between driving and driven parts has been broken at the cut. The frequency followed by the section below the transection is the intrinsic frequency of the muscle at the point of transection. When the segment below the cut is warmed a few degrees, the frequency at the point of warming rises and so does the frequency of as much as 75 cm of immediately distal intestine.

The frequency of the basic electric rhythm and of segmental contraction of the entire small intestine is higher than the intrinsic frequency of its component cells. In the duodenum there is some process whose nature is unknown that raises the frequency of that part above its intrinsic rate.

Short Propulsive Movements

Chyme is moved down the intestine by short, weak propulsive movements; it would be disadvantageous if long, vigorous ones drove chyme rapidly downward without allowing time for digestion and absorption. In human subjects observed by cineradiography an annular contraction originating in the upper duodenum may be seen to move as a stripping wave into the upper jejunum. Such a wave pushes chyme before it; and when the chyme is propelled into the jejunum, it rarely returns in any significant quantity into the duodenum. The wave travels with a mean velocity of 1.2 cm per sec for an average distance of 15 cm.

On fluoroscopic examination of subjects whose intestinal contents are mixed with barium, the shadow is often seen to swing back and forth like a pendulum. The to-and-fro movement is caused by sequential contractions, first above and then below. There is no specific "pendular movement" of the intestinal musculature.

Although motility of the small intestine is increased following feeding, its propulsive activity is actually reduced. In a dog, propulsive activity can be measured by allowing isotonic fluid at body temperature to flow into the upper small intestine at a pressure of about 1 mm Hg. The quantity of fluid transported downward in unit time reflects the ability of the intestine to move its contents. The rate of transport is reduced about 80% when the animal is fed. The reduction in flow is caused by increased contraction of the intestine, which narrows its lumen and increases resistance to flow. In man, the rate of flow of chyme past a point in the upper jejunum during digestion of a meal is only 2 ml per min. Extrinsic denervation abolishes this response to feeding.

On the other hand, progress of chyme may be abnormally rapid in the atonic bowel, for resistance to flow is low.

Peristalsis

Peristalsis is a progressive wave of contraction of successive rings of circular muscle moving a shorter or longer distance down the small intestine. Excitation is provided by the wave of depolarization passing along the longitudinal muscle, and it is transmitted to the circular muscle by electrotonic spread of current or by conduction through small bands of muscle connecting the two layers. Consequently, the peristaltic contraction moves down the intestine at the velocity of the electric control activity in the longitudinal muscle. The frequency with which peristaltic waves follow one another is likewise some submultiple of the frequency of the electric control activity. Whether or not circular muscle responds to the wave of depolarization in the longitudinal muscle depends upon the prevailing degree of excitability of the circular muscle.

In the normally innervated small intestine, the excitability of the circular muscle is strongly influenced by the intrinsic plexuses, which, in turn, are influenced by their extrinsic innervation.

Distention of the intestine elicits the peristaltic reflex, which consists of a compound wave of contraction above and relaxation below. First, the longitudinal muscle contracts, and then the circular muscle contracts. The two contractions have a phase difference of 90 degrees, for the circular muscle begins to contract when the contraction of the longitudinal muscle is half completed.

Although the reflex is unimpaired by extrinsic innervation, it is abolished by asphyxia and application of cocaine to the mucosa. Treatment with hexamethonium blocks conduction in the ganglia and abolishes contractions of the circular coat without affecting preceding contractions of the longitudinal one. However, if the reflex function of the intrinsic plexuses is completely blocked by hexamethonium, a peristaltic wave can be made to pass down the intestine by infusing a parasympathomimetic drug such as bethanechol. This increases the excitability of the circular layer so that it responds with action potentials to the electric control activity of the longitudinal layer.

The myenteric plexus contains a descending pathway, which is activated by distention; its last neuron passes to the circular muscle and releases an inhibitory transmitter. Because this pathway has short latency, the first effect of distention is to cause inhibition of the circular muscle below the point of distention. The plexus also contains a descending, excitatory pathway with long latency. Therefore the circular muscle is stimulated to contract after it has relaxed.

The Interdigestive Myoelectric Complex

Under surgical anesthesia, electrodes are sewn in a row on the serosal surface of the gastric antrum and along the entire length of the small intestine. Leads from the electrodes are brought to the surface of the body. A dog with implanted electrodes may remain in good health for many months, and records of the electric activity of the muscle of his

Fig 4–6.—The electric sign of the migrating interdigestive complex in the dog. Electric activity at electrodes along the small intestine of a dog were recorded on magnetic tape; then the tape was run at high speed so that only intense activity was displayed. (From Grivel, M-L., and Ruckebusch, Y.: J. Physiol. 227:611, 1972.)

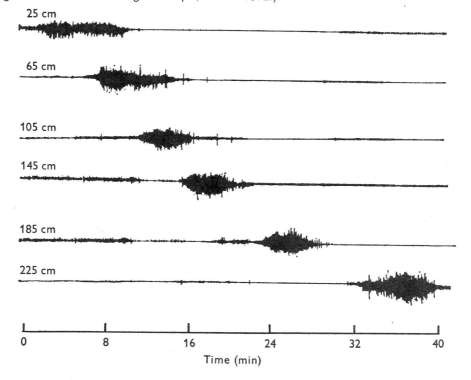

stomach and small intestine can be made for hours or days. Strain gauges may be sewn to the muscle to permit recording of movements, or intraluminal catheters may be inserted to measure pressures. Cinefluoroscopy can be used to demonstrate movement of contents of the bowel.

A characteristic sequence of activity, the interdigestive myoelectric complex, is found in a dog that has been fasted more than 12 hours (Figs 4–6 and 4–7). The cycle begins with action potentials in the stomach, these signal peristaltic waves passing over the gastric antrum. At the same time, bursts of action potentials, in phase with duodenal electric control activity, appear in the duodenum. They show that the duodenum is segmenting. Peristalsis dies out in the stomach, and a band of segmenting movements moves down the small intestine from duodenum to terminal ileum. As one complex finishes in the ileum, another begins in the stomach and duodenum (see Fig 4–7). At the end of a complex, a few peristaltic waves may sweep the small intestine.

The complex consists of four phases: At first, random action potentials begin to be associated with the electric control activity, and, as time passes, action potentials increase in incidence so that the electric record increases in intensity. This is a period of vigorous segmentation. This phase lasts 30–48 min. Then, suddenly, there is a large burst of action potentials coincident with each wave of electric control activity, and segmentation reaches a maximum. This lasts 12 min in the antrum, 8 in the duodenum and 6–7 in the jejunum and ileum. The phase of intense activity ends almost as abruptly as it begins, and it is followed by a period lasting 6–16 min in which action potentials die out. Then there is a period of quiet in which the regular waves of the electric control activity are followed by few or no action potentials.

Fig 4–7.—Migrating complexes recorded from the antrum, duodenum, jejunum and ileum of a fasted dog. The *stippled areas* represent the periods in which action potentials were beginning to appear and to increase in number in phase with the electric control activity. The *black areas* represent periods of intense activity. The *white areas* represent periods in which activity was rapidly dying out. (Adapted from Code, C. F., and Marlett, J. A.: J. Physiol. 246:289, 1975.)

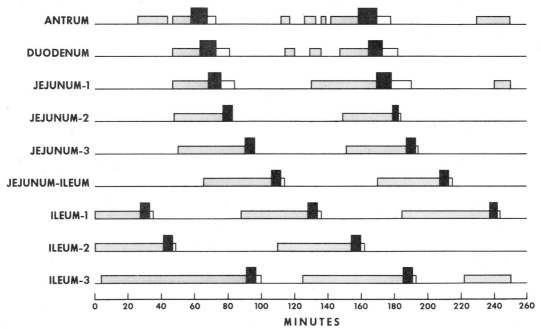

The period, propagation time and velocity are characteristic of the particular dog studied. The length of each cycle is from 90–114 min, and a complex migrates from stomach to terminal ileum in 105–134 min. The velocity of propagation is 6–12 cm per min in the upper part of the bowel, and it diminishes to 1–3 cm per min at the midpoint of the small intestine, remaining constant for the rest of the intestine's length.

The complex requires neither continuity of the bowel nor movement of contents. A piece of small intestine with its nerve and blood supply intact may be cut from the middle of the bowel. Each end is brought to the surface of the body so that the isolated loop of intestine may be irrigated. The continuity to the rest of the small intestine is established by end-to-end anastomosis. In a dog prepared in this way, the complex migrates from the stomach to the point of anastomosis. Then it begins in the loop at the end that originally had been uppermost and travels to the lower end. It begins again at the point of anastomosis and travels to the terminal ileum.

Feeding the dog interrupts any ongoing complex. Similar migrating complexes occur in the rabbit and the sheep, but, in those animals, feeding has no effect. Whether man resembles a dog more than a sheep has not been determined.

The Duodenal Bulb

The motility of the first part of the duodenum just distal to the pylorus, the duodenal bulb, differs from that of the stomach and lower duodenum. During gastric emptying the bulb may be filled during terminal antral contraction, and chyme spills over into the duodenum. In normal persons, there is little or no regurgitation of contents of the bulb into the stomach, for pressure in the pyloric canal is greater than that in the bulb. Most contractions of the bulb occur when the pyloric canal is firmly closed by terminal antral contraction. In patients with gastric ulcers, regurgitation is frequent.

The longitudinal muscle layer of the duodenal bulb has its own relatively rapid basic electric rhythm. The antral electric rhythm, whose frequency is much less than that of the duodenum, is conducted into the duodenal bulb by longitudinal muscle fibers, which cross the pyloric junction, chiefly along the lesser curvature. The two electric rhythms interact (see Fig 3–8) with the result that every fifth or so duodenal depolarization is augmented by depolarization conducted from the antrum; such augmentation can be detected, in the monkey, as far as 17 mm beyond the pyloric junction. Augmented duodenal depolarization is often followed by contraction. This coupling of antral and duodenal rhythms accounts for the fact that most duodenal contractions follow antral contractions.

The duodenal bulb is a sensitive receptor area, responding to qualities of the chyme entering it from the stomach; and osmotic equilibration, neutralization and intestinal digestion begin there and in the immediately succeeding few centimeters of duodenum.

Movements of the Mucosa and Villi

The mucosal surface of the small intestine, stomach and colon is thrown into a complex pattern of folds. An important characteristic is the rapidity with which the pattern shifts. In the jejunum, for example, the mucosa may at one moment exhibit a pattern of high longitudinal folds closely connected by transverse creases and, at another, a series of simple circular folds, which cut the lumen into a series of disk-shaped chambers. This movement is independent of contraction of the circular and longitudinal muscle of the gut wall, and folding of the mucosa is not a passive result of larger movements of the intestine. It is the result of contraction of the muscularis mucosae, occurring spontaneously or in response to local stimuli.

The muscularis mucosae of the small intestine of the dog contracts in an irregular rhythm of 3 or 4 times per min, and occasionally there is a slower rhythm on which the faster is imposed. Mechanical stimulation by gentle stroking or touching causes contraction lasting 30 sec to 1 min, which throws the mucosa into folds, ridges and pits. Solid matter in the intestine, lumps of food or tapeworms are stimuli for contraction. The chyme itself can be a sufficient stimulus, and stimulation by chyme accounts for the rapid shifts in pattern of folds seen roentgenoscopically in the barium-filled intestine.

Stimulation of the sympathetic nerves causes contraction of the muscularis mucosae and mucosal blanching. The nerves are adrenergic, and epinephrine given topically or intravenously also causes contraction; it does not inhibit the muscularis mucosae as it inhibits the major muscle of the gut wall. The muscularis mucosae also responds to parasympathomimetic agents (e.g., acetylcholine) with contraction. Vagal stimulation has little effect, although sometimes it may cause small contractions; this may mean that parasympathetic postganglionic fibers do not generally reach the muscularis mucosae. However, afferent fibers from the mucosa pass centrally in the vagus, and these carry a train of impulses whose rhythm corresponds to contractions of the muscularis mucosae. Since the muscularis mucosae contracts during digestion, the afferent impulses signal activity of the intestine, but their reflex significance is unknown. Afferent fibers in the splanchnic nerves also fire when the mucosa is irrigated with isotonic solutions of glucose or amino acids.

Villi of the intestinal mucosa move either by swaying to and fro or by abrupt contraction. Some of their movement results from contraction of the underlying muscularis mucosae. Partially purified extracts of human and dog intestinal mucosa cause villi to contract. The agent, assumed to be a hormone, is called villikinin. Individual villi contract at intervals, with no relation to the activity of their neighbors. The rhythm is irregular, being faster in the duodenum and jejunum and higher in fed, than in fasted, animals. Individual villi of the ileum seldom contract. The central lacteal of the villus is not completely emptied by its contraction, but lymph flow from the intestine is elevated during periods of increased activity of the villi.

Retrograde Movement and Effect of Reversal of Intestine

Segmentation and peristaltic movements carry most of the chyme downward, but with each contraction some fraction is transported upward. This accounts for the fact that some of any soot or other easily identifiable substance placed in the ileum can eventually be recovered from the stomach. Reverse peristalsis, which is rare or entirely absent from the human small intestine, need not be invoked to explain retrograde transport. During the spastic contraction of the whole duodenum accompanying nausea, its contents are squirted upward into the stomach, with the result that the vomitus is bile-stained, although reverse peristalsis may or may not have occurred. Surgical reversal of an intestinal segment is not necessarily fatal. Reversal changes the direction of peristalsis to one opposing downward movement; but if the reversed segment is short enough (under 20 cm) and if the diet is sufficiently bland, chyme can flow through the reversed segment. In fact, transplantation of the dog's pyloric antrum, with its direction reversed, to midjejunum does not seriously impede normal progression of intestinal contents. If the reversed segment is long or if the diet contains undigestible bits of solid, the reversed segment blocks movement, and fatal dilatation of the bowel at the upper end of the segment follows.

The Ileocecal Sphincter

The ileocecal sphincter, separating the terminal ileum from the colon, is normally closed. When an open-tip catheter is drawn from the dog's ileum into the colon, a zone of elevated pressure, about 1 cm long, is found in the cone-shaped segment comprising the ileocolic junction. In man, the zone is 4 cm long, and its resting pressure is about 20 mm Hg greater than colonic pressure. Action of the sphincter is partially governed by intrinsic mechanisms, for extrinsic denervation does not destroy its function. One factor keeping it closed is a myenteric reflex from the cecum. Mechanical stimulation of the cecal mucosa or distention of the cecum causes pressure in the sphincter to rise and delays passage of chyme through it. On the other hand, the ileocolic sphincter relaxes each time a propulsive wave passes along the last few centimeters of the ileum, and on such occasions a small amount of chyme may squirt through the sphincter. Distention of the terminal ileum in man or dog causes a fall in sphincter pressure (Fig 4–8). In man, the hormone gastrin in physiological doses causes increased segmentation in the terminal ileum at a frequency of 6– 9 per min, and at the same time it causes sphincter pressure to fall. Heightened activity of the ileum accompanies gastric secretion and emptying, and therefore emptying of the ileum through the sphincter is associated with eating. This relation is called the *gastroileal reflex*. The ileocecal sphincter has a valvelike action, which normally prevents regurgitation from cecum to ileum. It is not essential for this purpose; in carnivorous animals, such as the raccoon, the bear, the mink and the skunk, which have no ileocecal sphincter, regurgitation is kept at a minimum by the generally downward direction of intestinal movement.

The same gradient is adequate even in animals that have a sphincter, for its surgical removal in man and experimental animals is not followed by any significant regurgitation.

Long Intestinal Reflexes

The gastroileal reflex is one example of the influence of the activity of one part of the digestive tract on that of another, and some additional examples have been dignified with the name of reflex. Inhibition of gastric motility when the ileum is distended is called the *ileogastric reflex*. Immediate cessation of all intestinal movement (adynamic ileus) follows distention of an intestinal segment, rough handling of the intestine during abdominal surgery or peritoneal irritation. Adynamic ileus operates through three pathways: general sympathetic discharge, the peripheral reflex pathway through the celiac and mesenteric plexuses (see Fig 2–1, D) and the intramural plexuses. Gaseous distention follows paralysis, exacerbating and perpetuating the condition. Cessation of intestinal movements as the result of distention is called the *intestino-intestinal reflex*. Receptors are in the longitudinal muscle, and they are activated by stretch. Inhibition ceases as soon as distention is relieved, as by draining the obstructed segment through a tube. The

Fig 4–8. — Record of pressure measured by means of a 0.5 × 2 cm balloon having a volume of 0.35 ml placed in the cone-shaped junctional zone between the ileum and the colon of an unanesthetized dog. The zone, or ileocolic sphincter, displays rhythmic fluctuations of pressure. At the first *arrow* a balloon in the ileum was distended by injection of 1 ml of water; at the second *arrow* it was emptied. (From Kelley, M. L., Jr.; Gordon, E. A., and DeWeese, J. A.: Am. J. Physiol. 211:614, 1966.)

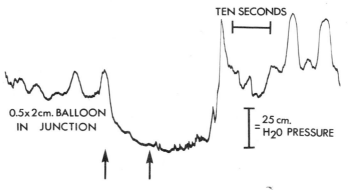

TEN SECONDS

0.5x2cm. BALLOON IN JUNCTION

25 cm. = H₂0 PRESSURE

intestino-intestinal reflex is both facilitated and inhibited by higher nervous centers. Many cardiovascular reflexes are affected by impulses arising in small-intestinal receptors.

Hormones also influence intestinal motility. Serotonin (5-HT) is probably a neurotransmitter between sensory and motor neurons in the peristaltic reflex. It is released into intestinal venous blood during peristalsis and when intraluminal pressure rises. Intravenous infusion of 5-HT into human subjects in a dose of 0.5 mg per min for 8 min stimulates upper and lower intestinal motility. Gastrin administered intravenously in a dose of $25-50$ μg to human subjects also increases upper intestinal contractions.

Intestinal Blood Flow

Mesenteric, renal and coronary blood flow, heart rate and arterial pressure have been measured by telemetry in unrestrained baboons at rest, during eating and after eating. The animals weighed between 22 and 26 kg, and the following figures are the means of the variables cited. At rest, mesenteric blood flow was 212 ml per min compared with renal blood flow of 150 and coronary blood flow of 25 ml per min. When the animals were given a meal of apples, oranges and bananas, their heart rate rose at once from $72-142$ beats a minute, and their mean arterial pressure went from $96-120$ mm Hg. At the same time mesenteric flow fell and mesenteric resistance rose. These immediate changes probably reflected the excitement of fasted animals upon receiving food. Within 15 min, heart rate and mean arterial pressure had returned nearly to normal, and mesenteric flow had risen to 373 ml per min. Mesenteric flow remained high for an hour and slowly declined in 4 hours to control level. Qualitatively similar results have been obtained in conscious dogs and in a chimpanzee who washed down three peanut butter sandwiches with a quart of milk.

No method for accurately measuring intestinal blood flow in normal human subjects has been devised. Measurements made in anesthetized patients at laparotomy have found intestinal blood flow to average 36 ml per minute in 100 gm of intestine; 75% of the flow went to the mucosa and the rest to the muscle.

It is possible that parasympathetic stimulation of the stomach, small intestine and colon has a direct vasodilator effect, but proof is incomplete. The increase in blood flow that follows parasympathetic stimulation is thought to be secondary to increased motor and secretory activity. The presence of acid, fat or amino acids in the intestine induces an increase in mucosal blood flow, and there is also active vasodilatation associated with an increase in oxygen consumption. Bile in contact with the mucosa causes prolonged vasodilatation. Reflexes within the intramural plexuses also control blood flow; gentle stroking of the jejunal mucosa of a cat causes a transient increase in flow through vessels of the villi. This reflex is blocked by local anesthetics, by tetrodotoxin and by antagonists of 5-HT but not by atropine or alpha- or beta-blockade.

Intra-arterial infusion of acetylcholine, bradykinin or histamine in the dog causes intestinal vasodilatation, and infusion of angiotensin II, norepinephrine or serotonin diminishes blood flow. Flow tends to return to base-line value during prolonged infusion of any of these agents. Similar autoregulatory escape occurs during stimulation of splanchnic vasoconstrictor nerves. Escape is in part myogenic, the result of relaxation of arteriolar smooth muscle in response to diminished stretch, and in part the result of accumulation of vasodilators during low blood flow. Villous vessels have a larger autoregulatory capacity than do vessels in the deeper part of the mucosa; this is important in maintaining oxygenation of the tips of the villi. However, during hemorrhagic shock, circulation through the villi is inadequate, and their tips become necrotic. When arterial pressure is low, venules draining the intestine contract as the result of an axon reflex in their sympathetic innervation; capillary pressure and capillary filtration are maintained.

Reactive hyperemia follows vasoconstriction. On the other hand, arteriolar resistance vessels usually contract when arterial pressure is high. When portal pressure rises, the common result is contraction of intestinal arterioles.

The flow of water across intestinal capillaries under the influence of a pressure gradient is at the rate of 1 ml per min·mm Hg·100 gm of tissue; this is far higher than in skeletal muscle and approaches the rate occurring in the renal glomerulus. If capillary pressure rises as the result of elevated venous pressure, tissue volume and interstitial fluid pressure increase as fluid is filtered from capillaries. In a few minutes equilibrium is reached when increased tissue pressure balances increased filtration pressure.

Blood vessels supplying intestinal villi form hairpin loops, and consequently they are the site of countercurrent exchanges. Each villus has a central arteriole, which arises from the submucosal network and which runs in the core of the villus without branching. This vessel has no muscular coat. At the tip of the villus the central vessel arborizes into a dense, subepithelial capillary network, which descends the villus and returns blood to venules at the base of the villus. The mean transit time of plasma in a villus is 4−8 sec at rest, about 1 sec during vasodilatation and 20−30 sec during low perfusion occurring in hemorrhagic shock. Plasma skimming occurs at the origin of the central arteriolar vessel, and the hematocrit of blood flowing through the villi is only 50−60% of that in major arterial vessels.

This anatomical arrangement permits countercurrent exchange of substances between blood flowing in opposite directions in the two limbs of the hairpin loop. The effect depends upon whether the substance enters the loop at the tip or the base. If an ion such as sodium is delivered to the blood at the tip of the loop by active absorption, it is in a higher concentration in the venous blood descending the loop than in the ascending arterial blood. The osmotic pressure of blood in the descending loop is also higher

than in the ascending arterial blood. Consequently, sodium tends to diffuse from the descending vessel into the ascending one, and its concentration in the blood flowing toward the tip of the villus increases (Fig 4−9). At the same time, water flows along its osmotic gradient from the ascending blood to the descending blood, and this also increases the sodium concentration and osmotic pressure of the blood reaching the tip of the villus. As long as active absorption of sodium continues, countercurrent multiplication keeps the concentration of sodium and the osmotic pressure high at the tip of the villus. The osmotic pressure in a cat's intestinal villus during active absorption of sodium may be as high as 1,200 mOsm at the tip.

Any substance entering the veins at the tip of the villus that can diffuse through the walls of the blood vessels undergoes the same process of multiplication. Oleic acid absorbed during fat digestion and carbon dioxide produced by metabolism in the epithelial cells are concentrated at the tips of the villi by countercurrent multiplication.

On the other hand, oxygen arriving at the arteriolar end of the vessels is short-circuited by diffusion into venous blood. If a slug of ^{51}Cr-labeled erythrocytes whose hemoglobin carries isotopically labeled oxygen is injected into a small artery serving a length of intestine, the isotopic oxygen appears in the venous blood draining the intestine 1−2 sec before the erythrocytes. Because oxygen can pass from arterial to venous blood without traversing the end-loops of the capillaries, the partial pressure of oxygen in the end-loops is low.

The effectiveness of countercurrent multiplication or dilution is a function of the rate of blood flow through the hairpin loops. An increase in the rate of flow tends to wash out the concentration gradients, and a decrease in flow exaggerates the gradients. In the resting state, enough oxygen reaches the end-loops to meet the needs of the epithelial cells, and when blood flow increases as the result of active hyperemia, the partial pressure of

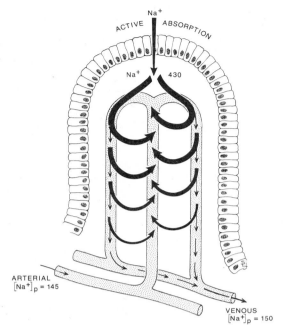

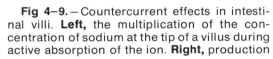

Fig 4–9. – Countercurrent effects in intestinal villi. **Left,** the multiplication of the concentration of sodium at the tip of a villus during active absorption of the ion. **Right,** production of a low partial pressure of oxygen at the tip of a villus as the result of short-circuit transfer of oxygen from the ascending arteriole to the descending venules.

oxygen in blood reaching the end-loops increases. When blood flow is substantially slowed, as in hemorrhagic shock, the partial pressure of oxygen at the tips is so low that epithelial cells die and slough off. This kind of necrosis of the tips of the villi can be prevented by perfusing the intestinal lumen with oxygenated fluid.

REFERENCES

Code, C. F., and Marlett, J. A.: The interdigestive myo-electric complex of the stomach and small bowel of dogs, J. Physiol. 246:289, 1975.

Code, C. F., Marlett, J. A., Szurszewski, J. H., Kelly, K. A., and Smith, I. B.: A concept of control of gastrointestinal motility, in Code, C. F. (ed.): *Handbook of Physiology:* Sec. 6. *Alimentary Canal,* Vol. V (Washington, D.C.: American Physiological Society, 1968), pp. 2881–2896.

Daniel, E. E., Gilbert, J. A. L., Schofield, B., Schnitka, T. K., and Scott, G. (eds.): *Proceedings of the Fourth International Symposium on Gastrointestinal Motility* (Mitchell Press: Vancouver, 1974).

Edwards, D. A. W., and Rowlands, E. N.: Physiology of the gastroduodenal junction, in Code, C. F. (ed.): *Handbook of Physiology:* Sec. 6. *Alimentary Canal,* Vol. IV (Washington, D.C.: American Physiological Society, 1968), pp. 1985–2000.

Furness, J. B., and Costa, M.: Adynamic ileus, its pathogenesis and treatment, Med. Biol. 52:82, 1974.

Haglund, U.: The small intestine in hypotension and hemorrhage. An experimental cardiovascular study in the cat, Acta Physiol. Scand. [Suppl.] 387:1, 1973.

Lundgren, O.: The circulation of the small bowel mucosa, Gut 15:1005, 1974.

Melville, J., Macagno, E., and Christensen, J.: Longitudinal contractions in the duodenum: Their fluid-mechanical function, Am. J. Physiol. 228:1887, 1975.

Vatner, S. F., Patrick, T. A., Higgins, C. B., and Franklin, D.: Regional circulatory adjustments to eating and digestion in conscious primates, J. Appl. Physiol. 36:524, 1974.

5

Movements of the Colon

THE COLON OF AN ADULT man daily receives about 1,500 ml of chyme from the terminal ileum. This contains undigested and unabsorbed residues of food. Man's usual diet contains some material, chiefly walls of plant cells, which is not easily digested higher in the tract. The residue accumulates and is to some extent digested by bacterial action in the cecum. Strictly carnivorous animals eat food leaving no such residues, and in them the cecum is small or absent. The diet of herbivorous animals, such as the rabbit, contains large amounts of cellulose; these animals have a large cecum in which vigorous fermentation takes place. The contents of the cecum and of the ascending and transverse colon are kneaded and turned over by churning movements, which are nonpropulsive. The contents are slowly dried by absorption of water until they are firm but not hard, and they are slowly moved toward the rectum by serial or systolic segmental contractions, mass propulsion and peristalsis. Net transport is increased after feeding, and the descending colon and rectum are frequently filled after a meal. Feces that arrive in the rectum distend it and arouse the defecation reflex whose nervous center is in the sacral segments of the spinal cord. If the reflex is facilitated, or at least not inhibited, by higher centers, the act of defecation, involving co-ordinated movement of both voluntary and involuntary muscles, follows. If defecation is inhibited, the rectum relaxes, or feces are returned to the upper part of the descending colon; tension in the wall of the rectum disappears, and defecation is postponed.

Structure and Innervation of the Colon

In man, the longitudinal muscle of the colon is concentrated in three bands, the taenia coli; and a thin layer of longitudinal muscle is present in the intervening spaces. The walls of the cecum and ascending colon are usually folded into sacs, or haustra, by contraction of the circular muscle. Haustra are not fixed structures. The proximal end of the colon is closed by the ileocecal sphincter, and the distal end by a thickening of the circular muscle, which forms the powerful internal anal sphincter. The taenia coli broaden at the sigmoid colon, and a thick layer of longitudinal muscle completely encircles the anal sphincter. The longitudinal muscle ends by fanning into muscle fascia and skin in the perineal region. Its contraction elevates and shortens the anal canal. This canal, distal to the internal sphincter, is surrounded by striated muscle fibers arranged in two groups, the superficial and the deep external anal sphincters, which overlap to a greater or lesser degree

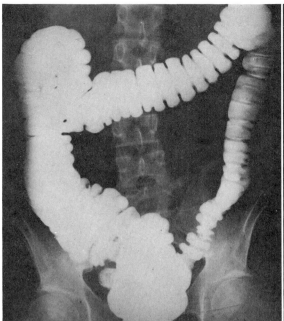 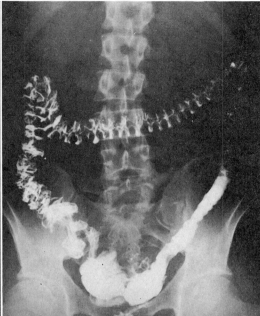

Fig 5–1.—X-ray photograph of a normal colon in an adult person. **Left,** colon filled with a barium enema. **Right,** after evacuation of the enema. (Courtesy of F. J. Hodges and J. N. Correa.)

with the smooth muscle of the internal sphincter. The puborectalis muscle arches around the uppermost part of the anal canal; it has no posterior attachment. It has the same innervation as does the external anal sphincter, and it is contracted except during defecation. It pulls the upper part of the anal canal forward so that the axis of the anal canal forms a right angle with the axis of the rectum. This angle is obliterated when the puborectalis muscle relaxes during defecation. The angle is also reduced when the hips are flexed more than 90 degrees, the usual position for defecation. The anus passes through and is supported by the pelvic diaphragm, striated muscle, which forms the pelvic floor. Both myenteric and submucous plexuses are present throughout the colon; their organization is similar to that in the rest of the bowel. Extrinsic parasympathetic innervation comes to the proximal colon by way of preganglionic fibers of the vagus.

Anatomists dispute how far distally vagal influence goes, but it perhaps reaches the first third of the transverse colon. Parasympathetic preganglionic fibers to the colon reach it by the pelvic nerves, which arise chiefly from the sacral segments of the cord; these overlap vagal innervation of the proximal colon. The preganglionic fibers are cholinergic, and they synapse with cells in the intrinsic plexuses, whose postganglionic fibers are likewise cholinergic. Therefore, parasympathomimetic drugs, anticholinesterases and acetylcholine congeners have the same effect on the colon as stimulation of these nerves. In the cat, which has a colon similar to man's, stimulation of vagal fibers to the colon increases rhythmic segmental mixing movements in the proximal colon only; the distal colon is unaffected. Stimulation of the pelvic nerves facilitates expulsive movements of the entire colon; contraction is not rhythmic but sustained. An exception is that the internal anal sphincter is relaxed. The pelvic nerves, but not the vagus, carry vaso-dilator fibers to the colonic mucosa.

Preganglionic cholinergic sympathetic fibers arise from lumbar roots and pass through the sympathetic chain to end in the inferior mesenteric ganglia, two to four in number, contained in the inferior mesenteric plexus. Postganglionic sympathetic fibers to the distal colon travel in the inferior mesenteric and hypogastric nerves. Postganglionic fibers also reach the colon from the splanchnic nerve. All postganglionic sympathetic fibers end on cell bodies of the intrinsic plexuses or on blood vessels, where they are vasoconstrictor; they do not directly innervate colonic muscle. The fibers are probably all adrenergic. Sympathetic stimulation antagonizes ongoing parasympathetic excitation; it does not directly inhibit the muscle. Blood-borne sympathomimetic drugs or epinephrine and norepinephrine released from the adrenal medulla are directly inhibitory. Splanchnic fibers to the colon mediate the intestinocolonic reflex by which the colon is inhibited when the intestine is distended.

Movements of Cecum and Appendix

The cecum contains only gas or the remains of previous meals when it is being slowly filled by the gastroileal reflex, which empties the terminal ileum. In the cat and the dog, distention of the wall of the cecum during filling initiates antiperistalsis. These waves begin at a tonic ring of contraction high in the ascending colon and pass slowly backward into the cecum. They are not preceded by a wave of relaxation. Six or so successive waves may be present at one time, and periods of antiperistaltic activity occur in cycles lasting 2–8 min. Their effect is to drive the contents of the colon into the cecum and to mix them thoroughly, but they never cause regurgitation through the ileocecal sphincter. Antiperistaltic waves are not prominent in man, and some experienced radiologists say they have never seen them.

Others report that antiperistaltic waves do occur in man, that they begin in the transverse colon close to the hepatic flexure and that they are weak and shallow. The vermiform appendix of man is filled from the cecum. When the cecal content contains contrast medium, the appendix throws a shadow, which goes through wormlike writhing movements. At other times, the shadow is divided into five or more beads. Then the appendix empties itself, presumably by peristalsis, only to be refilled a few moments later.

Types of Colonic Movement

When the human colon is studied by time-lapse cineradiography, the chief types of movement seen are haustral shuttling, segmental propulsion, multihaustral propulsion and peristalsis. The last two may produce mass propulsion of colonic contents. The frequency with which these forms occur in human subjects (each observed for a full hour) is given in Table 5–1.

Annular contractions in the transverse and descending colon fold the mucosa into sacs called haustra. The contractions form and reform at different sites, and the commonest form of movement detected in a subject fasting and at rest is a shuttling one in which liquid or semiliquid haustral contents are displaced short distances in both directions by apparently random segmental contractions of the circular muscle. The contractions are not progressive. There is little or no net movement of the fecal mass; it is slowly kneaded while water is absorbed. Similar contractions occur in the sigmoid colon, and they are responsible for the ovular shape of well-formed feces.

Segmental propulsion occurs when the contents of a single haustrum are displaced into the next segment, and thence into the one beyond, without being returned to the first. Subsequent contraction of the muscle surrounding the third and more distal haustra may expel their contents in both directions, but net movement of feces is in the aboral

TABLE 5-1.—PERCENTAGE OF SUBJECTS SHOWING
VARIOUS TYPES OF COLONIC MOVEMENTS OBSERVED
BY TIME-LAPSE CINERADIOGRAPHY FOR A FULL
HOUR'S FASTING, AFTER A MEAL OR AFTER
INTRAMUSCULAR INJECTION OF 0.25 MG CARBACHOL*

	FASTING	AFTER A MEAL	AFTER CARBACHOL
No. of subjects	101	63	29
Haustral shuttling	38%	13%	13%
Segmental propulsion	36%	57%	52%
Multihaustral propulsion	9%	17%	41%
Peristalsis	6%	8%	28%

*From Ritchie, J. A.: Gut 9:442, 1968.

direction. In one instance, such propulsive activity involved an 18-cm section of colon over a period of 5 min. Retropulsion by means of an identical but adoral displacement of contents through two or more haus-tra is about two thirds as common as orthograde propulsion.

Systolic multihaustral propulsion occurs when a number of adjacent segments contract more or less simultaneously. Part or

Fig 5–2.—Drawings made from cineradio-graphic studies of multihaustral propulsion in the normal human colon. At the time of the first observation *(upper left)*, barium-impregnated ileal contents *(arrow a)* were entering the cecum. Three minutes later, contraction of the cecum propelled a large part of its contents upward to distend the region of the hepatic flexure *(arrow b)*. After another 2 min, this section also contracted, and its contents were distributed over the proximal half of the transverse colon. Haustral markings disappeared over most of this length, and there was some narrowing of the mass in 2–3 inches of the middle section *(arrow c)*. Material forced out of the narrow section was accommodated by distention of the next four haustra *(arrow d)*. Four minutes later, as haustration was returning, most of the proximal half of the transverse colon between b and c also contracted. All the contents expelled from this section lodged in the distal half of the transverse colon. The conical outline at c in the fifth picture is typical of multihaustral contraction. (Adapted from Ritchie, J. A.: Gut 9:442, 1968.)

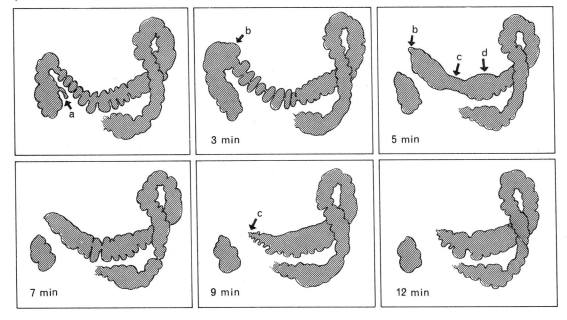

3 min

5 min

7 min

9 min

12 min

all of their contents is displaced into a neighboring length of bowel whose interhaustral folds are obliterated. The recipient segments may in turn contract in the same manner. A train of three such systolic multihaustral contractions occurring in about 10 min is shown in Figure 5–2. Infrequently, the recipient segment, instead of contracting as a unit, may undergo segmental haustration in which one after another of its original segments, beginning at the proximal end, recovers its original shape and size. The surplus contents slowly advance, distending still more distal parts of the colon.

Peristalsis in the colon consists of a progressive wave of contraction, which advances steadily, pushing the fecal mass ahead of it, at the rate of 1–2 cm per min. Muscular relaxation precedes the wave of contraction, and the relaxed segment is frequently filled with gas. The wave is followed by sustained contraction, which keeps the contracting segment of bowel closed and empty from 5 min to an hour. Reverse peristalsis does occur, but it is extremely rare.

Haustral shuttling is the most frequent type of movement in the fasting subject, but its occurrence diminishes after eating or when the colon is stimulated by a parasympathomimetic drug. Then segmental and multihaustral propulsion becomes more frequent. Persons having more than one bowel movement a day show 3 times as many propulsive movements as do those who habitually have only one daily evacuation.

Propulsive movements of a normal subject force the contents of the colon forward at a mean rate of 8 cm per hour. However, retropulsion forces the contents backward at 3 cm per hour, so that net forward movement is only 5 cm per hour. Net movements in a constipated subject, one having fewer than four bowel movements a week, is 1 cm per hour. After feeding, the forward rate of propulsion in a normal subject is 14 cm per hour. Retropulsion does not increase proportionately, and net propulsion becomes 10 cm per hour. After intramuscular injection of carbachol, net propulsion rises to 20 cm per hour.

Pressures in the Colon

Pressures within the lumen of the colon are measured by open-tip catheters or by a catheter whose opening is covered by a small balloon. The open-tip system accurately records pressure, but it gives no information about muscular contraction. The small balloon records pressure, but it also responds to mechanical deformation produced by contraction, making no distinction between the two. Since contraction can occur without change in pressure, and change in pressure can be produced by distant contraction, a record obtained from a balloon alone may be misleading. Combined pressure recording (preferably by means of an open-tip catheter) and cineradiography give more readily intelligible information. The following paragraphs are based on such studies.

Extensive shifts of colonic contents often occur with very little change in intraluminal pressure, and large variations in pressure may not be accompanied by transport of feces. Pressure within the lumen of the colon is generated by a combination of contraction and resistance. If adjacent segments of bowel are patent, their resistance is low, and contraction in one segment pushes its semiliquid contents into the area of low resistance with a pressure gradient of only a few mm Hg. If interhaustral contractions completely or nearly completely occlude the lumen, contraction can raise intraluminal pressure 70–80 mm Hg without displacing the contents of the haustrum.

Very slow changes in pressure, amounting to 2–3 mm Hg, occur over a period of 2–5 min in the human descending and sigmoid colon. Local contractions are isometric—i.e., without change in volume—if the colonic segment is closed at both ends. Although a high pressure in a single segment may be attained, the subject experiences no discomfort. Concurrent contractions of many segments are often painful. Contractions may be isotonic—i.e., without change in pressure—if the contents are free to move. These local contractions are often irregularly rhythmic, one frequency being about 12 per min and another 2 per min. In a normal subject at rest, 14 pressure peaks, averaging 27 mm Hg, appear each hour and pressure is elevated above base line 10–19% of the time. If the subject is engaged in an emotionally charged conversation about any topic likely to arouse anxiety, the frequency of contractions rises to an average of 34 per hour, and pressure is elevated 22–45% of the time.

Propulsive movements, consisting of waves of contraction moving from one segment to another in either direction, are preceded by a zone of raised pressure as gas is forced ahead of them. The pressure reached depends more on resistance than on contractile force. On the other hand, peristaltic contractions are preceded by a zone of low pressure as the wall of the colon relaxes ahead of the advancing wave of contraction. Solid feces, scybala, appear to be propelled by peristalsis only when resistance is drastically reduced.

Propulsive movement usually distends the colon ahead of it, and if the segment distended has been quiet, distention evokes a secondary contraction. After a latent period of 3–4 sec, the longitudinal muscle contracts, and within the next 4 sec, contraction of the circular muscle follows. If contraction is strong, the feces are pushed back.

The slowness of secondary contraction explains why gas moves through the colon so much faster than feces. Liquids or semisolids fill each haustral pocket before moving on to the next, and they are stopped by secondary contraction. Gas, however, rapidly passes through the interhaustral constrictions; although it distends each segment as it passes, it has blown by before a secondary contraction can stop it.

Intraluminal pressure changes in children or adults with uncomplicated diarrhea are actually subnormal, but intraluminal pressure waves in persons with constipation are more frequent and stronger than those in normal subjects. The explanation of this apparent paradox is that pressure is generated by contraction plus resistance. In diarrhea, no resistance is offered by local segmental contractions, and feces trickle along until they reach the rectosigmoid area, where the defecation reflex is aroused. In constipation, strong segmental contractions offer resistance, and they impede movement of feces.

Mass Propulsion

Colonic motility is strongly influenced by the rate of ileal outflow, and this in turn is affected by eating. In one series of cineradiographic observations of human subjects, ileal outflow substantially increased in the hour after lunch in 8 of 37 subjects; and in the whole group, net propulsion rose on the average from 3–10 cm per hour. In those subjects with the greatest ileal outflow, net propulsion was 34 cm per hour.

The increased activity in the right colon that follows eating, or sometimes discussion of food or listening to a lecture on defecation, stimulates mass propulsion. When primed by movement of contents from the right colon, mass propulsion often begins in the middle of the transverse colon as a series of multihaustral movements or as peristalsis (see Fig 5–2). Contents advance into a narrow, tubular section of bowel at least 20 cm long. Bowel contents are transported at various speeds, commonly 2–5 cm per min, as far as the pelvic colon. On some occasions, mass propulsion may originate in the cecum when ileal outflow is increased by feeding; then rapid transfer of some cecal contents fills the sigmoid colon and rectum. This extensive process, called mass movement or mass peristalsis, is described in Figure 5–3. Mass peristalsis is stimulated when the cathartic drug, oxyphenisatin, comes into contact with the colonic mucosal surface.

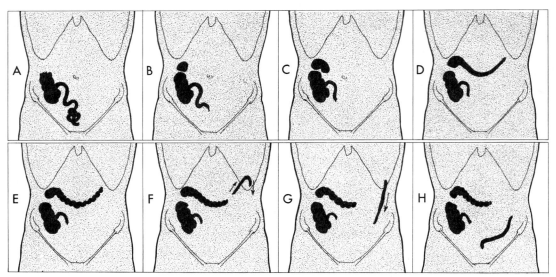

Fig 5–3.—The original description of a deliberately evoked mass movement reads: "A normal man had an ordinary breakfast with two ounces of barium sulfate at 7 A.M. At 12 noon the shadow of the end of the ileum, the cecum and the ascending colon was visible **(A)**. He then had an ordinary luncheon consisting of meat, vegetables and pudding. During the meal the cecum and ascending colon became more filled owing to the rapid emptying of the end of the ileum, and toward the end of the meal a large, round mass at the hepatic flexure became cut off from the rest of the ascending colon **(B)**. Immediately after the meal was finished some of this was seen to move slowly round the hepatic flexure **(C)**; the diameter of the separated portion then became suddenly much smaller, the large round shadow being replaced by a long narrow one, which extended from the hepatic flexure almost to the splenic flexure **(D)**. The shadow was at first uniform, but in a few seconds haustral segmentation developed **(E)**. About 5 minutes later the shadow suddenly became still more prolonged and passed round the splenic flexure **(F),** down the descending colon **(G)** to the beginning of the pelvic colon **(H).''** (Adapted from Hertz, A. F., and Newton, A.: J. Physiol. 47:57, 1913.)

Internal and External Anal Sphincters

The thick terminal portion of the circular smooth muscle of the rectum, 2.5 – 3 cm long, forms the internal anal sphincter surrounding the anal canal. It is overlapped distally by the striated muscle of the external anal sphincter.

The external anal sphincter is normally in a state of tonic contraction resulting from reflex activation through dorsal roots in the sacral segments, and the sphincter remains contracted even in sleep. However, the usual degree of contraction is minimal; stronger contraction occurs during distention of the rectum or when the urge to defecate is voluntarily resisted. Maximal contraction can be sustained for only 50 sec. The puborectalis muscle, which has the same innervation as the external anal sphincter, is also tonically contracted. The receptors from which many of the afferent impulses arise are in the sphincter itself. They operate like muscle spindle systems of other skeletal muscles, and proprioceptive feedback from them maintains discharge of motoneurons whose discharge over ventral root fibers excites • contraction of the sphincter. When the dorsal root fibers are destroyed, as in tabes dorsalis, resting contraction of the sphincter, but not voluntary contraction, disappears. Stimulation of the receptors by stretch accounts for the strong and sometimes painful contraction of the external anal sphincter that follows sudden distention of the anus. Such sudden distention is followed by a gasp and stimulation of respiration, a fact useful to those dealing with respiratory depression. Whereas the internal anal sphincter is not under voluntary control, the external sphincter is; reflex contraction can be modified by voluntary effort.

The internal anal sphincter is usually in a state of nearly maximal contraction; its major reflex response is to relax. Because its muscle is continuous with the circular muscle of the rectum, the intrinsic plexuses are its most important innervation. Action of the plexuses is augmented by the extrinsic motor supply of the sphincter, which arrives in the hypogastric nerves. Impulses in sympathetic fibers cause contraction, and impulses in parasympathetic fibers cause relaxation.

Because it is voluntary muscle, the external anal sphincter is paralyzed after destruction of the lower spinal cord, but the internal anal sphincter, being smooth muscle, is not.

Responses of the sphincters have been

measured by placing small balloons within them. When the rectosigmoid portion of the colon or the rectum is distended, the internal anal sphincter relaxes and the external anal sphincter contracts (Fig 5–4). The average peak pressure found 2 cm from the anal verge is between 25 and 120 mm Hg in a normal adult at rest. The internal anal sphincter is responsible for 85% of this pressure. Relaxation of the internal anal sphincter occurs whether or not the colon and rectum themselves contract following distention. If distention is transitory, relaxation is likewise brief; but during prolonged distention of the rectum, the internal anal sphincter usually returns to its contracted state. This reflex relaxation does not require extrinsic innervation, for it is found in patients whose sacral nerve roots have been destroyed. Contraction of the external anal sphincter does require integrity of extrinsic innervation. Reflex contraction occurring when the

Fig 5–4.—Response of the rectum, internal anal sphincter and external anal sphincter to distention of the rectum. Mild distention of the rectum stretches its wall and causes a passive increase in pressure. More distention is followed by further increase in pressure resulting from active contraction. Still more distention is followed by greater active contraction, and subsequent contractions may occur rhythmically at 20-sec intervals. Each increase in pressure in the rectum is accompanied by a decrease in pressure in the internal anal sphincter and an increase in pressure in the external anal sphincter. (From Davenport, H. W.: *A Digest of Digestion* [Chicago: Year Book Medical Publishers, Inc., 1975]. Adapted from Denny-Brown, D., and Robertson E. G.: Brain 58:256, 1935; and from Schuster, M. M., et al.: Bull. Johns Hopkins Hosp. 116:79, 1965.)

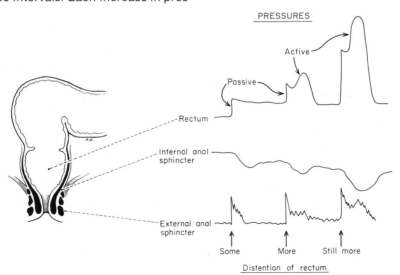

rectum is moderately distended, as with 50 cc of air, is a major factor in maintaining continence. In incontinent patients, the reflex is diminished or absent. However, the external anal sphincter relaxes when the rectum is grossly distended with 200 cc at a pressure of 45−55 mm Hg. Any activity (except defecation and micturition) that increases intra-abdominal pressure increases external anal sphincter contraction. Micturition is accompanied by sudden and complete relaxation of the external anal sphincter, and gas is frequently passed.

Distention of the rectum in a patient with anal fissures causes relaxation of the internal anal sphincter as in a normal person, but relaxation is followed by reflex contraction, which elevates pressure in the anal canal by 20−30 mm Hg above resting pressure.

Defecation

The rectum is usually empty and collapsed; the angulation caused by the contraction of the puborectalis muscle and helical folds of the mucosa resist progress of feces into it. Because the rectum contracts more than does the sigmoid colon, there is a reversed gradient of motility; this gradient empties contents of the rectum, including suppositories, upward.

When contractions of the upper parts of the colon force feces into the rectum, the rectum contracts and the internal anal sphincter relaxes. In paraplegics, whose external anal sphincter is paralyzed, this is enough to produce automatic defecation. In normal persons, contraction of the external anal sphincter prevents expulsion of feces and permits build up of volume in the rectum. As volume increases, pressure increases; and when intrarectal volume is 150−200 cc and pressure 55 mm Hg, there is complete relaxation of both internal and external anal sphincters, and the contents of the rectum are expelled.

Distention of the rectum brings the urge to defecate. However, as anyone who has had diarrhea knows, fullness of the rectum is not an essential part of urgency. The massive and strong contractions of the hyperexcitable colon are enough; and in persons who have strong rhythmic contractions, a sense of fullness of the rectum and a desire to defecate accompany each contraction, despite emptiness of the rectum.

When the rectum is filled to about 25% of its capacity by 100 ml of feces and when its pressure has increased by about 18 mm Hg, the urge to defecate is experienced. Volumes and pressures in the rectum can be trebled before the need becomes imperative. During each increment of filling there is a transitory relaxation of the internal anal sphincter and contraction of the external anal sphincter. However, as the urge to defecate increases, the external anal sphincter and the puborectalis muscle relax, and relaxation is facilitated by centers above the spinal cord. The complex act of defecation follows. Contraction of the longitudinal muscles of the rectum and distal colon shortens the rectum, and the angle between the distal colon and rectum is straightened. Pressure rises within the rectum. These changes are enough in themselves to expel feces at rates ranging from stately slowness to explosive abruptness. They frequently empty the distal colon as high as the splenic flexure.

As the rectum empties, there is rebound contraction of both the internal and external anal sphincters, but voluntary facilitation of defecation can maintain relaxation of the sphincters until the rectum is completely empty.

In a normal person, evacuation is assisted by descent of the diaphragm to its extreme inspiratory position; at the same time the glottis is closed and contraction of the chest muscles on the full lungs raises both intra-abdominal and intrathoracic pressures. The hemodynamic consequences of this maneuver are an abrupt rise in arterial pressure as increased intrathoracic pressure is transmitted across the wall of the heart, stoppage of venous return with a rise in peripheral venous pressure, a fall in stroke output and a subsequent fall in arterial pressure. Death while straining at stool can result from cerebral vascular accidents due to the rise in intracranial pressure, from ventricular fibrillation or coronary occlusion, possibly related to the decreased stroke output, from dissecting aneurysm and from pulmonary emboli caused by mural clots that have been dislodged by changes in cardiac transmural pressure. Intra-abdominal pressure is further increased to 100–200 mm Hg by powerful contractions of the abdominal muscles. Greatest pressures with easiest evacuation occur in the squatting position. Persons sitting on high modern water closets bend forward to assist abdominal contraction to raise intra-abdominal pressure. The greatest straining, measured both by duration and by intra-abdominal pressure reached, occurs with the smallest stool; the greater the fecal mass, the less the effort to evacuate it. During defecation, the abdominal contents are supported by the pelvic diaphragm, which draws the anus up over the fecal mass and resists prolapse of the anus and rectum.

The defecation reflex is further augmented by tactile stimuli of the skin and anus so that actual passage of the feces reinforces the reflex. Arousal of the micturition reflex by a full bladder also facilitates defecation. Although micturition may precede defecation, the stream of urine is usually interrupted during defecation while the stool is actually passing; then, after the bulk of the feces is evacuated, micturition resumes and the bladder is emptied. The final emptying of the bladder may be accompanied by contractions of the empty rectum.

The efferent pathways of defecation reflex are parasympathetic and cholinergic; therefore the reflex is made more vigorous by parasympathomimetic drugs. Prolonged administration of the destructive anticholinesterase diisopropylfluorophosphate results in diarrhea, which eventually becomes bloody as the mucosa is rubbed off by repeated vigorous contractions of the muscle of the colon. Defecation is entirely normal in the complete absence of sympathetic innervation. In man, transection of the cord above the lumbosacral region results in early in-

continence of feces, but the reflex soon returns, so that autonomous evacuation follows a mass movement in the proximal colon.

The fact that voluntary facilitation or inhibition of defecation exists shows that nervous activity of the highest centers projects on the medullary and sacral centers. Voluntary inhibition of defecation is expressed in strong contraction of the striated muscles of the pelvic diaphragm and external anal sphincter. Depression of motility of the whole colon also occurs, for the movements of colonic segments viewed through a colostomy have been seen to cease when the subject was attempting to restrain defecation. When voluntary closure of the external sphincter is effective, feces may remain in the rectum, and subsequent relaxation of the rectum, reducing tension in its wall, removes the stimulus for defecation. Likewise, when the urge to defecate following inflation of a balloon in the rectum is successfully defeated, the pressure in the rectum and tension in its wall drop and the urge passes away. Then, if the balloon is further inflated, pressure and tension rise and the urge to defecate returns. Consequently, the rectum may be filled with feces that fail to stimulate defecation until a subsequent mass movement increases their volume. The defecation reflex is also inhibited by pain or fear of pain, especially the pain aroused, as with hemorrhoids, by defecation itself.

When the call to stool is refused, contents of the rectum are often returned to the descending colon for safekeeping. This is accomplished by a series of retroperistaltic progressive movements traveling at 0.5–1 cm per min.

Megacolon

The underlying physiological disturbance in the disease of congenital megacolon is similar to that in achalasia of the esophagus. In fact, megacolon, megaureter and esophageal achalasia have occurred in the same person. In megacolon, the colon is enormously enlarged and hypertrophied above a narrow segment, which is usually near the rectosigmoid junction. Dilatation and hypertrophy are entirely the result of failure of the feces to pass through the constriction, for they disappear following colostomy or excision of the segment. Failure of feces to pass is the result of lack of coordination of movement in the segment with that of the rest of the colon. Although the constricted part may contract rhythmically, its contractions are not propulsive.

Ganglion cells are completely absent from the affected segment, and the internal anal sphincter fails to relax when the part above it is distended. Excision of the aganglionic segment usually allows establishment of essentially normal defecation, followed by regression of the colonic enlargement.

Megacolon can be produced experimentally in the dog. Ganglion cells of the myenteric plexus are killed by ischemia when a segment corresponding to the rectosigmoid junction is perfused for 4 hours with physiological salt solution. After blood flow is re-established, the perfused segment remains persistently constricted, often with dilatation of the colon proximal to the site.

Psychosomatic Factors

Alterations in function of the colon frequently accompany emotional disturbances. The acute reaction to extreme fright is known to everyone and is described in familiar phrases of the common speech. Medical students facing a serious examination frequently have diarrhea. Chronic dysfunction may be expressed either as depression or as enhancement of activity. In general, reactions to pain, fear and anxiety produce pallor of the mucosa, reduced secretion of mucus and inhibition of motility.

If, in response to his troubles, the subject is "grimly hanging on" not only to his life but to his fecal mass as well, he is likely to be constipated. On the other hand, anger, resentment and hostility, overt or subconscious, are generally associated with hyper-

emia, engorgement and increased motility. When extreme, these responses produce the *irritable* or *spastic colon* characterized by uncoordinated, abnormal motor function, abdominal pain, flatulence and either constipation or diarrhea. As the result of irregular spasms, the fecal masses may be retained for a long time while water is absorbed; they are eventually evacuated as hard, dry balls or, as the result of spasm of the anus, rectum and sigmoid colon, as pencil-thin ribbons. This state may alternate with diarrhea.

A patient with irritable colon has hyperalgesia; a degree of tension in the wall of his colon that would not bother a normal person causes him to feel pain in the hypogastrium, in the iliac fossa and in the anorectal region. During irregular spasms of the sigmoid colon and rectum, the internal anal sphincter relaxes; this, together with the hyperalgesia, produces a sense of urgency without the ability to defecate.

When the behavior of normal subjects or of patients with irritable colon is observed in the laboratory, functional changes are often found to coincide with mood swings. Depression of motility may accompany self-reproach and a feeling of helplessness; hypermotility may occur during expression of hostility. Hyperemia without motility changes occurs during embarrassment; the colon of a male medical student being examined through a sigmoidoscope blushed when he learned that a young woman was peering through the instrument. However, it is impossible to predict what emotional states will have an objective correlative in colonic activity; there may be no relation between the two; or the relation may be a quixotic one. The colon of one woman thoroughly studied on four occasions had a very similar pattern of motility when she was calm and when she was weeping freely during discussion of distressing life experiences. Pressures in the colon of a 58-year-old patriotic Briton were measured when he was recovering from a slight stroke. There were no changes when his physician probed the patient's responses to his hemiplegia, the high probability of future disablement and the likelihood of his being thrown on the parish, but his colon rose to a perfect storm of activity when talk was turned to the Royal Family.

REFERENCES

Christensen, J.: Myoelectric control of the colon, Gastroenterology 68:601, 1975.

Connell, A. M.: Motor action of the large bowel, in Code, C. F. (ed.): *Handbook of Physiology:* Sec. 6. *Alimentary Canal,* Vol. IV (Washington, D.C.: American Physiological Society, 1968), pp. 2075–2092.

Daniel, E. E., Bennett, A., Misiewicz, J. J., Edmonds, C. J., Hill, M. J., and Cummings, J. H.: Symposium on colon function, Gut 16:298, 1975.

Duthie, H. L.: Anal continence, Gut 12:844, 1971.

Hulten, L.: Extrinsic nervous control of colonic motility and blood flow: An experimental study in the cat, Acta Physio. Scand. [Suppl.] 335:1, 1969.

Ihre, T.: Studies on anal function in continent and incontinent patients, Scand. J. Gastroenterol. [Suppl.] 9:1, 1974.

Mendeloff, A. I.: Defecation, in Code, C. F. (ed.): *Handbook of Physiology:* Sec. 6. *Alimentary Canal,* Vol. IV (Washington, D.C.: American Physiological Society, 1968), pp. 2140–2146.

Patel, P. D., Picologlou, B. F., and Lykoudis, P. S.: Biorheological aspects of colonic activity. II. Experimental investigation of the rheological behavior of human feces, Biorheology 10: 441, 1973.

Schuster, M. M.: Motor action of rectum and anal sphincters in continence and defecation, in Code, C. F. (ed.): *Handbook of Physiology:* Sec. 6. *Alimentary Canal,* Vol. IV (Washington, D.C.: American Physiological Society, 1968), pp. 2121–2139.

Schuster, M. M.: The riddle of the sphincters, Gastroenterology 69:249, 1975.

Truelove, S. C.: Movements of the large intestine, Physiol. Rev. 46:457, 1966.

Young, S. J., Alpers, D. H., Norland, C. C., and Woodruff, R. A., Jr.: Psychiatric illness and the irritable bowel syndrome, Gastroenterology 70:162, 1976.

6

Vomiting

VOMITING, OR EMESIS, is the forceful expulsion of gastric and intestinal contents through the mouth; being forceful, it is distinguished from the gentle regurgitation occurring in infants, whose lower esophageal sphincter is inadequately closed. Preceding or accompanying vomiting are tachypnea, copious salivation, dilatation of the pupils, sweating and pallor, and rapid or irregular heartbeat—all signs of widespread autonomic discharge. In man, these are usually associated with the psychic experience of nausea. It is impossible to tell whether an animal showing the same signs is nauseated; and because all the signs can occur in a decerebrate man or animal, they cannot be equated with the experience of nausea. Labored breathing with the mouth partly or entirely closed, and retching, the involuntary movements of vomiting without actual expulsion of vomitus, may precede vomiting. The involuntary muscles of the esophagus, stomach and small intestine participate with movements that assist in expelling vomitus. These five components may be mixed in any proportion: (1) actual expulsion, (2) associated autonomic discharge, (3) the sensation of nausea, (4) retching and (5) other movements of the digestive tract.

Mechanics of Vomiting

Vomiting is usually preceded by retching, which defeats the normal antireflux mechanisms. Retching begins with a deep inspiration. The glottis is closed, and intrathoracic pressure falls far below atmospheric pressure. At the same time, strong contractions of the abdominal muscles raise intra-abdominal pressure, and the pressure gradient from abdomen to thorax may be as great as 200 mm Hg. The abdominal portion of the esophagus and the cardiac portion of the stomach are aspirated into the thorax; a sliding hiatal hernia occurs at each retch. At the same time, a strong annular contraction at the angular notch nearly divides the body of the stomach from the antrum. While the body of the stomach is flaccid, the antrum contracts so that gastric contents are shifted into the upper part of the stomach (Fig 6–1, B), and then gastric contents are emptied, as through a funnel, into the relaxed esophagus (Fig 6–1, C). Because the pharyngoesophageal sphincter is closed, none of the gastric contents enters the pharynx or mouth. As abdominal contraction slows and stops, the abdominal wall moves outward; at the same time the chest wall moves inward and the diaphragm downward. The stomach refills from the esophagus as rapidly as it empties. The esophagus remains dilated, and there is no sign of esophageal peristalsis. With return of most of the esophageal contents into the stomach, the esophagus quickly collapses symmetrically along its entire length; but the lower esophageal sphincter remains open. The retching reaction does not hesitate at this point, but proceeds with immediate repetition of the cycle.

During retching, secretion of mucus by the stomach increases, but secretion of acid falls. Segmental movements in the duodenum become violent, and they may be converted into spasm. Reverse peristalsis in the duodenum has been observed in vomiting cats and dogs; whether it occurs in man is unknown. Gastric relaxation and duodenal contraction reverse the usual pressure gradient so that duodenal contents are forced into the stom-

ach. Contents of the lower intestine may be regurgitated during prolonged vomiting. The most distal parts of the intestine are quiet during vomiting, but the colon often becomes active, and defecation follows.

The act of vomiting builds up on a developed retch. While the esophagus is full of gastric contents, there is a sudden rise in intrathoracic pressure. Strong contractions of abdominal muscles increase intra-abdominal pressure and force the diaphragm high into the thorax. The larynx and hyoid bone are drawn forward, and high intrathoracic pressure, which may be 100 mm Hg above atmospheric pressure, forces the contents of the esophagus through the pharyngoesophageal sphincter and out of the mouth. Because abdominal contraction is maintained, there is limited reflux from the esophagus into the stomach, and at the end of expulsion the esophagus is still full. After several seconds,

Fig 6–1.—Radiographs taken of an erect human subject. **A,** before retching began. The duodenal bulb *(DB)* is distended with barium, and the stomach has not yet begun to contract. **B,** taken during retching. The duodenal bulb *(DB)* has emptied, and its contents appear to have passed back into the stomach. The proximal part of the gastric antrum *(GA)* is tightly contracted, and there has been a shift of barium from the lower to the upper part of the stomach. **C,** taken during emesis. The cardia *(CA)* is elevated and opened widely. Barium is being forced out of the stomach and up the esophagus. There is a long contracted segment in the lower part of the stomach. (From Lumsden, K., and Holden, W. S.: Gut 10:173, 1969.)

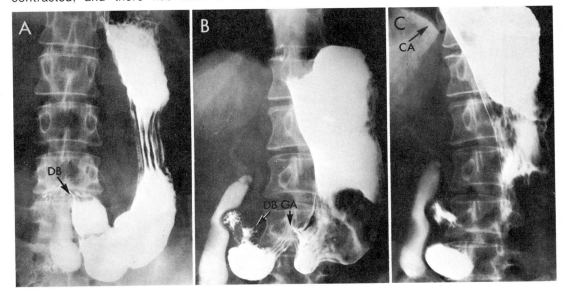

peristalsis begins in the upper esophagus; the esophagus empties into the stomach, and the lower esophageal sphincter closes. If a significant volume remains in the stomach after several seconds, a second episode of cyclic filling and emptying of the esophagus begins, culminating in further expulsion of vomitus. Finally, the voluntary muscles relax and normal respiration resumes.

The Vomiting Center

Straightforward vomiting unaccompanied by retching, called *projectile* vomiting, is governed by a vomiting center in the dorsolateral border of the lateral reticular formation of the medulla, lying ventral to the solitary tract and its nucleus. Projectile vomiting follows electric stimulation of this region in experimental animals with a latent period no longer than is required for the initial inspiration, and it ceases abruptly at the end of stimulation. The center in the lateral reticular formation is closely associated anatomically and functionally with the respiratory center, vasomotor nuclei and a locus whose stimulation produces retching without vomiting. The vomiting center coordinates activities of these neighboring structures to produce the complex response. The heavy, rhythmic respiratory movements of retching may precede or follow expulsion of vomitus, but retching is distinct from vomiting and may occur for many hours without actual vomiting. In the cat, electric stimulation of the vestibular nucleus has occasionally produced retching that could not be converted to vomiting even by filling the animal's stomach with milk.

Afferent Pathways

The vomiting center is activated only by afferent impulses, which arise from many parts of the body; it is not directly stimulated by emetic substances carried to it by the blood. Among the effective stimuli exciting afferent impulses are: tactile stimulation to the back of the throat; distention of the stomach or duodenum to a pressure of about 20 mm Hg; distention or injury of the uterus, renal pelvis or bladder; a rise in intracranial pressure; rotation or unequal stimulation of the labyrinths; acceleration of the head in any direction; and many kinds of pain, such as that attending injury to the testicles. Many young children can vomit voluntarily to express disapproval of parental actions, and some adults may do so, often by thinking about nauseating experiences. Persistent psychogenic vomiting in adults occurs in persons locked in an inescapable hostile relationship within the family. Such persons may become severely hypokalemic.

There are two general pathways by which emetic substances or chemical changes in body fluids affect the vomiting center. The first is by way of the chemoreceptor trigger zone consisting of a restricted group of receptor cells in the area postrema, on the floor of the fourth ventricle. Stimulation of the chemoreceptor trigger zone by emetics in the blood or cerebrospinal fluid elicits vomiting. Destruction of the trigger zone eliminates the response to centrally applied emetics and also the vomiting accompanying uremia, radiation sickness and motion sickness. Other pathways lie in many afferent nerves, especially from the digestive tract, which are activated by drugs or noxious substances. The most sensitive receptors are in the first part of the duodenum. Ipecac placed in the human stomach through a fistula has no effect until it passes through the pylorus; then it immediately induces vomiting. Copper sulfate stimulates receptors whose fibers run centrally in both vagus and sympathetic nerves; but after complete denervation of the gut, absorbed copper sulfate still causes vomiting by stimulating the chemoreceptor trigger zone. After ablation of the trigger zone, even a lethal dose of copper sulfate will not cause vomiting.

REFERENCES

Borison, H. L., and Wang, S. C.: Physiology and pharmacology of vomiting, Pharmacol. Rev. 5: 193, 1953.

Hill, O. W.: Psychogenic vomiting, Gut 9:348, 1968.

Lumsden, K., and Holden, W. S.: The act of vomiting in man, Gut 10:173, 1969.

McCarthy, L. E., Borison, H. L., Spiegel, P.K., and Friedlander, R. M.: Vomiting: Radiographic and oscillographic correlates in the decerebrate cat, Gastroenterology 67:1126, 1974.

Smith, C. C., and Brizee, K. R.: Cineradiographic analysis of vomiting in the cat: I. Lower esophagus, stomach and small intestine, Gastroenterology 40:654, 1961.

PART II
SECRETION

7

Salivary Secretion

THE SALIVARY GLANDS of man secrete 1–2 liters of saliva a day, the rate of secretion ranging from barely perceptible to as high as 4 ml per min on maximal stimulation. In keeping the mouth wet, saliva facilitates speech and lubricates food for swallowing. It is essential for dental health, for caries occurs in the absence of secretion. Decreased secretion in dehydration makes the mouth dry, and this contributes to the sensation of thirst. Saliva dissolves sapid substances, making them available for taste. Human saliva contains the α-amylase ptyalin, which, when mixed with food by chewing, begins the digestion of starch. Saliva secreted in response to noxious or unpleasant substances dilutes them and helps to cleanse the mouth, and its bicarbonate content neutralizes acids.

Structure and Innervation of Salivary Glands

The mouth contains numerous small buccal glands, together with some purely mucous glands on the inferior surface and margin of the tongue. These glands do not secrete enough saliva to keep the mouth comfortably moist; and so, a man having no functioning major salivary glands must frequently rinse his mouth with small drinks of water. Most saliva is secreted by three pairs of large glands: the parotid, the sub-maxillary and the sublingual. The parotid glands are "serous" glands, for their acini contain only one type of cell, which secretes fluid devoid of mucin. Their secretion is therefore watery as compared with the mucin-containing saliva from the mixed sub-maxillary and sublingual glands, whose acini contain cells capable of secreting muco-proteins. The acini lead to short intercalated tubules, which empty into interlobular tubules and then into the excretory ducts. The cells of the acini and the two kinds of tubules have ultrastructure similar to that of other secretory cells, and the physiological evidence shows that all three structures participate in formation of saliva. There is a sharp transition at the point where the tubules become excretory ducts.

All three pairs of glands receive sympathetic innervation from the superior cervical ganglion. These fibers are all adrenergic, liberating norepinephrine, and they are distributed to blood vessels and secretory cells. Stimulation of sympathetic fibers to all glands causes vasoconstriction; in man, stimulation of the sympathetic trunk in the neck or injection of epinephrine causes secretion by the submaxillary but not by the parotid glands. Preganglionic parasympathetic innervation reaches the glands from the cranial outflow by way of the glosso-pharyngeal nerve (IX), the chorda tympani branch of the facial nerves (VII) and the hypoglossal nerve (XII). These synapse in, or close to, the glands; their postganglionic fibers are distributed to all secretory cells. Postganglionic fibers innervated by 5–10 preganglionic fibers converge on each secretory cell, and their stimulation causes copious secretion. The mediator is acetylcholine, and injection of parasympathomimetic drugs produces equal but no greater response. Atropine administration causes dry mouth by blocking the effect of acetylcholine on secretory cells. All salivary secretion in man, with the exception of paralytic secretion, is a response to nerve impulses; there is no humoral control.

Salivary glands atrophy when they are not used, and a decrease in the flow of nerve impulses to the glands in both the sympathetic and parasympathetic fibers is responsible for the atrophy. The rat is a nocturnal animal; when it is fed solid chow, there is a diurnal rhythm of protein synthesis in its salivary glands. Synthesis is decreased by day and increased by night. If the solid food is replaced by a liquid diet, the glands atrophy, their store of secretory products falls to a low level, and there is no diurnal pattern of protein synthesis. Increased stimulation causes hypertrophy of the salivary glands.

Composition of Saliva

Figure 7–1 shows representative values for the composition of human parotid saliva

secreted in response to a parasympathomimetic drug. The saliva used for analysis was being secreted at a steady rate in response to a graded dose of the drug, and the graph was constructed from many observations made at rates of secretion covering the range of 0.1–4 ml per min. At high secretory rates, the sodium and chloride concentrations are greater than at low rates, but they never equal their concentrations in plasma. The total osmotic pressure is always hypotonic in man, being, at maximal secretory rates, about two thirds of the plasma value. Dehydration and repletion lead to corresponding changes in saliva osmolarity. Phosphate concentration is constant at 6 mEq per liter at rates above 1 ml per min, and calcium concentration is likewise constant at 3 mEq per liter. In the same range of secretion rates, the concentration of magnesium is only 0.06 mEq per liter.

The pH of saliva secreted by the unstimulated human parotid gland ranges from 5.45 to 6.06. On stimulation, the pH of parotid saliva rises by two pH units to a maximum of 7.8. After secretion, saliva becomes more alkaline in the mouth as the result of loss of dissolved carbon dioxide. The reason for the pH rise is that the partial pressure of carbon dioxide remains approximately constant at all secretion rates, but the bicarbonate concentration rises as rate of secretion increases.

Parotid glands on opposite sides may secrete at different rates in response to the

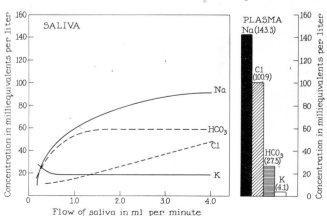

Fig 7–1.—Average composition of the parotid saliva of three young women as a function of the rate of secretion. (From Thaysen, J. H.; Thorn, N. A., and Schwartz, I. L.: Am. J. Physiol. 178:155, 1954.)

same stimulus, but the same relation between rate and electrolyte secretion obtains for each. There are detectable differences between individuals, and a small day-to-day variation occurs in any one person. Human submaxillary and sublingual salivas have the same general relation between rate of secretion and sodium and potassium concentrations.

Reasons for Variations in Composition of Saliva

Systematic variation in the composition of juice according to rate of secretion is characteristic of all digestive glands. Three general types of explanation are given:

1. The composition of the juice as it is extruded from the secreting cells may itself be variable. This explanation is usually intuitively rejected, apparently on the ground that it violates some presumed constancy of nature.

2. The juice as collected may be a mixture of two or more juices, each secreted at constant composition but at different rates by distinct cells. This idea explains why, in the case of mixed glands such as the submaxillary, samples of saliva may contain greatly differing amounts of mucin, for mucin is secreted by one type of cell, while its aqueous menstruum is secreted by another. This principle accounts for much of the variability of gastric secretion, and it will be more fully discussed when the stomach is considered.

3. A secretion extruded from cells at constant (or perhaps variable) composition may subsequently be acted upon by other cells,

Fig 7–2.— A scheme of electrolyte exchanges during secretion of saliva by the parotid gland.

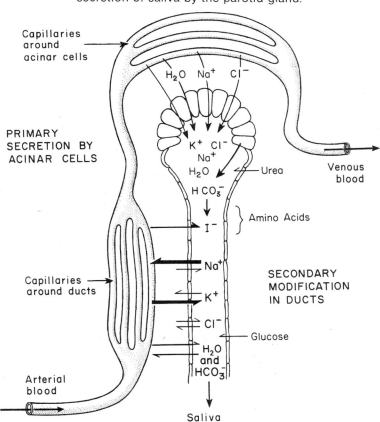

such as those of the gland's ducts. This theory of salivary secretion, which asserts that a primary secretion from acinar cells undergoes secondary changes while passing down the ducts, is the most coherent, if still fragmentary, explanation of salivary secretion. The theory is shown diagrammatically in Figure 7–2.

Primary Secretion by Acinar Cells

Fluid has been obtained by micropuncture from the acini and intercalated ducts of salivary glands of several species. The concentrations of sodium, chloride and bicarbonate in the fluid are approximately equal to their concentrations in plasma water. The concentration of potassium in fluid obtained from nonsecreting glands is higher than the plasma concentration, but it falls as the glands secrete. The fluid is usually slightly hypertonic. Thus, the fluid that begins its journey down the ducts is similar to, if not identical with, an ultrafiltrate of plasma.

As a microelectrode passes through the basal membrane of a resting acinar cell of the cat's sublingual gland, it detects a potential difference of about -31 mV, the inside of the cell being negative to the outside. A value of -95 mV would be predicted on the basis of the potassium concentration gradient, and the measured value obviously does not agree with the predicted one. Neither does it agree with the calculated sodium potential ($+29$ mV) or the chloride potential (-12 mV). None of these ions appears to be in electrochemical equilibrium across the membrane. Since the concentration of potassium is high in the resting cell, potassium may be continuously accumulated, and perhaps sodium is extruded. According to this view, the resting potential is established by the processes continuously transporting the ions.

A similar and almost identical potential difference exists across the apical membrane of the cell. The membrane is electrically inexcitable, and it does not show a propagated action potential when it is stimulated.

Two seconds after the cell is stimulated through its parasympathetic nerve supply, the basal potential difference rises to -56 mV, and later the apical potential may rise to -43 mV. These secretory potentials may also be produced by the injection of acetylcholine into the gland's arterial supply. They are not all-or-none potential changes, such as the action potentials of nerve and muscle, for they may be graded by varying the strength of stimulation. They are likewise independent of the resting potential of the membrane. The resting potential may be set at any value by passing a current through the membrane. When this is done, secretory potentials of the usual magnitude still follow stimulation.

Most or all of the water contained in saliva is secreted by the acinar cells; although modification of the composition of acinar fluid occurs in the ducts, the ducts have little effect upon the volume secreted. Water flow from interstitial fluid through the cells into acinar fluid is probably secondary to secretion of electrolytes.

Blood Flow and Tubular Exchanges

Most or all of the blood supplying the salivary glands flows first (probably in a countercurrent direction) through capillaries surrounding the main excretory ducts or tubules, and exchanges occur between the blood and the saliva contained within the ducts.

When radioactive iodide, $^{131}I^-$, is present in blood, the cells of the ducts accumulate it and secrete it in high concentration into the saliva. Accumulation of ^{131}I within the tubular cells can be demonstrated by radioautography. As much as 88% of the ^{131}I present in arterial blood is extracted as it passes through the gland. Since ^{131}I is secreted only by the tubular cells, most of the blood must come in contact with the tubules. Because the volume of blood cleared of iodide is so high, much of the blood passing the tubules must come in contact with the acini,

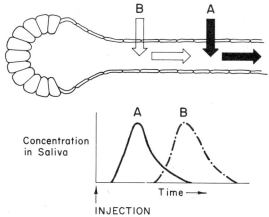

Fig 7–3.—Method of locating the place at which a substance crosses the salivary tubules to enter the saliva. The gland is stimulated to secrete at a constant rate, and the saliva is collected in small, serial samples. A small volume of solution containing two isotopically labeled substances, *A* and *B*, is quickly injected into an artery close to the gland. Substance *A*, because it crosses a more distal portion of the tubule, appears in the saliva and reaches a peak concentration earlier than substance *B*, which enters a more proximal part of the tubule. The relative positions of entry of iodide, sodium, potassium, chloride, water and bicarbonate, shown in Figure 7–2, were determined in this manner. (Adapted from Burgen, A. S. V., and Emmelin, N. G.: *Physiology of the Salivary Glands* [London: Edward Arnold & Co., 1961].)

and therefore the two structures are serially perfused. The question remains: Is flow concurrent or countercurrent? If a salivary gland is loaded with an isotopically labeled substance that can move across the epithelium of the tubules, and if the gland is then perfused with arterial blood in which the isotope's concentration is zero, a distinction can be made. If flow is concurrent, then minimal specific activity (the ratio of the concentration of the isotope to total concentration of the labeled substance) of the substance in saliva must be that of venous blood. If flow is countercurrent, the specific activity of the substance in saliva approaches the zero specific activity of arterial blood and is lower than that of venous blood. In such experiments, the specific activity of radioactive sodium or radioactive urea is frequently lower in saliva than in venous blood; this argues for countercurrent flow as depicted in Figure 7–2.

If small amounts of two tracer substances, *A* and *B*, are injected into the arterial supply of a gland secreting at constant rate, their patterns of efflux into the saliva show the relative positions at which they crossed tubules (Fig 7–3). Since substance *A* can penetrate at a more distal site, it appears in the saliva first. Such experiments have shown that urea enters the tubular fluid soon after it leaves the acini. Iodide is next and is followed by sodium. Potassium (and rubidium and cesium) enter midway. Chloride (and bromide) cross still farther down, and water and bicarbonate exchange at the most distal region. Amino acids and glucose enter saliva at the positions indicated in Figure 7–2.

Movement of Sodium, Potassium and Water

Samples obtained by micropuncture at the beginning of the main excretory duct contain sodium at 117 ± 32 mEq per liter, whereas samples collected at the duct opening contain sodium at 21 ± 15 mEq per liter. The potential difference across the duct wall is such that the lumen is negative with respect to the blood, and therefore sodium must be actively transported by the cells of the duct against sodium's electrochemical gradient. Net transport of sodium out of the lumen is actually the resultant of two opposing unidirectional fluxes. If radioactive $^{24}Na^+$ is injected into the carotid artery supplying an actively secreting parotid gland, it appears in saliva within 5 sec. After subtracting circulation time of $1-2$ sec and deadspace time of $1-2$ sec, the actual transfer time from blood to saliva is less than 3 sec. Considering the dimensions of the salivary glands, this could be achieved in such short time only if some sodium were transferred from blood to saliva without passing through the acinar cells. The simplest explanation is that sodium moves across the walls of the ducts from blood to lumen despite the fact that net flow of sodium occurs in the opposite direction.

Reabsorption of sodium from the ducts accounts for the relation between salivary sodium concentration and rate of flow of saliva shown in Figure $7-1$. At low rates of flow, saliva is present in the ducts for a long time, and sodium reabsorption can be nearly complete. At the lowest rates of secretion, salivary sodium content may be as low as $1-2$ mEq per liter. As flow rate increases, reabsorption of sodium becomes less complete, and salivary sodium concentration rises.

Potassium enters saliva from two sources: some of it from the acinar cells and most from the cells of the main excretory ducts. If fluid containing no potassium but having an initial sodium chloride concentration of 150 mEq per liter is placed between two oil drops in the main excretory duct, at the end of 9 min its sodium concentration has been reduced to 60 mEq per liter and its potassium concentration has become 75 mEq per liter. Secretion of potassium into the luminal fluid appears to be coupled with sodium reabsorption.

Although measurement of the movement of tritiated water between blood and saliva shows that water molecules can move in both directions across the cells forming the gland's ducts, the ducts are relatively impervious to bulk flow of water along an osmotic gradient. Active reabsorption of sodium and an accompanying anion reduces the osmotic pressure of fluid within the ducts, and the epithelial cells of the ducts can establish and maintain an osmotic gradient of $40-60$ mOsm per liter. The permeability of the tubules to water and to nonelectrolytes increases as the rate of flow of saliva increases. The permeability to urea increases about 5 times as the rate of secretion changes from lowest to highest rates. Because sodium reabsorption is more complete at low rates of flow and osmotic permeability is low, saliva secreted at a low rate has a lower osmotic pressure than that secreted at a high rate.

Stimulation of sympathetic nerves to the salivary glands (or administration of epinephrine) increases their permeability; compounds normally excluded, glucose for example, then enter the saliva. In fact, compounds having a molecular radius as large as 0.81 mμ diffuse into the saliva after sympathetic stimulation; but inulin, whose molecular radius is 1.47 mμ, does not. Sympathetic stimulation appears to open pores. Because this stimulation allows sucrose to move from plasma into the saliva but not to enter intracellular water, the pores opened by sympathetic stimulation are probably between the cells of the glands.

Hormones and Composition of Saliva

The sublingual glands of man and the cat and the parotid glands of the rat and of sheep secrete isotonic or slightly hypertonic saliva. A single 13-gm gland of a sheep secretes $1-4$ liters per day; this provides a large volume of alkaline juice, which maintains a neutral pH and fluid consistency of the fermenting contents of the animal's rumen. Sheep parotid saliva contains sodium and potassium in a ratio of about $180:10$. If a fistula is made so that the juice of a single parotid gland is lost, there is rapid depletion of $400-600$ mEq of sodium and a collapse of the extracellular fluid volume. As this occurs, the ratio of sodium to potassium in the saliva reverses, and it may become 10:180. The major part of this reversal is mediated by products of the adrenal cortex, for if these are removed, the animal, despite its severe sodium depletion, continues to secrete saliva with a high ratio of sodium to potassium. In such a depleted, adrenalectomized sheep, the administration of adrenal mineralocorticoids is followed by a fall in salivary sodium and a rise in potassium. Aldosterone acts directly upon the ducts to increase reabsorption of sodium and secretion of potassium.

The generalization that the effect of adrenal mineralocorticoids is to lower salivary sodium and raise potassium concentrations applies to man as well. Administration of aldosterone reduces the salivary Na:K ratio; and excess of endogenous aldosterone, occurring in primary or secondary aldosteronism, likewise reduces the ratio, which rises after the cause of excess hormone secretion has been corrected. On the other hand, patients with adrenal cortical insufficiency secrete saliva with a high Na:K ratio, and the ratio falls during appropriate mineralocorticoid treatment. The concentration of 17-hydroxycorticosteroids in human parotid saliva is directly correlated with the concentration of free, not protein-bound, cortisol in the plasma, and it is a measure of cortisol available to cells.

Salivary Iodide Secretion

Although only the most minute amounts of iodide are normally present in plasma, the tubules of salivary glands of some species are capable of accumulating and secreting iodide. The ion can be secreted into saliva at a concentration 10 or more times the plasma level. Glands capable of secreting iodide at high ratios of saliva to plasma include the parotid glands in man, the dog and the guinea pig and the submaxillary glands in the mouse and man. Accumulation by tubular cells has been demonstrated by radioautography. When ^{131}I is injected into the carotid artery, it appears in saliva within 3 sec, a time too short for secretion by the acinar cells. Iodide

secretion shows self-depression, for the ratio of salivary to plasma concentration falls toward unity when the plasma iodide concentration is raised. Furthermore, the ability of the glands to concentrate iodide is reduced by thiocyanate, perchlorate and nitrate, just as the ability of the thyroid gland to accumulate iodide is reduced. (A similar situation obtains in the stomach.) Thiocyanate itself may be a normally secreted constituent, for saliva gives the color reaction for thiocyanate with ferric ions, and the intensity of color increases when food containing thiocyanate is eaten or tobacco smoke is inhaled.

No organic iodine is present in human saliva, but the dog parotid saliva (and gastric juice and bile) contains iodotyrosine linked in peptide bonds in protein.

Salivary Proteins

Cells of the resting salivary glands contain histologically demonstrable granules whose number diminishes after prolonged secretion. They are called zymogen granules because they are thought to be the locus of storage of enzymes about to be secreted. The major enzyme of saliva is the α-amylase ptyalin, which is secreted by man, apes, pigs, rats, mice, guinea pigs and squirrels but not by horses, cats or dogs. It is a hydrolytic enzyme which splits α-1,4-glucosidic linkages, a reaction that is practically irreversible because its equilibrium is far in the direction of hydrolysis. The enzyme's pH optimum is 6.9, and the enzyme is stable between pH 4 and 11. It is inactive without chloride ions, with full activity occurring in 0.01 N Cl$^-$. Native starch is a mixture of two glucose polymers: (1) unbranched amylose, which is a chain of glucose residues linked through α-1,4-glucosidic bonds, and (2) branched amylopectin, which contains, in addition to straight chains formed by α-1,4-glucosidic bonds, 3–5% branching linkages formed by α-1,6-glucosidic bonds. Salivary amylase hydrolyzes only α-1,4-linkages within the chain; it does not attack the terminal α-1,4-linkages, nor does it break the α-1,6-branching linkages. Consequently, the products of salivary amylytic digestion of amylopectin are maltose (glucose-glucose; α-1,4-linkage), maltotriose (glucose-glucose-glucose; α-1,4-linkage) and a mixture of dextrins containing α-1,6-branches and averaging six glucose residues per molecule. When the enzyme acts on starch paste, the immediate rupture of only 0.1% of the glucoside linkages produces particles 1/100th the original size, and there is a large drop in the viscosity of the suspension.

The physical properties of saliva depend on the kind of protein it contains, not on protein concentration. Parotid saliva, whose viscosity is only slightly greater than that of water, may contain more protein than sticky, stringy submaxillary saliva. The juice of the submaxillary and sublingual glands contains mucin, which is responsible for the lubricating action of the saliva. There are two principal types: those rich in neuraminic acids and N-acetylgalactosamine but devoid of blood-group activity, and others with blood-group activity containing much L-fucose.

The two processes of synthesis of proteins and their secretion are distinct. The formation of new protein in the glands occurs rapidly; the rat submaxillary gland increases its content of α-amylase by 10 times when it is stimulated by feeding or by injection of acetylcholine. Such synthesis can be blocked by atropine. When slices of rat submaxillary gland are incubated in vitro, the addition of acetylcholine or epinephrine doubles or triples the amount of mucin extruded from the cells into the medium. Acetylcholine and other parasympathomimetic drugs also stimulate secretion of amylase by slices of guinea pig and rabbit parotid glands. In both instances, stimulation is accompanied by stimulation of the turnover, but not new synthesis, of the phospholipids of the gland cells. When phosphate containing the radioactive isotope ^{32}P is present, stimulation of secretion is accompanied by a large increase in the specific activity of the phospholipids; there is a 400% increase in that of phosphoinositide and of phosphatidic acid and a much smaller

increase in that of phosphatides containing choline or ethanolamine. At the same time, there is only a small increase in the turnover of the fatty acid and glycerol parts of the phospholipid molecules. The phospholipids are probably part of the carrier mechanism by which secreted protein is transported across the cell membrane.

Nervous Control of Salivary Secretion

The neural apparatus governing salivary secretion consists of secretory centers in the medulla, which receive afferent innervation from the mouth, pharynx and olfactory area and which supply efferent innervation to the glands by way of both parasympathetic and sympathetic outflows. In the cat, unilateral stimulation of the medulla results in either ipsilateral or bilateral secretion by parotid and submaxillary glands. When bilateral responses occur, ipsilateral is greater than contralateral secretion. The responsive centers are in the dorsolateral region of the reticular formation, dorsomedial to the spinal trigeminal nucleus and dorsal to, and at the level of, the facial nucleus. From there, fibers exit in the ventrolateral portion of the medulla by way of the facial nerve to the sublingual and submaxillary glands and by way of the glossopharyngeal nerve to the parotid glands. These parasympathetic fibers are preganglionic, and they synapse with postganglionic fibers in or near the glands themselves. Despite differences in efferent pathways, there is no sharp division between centers. Rostral portions of the center supply the submaxillary, and caudal portions the parotid gland; but there is an intermediate portion supplying both.

Sympathetic innervation supplies all the glands by way of preganglionic fibers in the cervical sympathetic trunk that synapse in the superior cervical ganglion. Centers for the sympathetic outflow have not been clearly delimited. Afferent innervation is included in the intramedullary oral afferent systems, such as the solitary tract and portions of the spinal trigeminal nucleus and tracts; but, as with all medullary centers, additional afferent systems converge on the final common path. For example, salivation is associated with swallowing, licking and chewing movements and with affective reactions.

Sympathetic stimulation acts through the beta-receptors on the secretory cells, and therefore isoproterenol stimulates salivary secretion. Saliva obtained during such stimulation has a very high concentration of potassium and bicarbonate. The reason is that isoproterenol causes only a little secretion of primary fluid by acinar cells, and the cells of the ducts have the opportunity to remove most of the sodium and to secrete much potassium and bicarbonate into the small volume of fluid passing through them.

Stimulation of the parasympathetic efferent fibers causes secretion of a large volume of serous juice by the parotid glands, secretion of a large volume of juice and mucoprotein by the submaxillary gland and vasodilatation in both glands. Sympathetic stimulation causes contraction of myoepithelial cells on the ducts, and their contraction helps to empty viscous saliva into the mouth.

Paralytic Secretion of Saliva

If one chorda tympani nerve in a dog or a cat is cut, the submaxillary gland on the cut side shows paralytic secretion. This secretion by the denervated gland ranges from scanty to copious, begins 3 days after denervation and occurs in circumstances when the adrenal medulla is stimulated. Paralytic secretion thus appears to be similar to dilatation of the denervated pupil, another example of Cannon's law that denervated structures become exquisitely sensitive to the chemical mediator that had been their stimulus when the nerves were intact. The denervated gland is, in fact, much more sensitive to injected epinephrine or norepinephrine than is a normally innervated one.

Stimuli for Salivary Secretion in Man

Secretion of saliva is almost abolished during sleep. When saliva is collected by spitting out the contents of the mouth every 2 min, a procedure in itself minimally stimulating, the basal flow occurring in the absence of any obvious stimuli is about 0.5 ml per min. The submaxillary glands contribute 69% of resting flow, the parotid glands 26% and the sublingual glands 5%. Upon stimulation by acid in the mouth, the rate of secretion by the parotid glands increases proportionately more than that of the other glands. Resting flow is reduced during dehydration, anxiety, fear or severe mental effort. Chewing tasteless wax at the normal rate of 40–80 strokes per min raises secretion to an average of 2.3 ml per min. When chewing is unilateral, the glands on the active side secrete vigorously while those on the inactive side secrete only a little. Secretion increases with the size of the bolus and with the pressure required to chew it. The most effective stimuli are substances arousing sensations of taste, and acid lemon drops are among the most potent. When the mouth is rinsed with a sapid solution in order to stimulate the whole gustatory receptor field, the secretory response is roughly proportional to the log of the stimulus strength. Maximal secretion averaging 7.4 ml per min follows rinsing of the mouth with 0.5 M citric acid. Smells also stimulate, amyl acetate calling forth secretion at about twice the basal rate.

Our knowledge of conditioned reflexes is built on the observation that a dog's salivary glands secrete in circumstances associated with feeding when food is not actually placed in its mouth. An unconditioned reflex is one that does not depend on previous experience. The secretion of saliva in response to food in the mouth is an unconditioned reflex; the stimulus evoking it is the unconditioned stimulus. If the unconditioned stimulus is presented together with the conditioned stimulus, which in itself would not evoke salivary secretion, a conditioned reflex is eventually established. Then, presentation of the conditioned stimulus alone calls forth salivary secretion. Thus it happens that sound, associated naturally or experimentally with food, comes to be a sufficient stimulus for salivation, and, when presented, the dog's mouth waters.

The universal impression that man's mouth waters at the thought of savory food, the clink of the cocktail shaker or the sight of a sizzling steak supports the notion that conditioned reflexes are also important for human salivation. Unfortunately, careful studies do not confirm this belief. Although a lemon, as a child with mumps is told, is believed to be a strong conditioned stimulus, one subject, whose parotid ducts were cannulated so that secretion could be accurately measured, experienced no increased secretion whatever when he saw a lemon and watched it being cut, squeezed and sucked by a colleague. When he himself drank 2 ml of the juice, his secretory rate increased 12 times. In another experiment, bacon and eggs were cooked within 4 ft of three hungry subjects. In one, there was no increased secretion, and in the other two, saliva flowed at less than twice the resting rate. One subject reported his mouth watering when objectively there was no increase in his rate of secretion. Conditioned reflexes, at least when tested by these homely means, are weak in man, whatever they may be in the dog. We can account for the notion that the human mouth waters by observing that in the waking state the mouth always contains saliva, although usually we are not aware of it; thinking about food or participating in its preparation makes us conscious of the saliva, and we conclude that the saliva has just been secreted.

Salivary Blood Flow

Blood supplying the salivary glands flows first through a dense capillary network surrounding the intralobular ducts and then through a sparser capillary bed serving the secretory cells of the acini. This arrangement reflects the embryologic development of the glands, for the ducts and their blood vessels differentiate before the acini. Venous drainage follows the same general route.

Normal resting blood flow through the dog's submaxillary gland ranges from 0.1 – 0.6 ml per gm of gland per min. This is about 20 times the flow through resting muscle. During maximal stimulation, blood flow increases to 4 – 6 ml per gm of gland per min, or 10 times that through active skeletal muscle. The capillaries in salivary glands have an enormous filtration capacity, the value being 40 times that of muscle capillaries. Two mechanisms, and perhaps three, mediate active hyperemia. Parasympathetic nerves to the glands liberate acetylcholine, and this mediator is responsible for the rapid vasodilatation that occurs when the glands are first stimulated. Stimulation of secretory fibers also releases an enzyme, kallikrein, from glandular tissue into the interstitial fluid. This enzyme acts on a protein substrate in plasma to produce a powerful octapeptide vasodilator, bradykinin, and bradykinin is responsible for the slow, sustained vasodilatation that follows stimulation. However, another process must also be responsible for slow dilatation, for it occurs when a submaxillary gland is perfused with a solution containing no bradykinin precursor. There is no necessary relation between increased blood flow and secretion. When secretion in response to chorda tympani stimulation is inhibited by appropriate small doses of atropine, nerve stimulation still results in vasodilatation through bradykinin formation. On the other hand, increase of the blood flow sevenfold by injection of yohimbine does not cause secretion.

Salivary Metabolism

Resting oxygen consumption is 0.01 – 0.05 cc per gm of gland per min, which is about 6 times the rate of oxygen uptake by resting muscle. When the gland is stimulated, oxygen consumption increases. There is an essentially linear relation between extra oxygen consumed and salivary flow; 1 ml of saliva requires 0.3 ml of oxygen. At the beginning of stimulation, the gland's content of glycogen and phosphocreatine diminishes, and a small oxygen debt is accumulated. Resting submaxillary glands of dogs consume 0.8 – 2.9 (mean 2.1) mg of glucose per gm of gland per hour; and when stimulated, they consume an additional 0.7 – 2.7 (mean 1.5) mg per ml of saliva secreted. If these figures can be transposed to human salivary glands, the secretion of 1 liter of saliva per day calls for 1.5 gm of glucose above the resting requirements of the gland. Energy cost is 6 calories per liter.

REFERENCES

Babkin, B. P.: *Secretory Mechanism of the Digestive Glands* (2d ed.; New York: Paul B. Hoeber, Inc., 1950).

Blair-West, J. R, Coghlan, J. P, Denton, D. A., and Wright, R. D.: Effects of endocrines on salivary glands, in Code, C. F. (ed.): *Handbook of Physiology:* Sec. 6. *Alimentary Canal,* Vol. II (Washington, D.C.: American Physiological Society, 1967), pp. 633 – 664.

Burgen, A. S. V.: Secretory processes in salivary glands, in Code, C. F. (ed.): *Handbook of Physiology:* Sec. 6. *Alimentary Canal,* Vol. II (Washington, D.C.: American Physiological Society, 1967), pp. 561 – 580.

Emmelin, N.: Nervous control of salivary glands, in Code, C. F. (ed.): *Handbook of Physiology:* Sec. 6. *Alimentary Canal,* Vol. II (Washington, D.C.: American Physiological Society, 1967), pp. 595 – 632.

Gautvik, K.: Studies on kinin formation in functional vasodilatation of the submandibular salivary gland in cats, Acta Physiol. Scand. 79:174, 188, 204, 1970.

Kreusser, W., Heidland, A., Hennemann, H., Wigand, M. E., and Knauf, H.: Mono- and divalent electrolyte patterns, Pco_2 and pH related to flow rate in the normal human parotid saliva, Eur. J. Clin. Invest. 2:398, 1972.

Mangos, J. A., McSherry, N. R., Irwin, K., and Hong, R.: Handling of water and electrolytes by rabbit parotid and submaxillary glands, Am. J. Physiol. 225:450, 1973.

Martinez, J. R.: Water and electrolyte secretion by the submaxillary gland, in Botelho, S. Y., Brooks, F. P., and Shelley, W. B. (eds.): *Exocrine Glands* (Philadelphia: University of Pennsylvania Press, 1969), pp. 20 – 29.

Petersen, O. H.: Acetylcholine-induced ion transports involved in the formation of saliva, Acta Physiol. Scand.[Suppl.] 381:1, 1972.

Petersen, O. H., and Poulsen, J. H.: Secretory transmembrane potentials in salivary glands, in Botelho, S. Y., Brooks, F. P., and Shelley, W. B. (eds.): *Exocrine Glands* (Philadelphia: University of Pennsylvania Press, 1969), pp. 3 – 19.

Schneyer, L. H., and Schneyer, C. A.: Inorganic composition of saliva, in Code, C. F. (ed.): *Handbook of Physiology:* Sec. 6. *Alimentary Canal,* Vol. II (Washington, D.C.: American Physiological Society, 1967), pp. 497 – 530.

Schneyer, L. H., Young, J. A., and Schneyer, C. A.: Salivary secretion of electrolytes, Physiol. Rev. 52:720, 1972.

8

Gastric Secretion

THE STOMACH RECEIVES diverse amounts of food of heterogeneous composition and consistency at irregular intervals. Stimulated by the act of eating and by the presence of food in the stomach, reflexes acting through nervous and humoral channels cause the oxyntic glandular mucosa to secrete pepsinogens and hydrochloric acid. The concentration of acid in gastric contents is determined by the rate at which it is secreted, by neutralization and dilution by other digestive secretions and food and by back-diffusion of acid into the mucosa itself. The cells of the mucosa are in a dynamic state of growth, migration and desquamation, and most injury to the mucosa is rapidly repaired. The property of the mucosa that normally resists penetration by acid may be impaired; then acid diffusing rapidly into the mucosa may cause it to shed large amounts of fluid and to bleed.

Structure of the Gastric Mucosa

The gastric mucosa is divided, approximately at the notch, into the oxyntic glandular mucosa of the body and that of the pyloric gland area. Separation of the two types of mucosa is not precisely at the notch; generally, the pyloric type of mucosa extends above the notch into the body, particularly along the lesser curvature. The mucosa of the body is covered with simple columnar epithelial cells that contain and secrete mucus and an alkaline fluid. The surface is studded with depressions forming gastric pits, into each of which the lumina of three to seven glands empty. The area occupied by the pits is 50% of the total surface. The distance between centers of the pits averages 0.1 mm, and there are about 100 pits per mm^2. The cells at the transition between surface epithelial cells and the glands are neck chief cells, whose cytoplasm stains for mucin. The chief cells, containing and secreting pepsinogens, are the most numerous of the several types of cells in the glands. On the borders of the glands, and especially in the neck region, are the cells responsible for acid secretion. Because of their location they are frequently called parietal cells, but their function of secreting acid is more appropriately described by the name oxyntic cells. They contain intracellular canaliculi, which communicate with the lumen by ducts passing between the chief cells. The surface of the pyloric gland area is also covered with epithelial cells. The glands in this region are composed of cells resembling neck chief cells, but oxyntic cells are almost or entirely absent. Chief cells and oxyntic cells are absent from the cardiac gland area, a ring of mucus-secreting glands 1 – 4 cm wide at the junction of the esophagus with the stomach.

Secretion of Surface and Neck Chief Cells

The approximate composition of the alkaline juice secreted by surface cells is given in Table 8 – 1. The estimated values for the dog in column 2 were calculated from analyses of samples of juice being secreted at different rates in response to graded doses of histamine. It was assumed that these samples consisted of different proportions of acid and nonacid juices. On the further assumption that variations in the rate of secretion were caused only by variation in the proportion of acid juice of fixed composition, an extrapolation to a hypothetical sample containing no acid was made. The values for man were calculated by applying the same assumptions to data obtained from numerous samples of gastric juice stimulated either by histamine or by insulin hypoglycemia. Data for the dog in column 3 were obtained from samples secreted by an exposed flap of gastric mucosa stimulated by high doses of acetylcholine. The ionic composition of the secretion is close to that of an ultrafiltrate of plasma to

TABLE 8 – 1.—COMPOSITION OF JUICE SECRETED BY SURFACE EPITHELIAL CELLS OF THE STOMACH

	MAN, EST.* RANGE (mM)	DOG, EST.† MEAN (mM)	DOG, BY ANALYSIS‡ MEAN (mM)
Na^+	150 – 160	155	138
K^+	10 – 20	7	4
Ca^{2+}	3 – 4	4	5
Cl^-	125	133	117
HCO_3^-	45	33	23
pH	(7.67)§	(7.54)§	7.42
Protein	. .	. .	10 – 11 gm/L

*Equals nonparietal secretion, calculated by Hunt, J. N.: Physiol. Rev. 39:491, 1959.

†Gray, J. S., and Bucher, G. R.: Am. J. Physiol. 133:542, 1941.

‡Altamirano, M.: J. Physiol. 168:787, 1963.

§Calculated from bicarbonate concentration and assumed Pco_2 of 40 mm Hg.

which has been added mucus containing 3 mEq of sodium and potassium per liter.

The major organic secretion of the surface cells is visible mucus. Although desquamation frequently accompanies mucus secretion, it is not a necessary part of secretion, for visible mucus can be collected free from cells. Mucus is secreted as a gel whose viscosity is 30 – 260 times that of water; and, as such, it entraps the alkaline fluid, with the result that in the resting state the surface of the mucosa is covered by a tenacious, slimy alkaline coat. On contact with acid, visible mucus precipitates to shreds or clumps. Synthesis and discharge of mucus can be followed by radioautography, for intravenously administered $^{35}SO_4^=$ is quickly incorporated into gastric mucus by ester sulfate linkages. The greatest secretion of tagged mucus is at the base of the pits. The secretion of the surface cells also contains the mucolytic enzyme lysozyme (which does not attack gastric mucus) and some dissolved mucoproteose. Secretion of alkaline fluid and mucus is spontaneous in the sense that, in the absence of other stimuli, the resting stomach remains covered by them, and if they are gently removed they quickly reappear. Mechanical irritation of the surface, as by rubbing with a glass rod, stimulates mucus secretion; and more severe stimulation by substances that tend to injure the mucosa causes copious secretion. Splanchnic nerve stimulation at low rates within physiological range causes secretion of mucus, and so does stimulation of the vagus nerve. Mucus secretion is probably under the same reflex control as the other gastric secretions. Secretion is not stimulated by histamine, but it is by acetylcholine and parasympathomimetic drugs.

Neck chief cells secrete glandular mucoprotein, or soluble mucus, which is chemically and physically distinct from the secretion of the surface cells. It is absent from gastric juice in fasting, but it is secreted in response to parasympathomimetic or to vagal excitation by sham feeding or insulin hypoglycemia.

Cardiac and Pyloric Glands

In man, a small zone of gastric mucosa (0.5–4 cm in radius) surrounding the esophagus is lined with cardiac glands. These tubular glands are composed of mucous cells and have few or no chief or oxyntic cells. Oxyntic cells also are rare in the mucosa lining the pyloric gland area of the stomach. The mucosa in this area is thin (1–1.5 mm) and composed of cells resembling mucous neck cells. It produces a scanty, highly viscid secretion, which does contain some pepsinogen. In the dog, the rate of secretion is 0.5–5 ml per hour; and the rate is unaffected by feeding, histamine injection or parasympathomimetic drugs. The secretion's average composition is: sodium, 151 mN; potassium, 9; chloride, 145; and bicarbonate, 8. The acid-neutralizing property of this juice is slight, but its mucus lubricates the surface over which a large volume of chyme moves back and forth during gastric digestion.

Pepsinogens and Pepsins

Pepsinogens are zymogens secreted by gastric mucosa, and they are converted to active, proteolytic enzymes by acid. They can be separated electrophoretically into two groups. Group I contains seven rapidly migrating proteins, and Group II contains one or more slowly migrating ones. Group I pepsinogens occur only in the proximal stomach, but Group II are found throughout the stomach and duodenum. Both groups occur in high concentration in the chief cells of the oxyntic glandular mucosa and in lower concentration in mucous neck cells. Group II pepsinogens are abundant in the pyloric glandular mucosa and in Brunner's glands of the duodenum. Different fractions of pepsinogen, when activated to pepsin, have different pH optima and substrate specificity, and they may also be functionally heterogeneous. Nevertheless, all the zymogens and enzymes will be referred to simply as pepsinogen and pepsin. Little or no pepsinogen can be found in the chief cells 3 hours after feeding, although pepsinogen secretion continues. The granules are not obligatory intermediates in the path of synthesis and secretion. The granules begin to reappear after 6 hours, and they are maximally concentrated after 72 hours.

Pepsinogen is converted to proteolytically active pepsin when it mixes with acid secreted by the oxyntic cells. Activation occurs more rapidly the lower the pH and is almost instantaneous at pH 2.0. Pepsin itself activates pepsinogen, and in the process of activation a small polypeptide is cleaved from the zymogen. Pepsin is maximally active as a proteolytic enzyme between pH 1.0 and 3.0. Although pepsinogen is stable in neutral solution, pepsin rapidly and completely loses its catalytic activity when the solution in which it is dissolved is neutralized. Under some experimental conditions, denaturation of pepsin can be reversed; but for all practical purposes, peptic activity of gastric juice is destroyed by neutralization.

Stimuli for Pepsinogen Secretion

There is a small continuous basal secretion of pepsinogen in man. The most powerful stimulus for additional secretion is vagal stimulation; and, therefore, insulin hypoglycemia, acetylcholine and its congeners and the direct vagal phase of stimulation by feeding cause flow of juice with high pepsin concentration. This effect is mediated by acetylcholine liberated in the neighborhood of the chief cells. Pepsinogen secretion is stimulated by the hormone secretin. Pepsinogen secretion is also increased by the hormone gastrin, but this may not be entirely the result of the action of the hormone on chief cells. Gastrin stimulates acid secretion, and acid in contact with the surface of the gastric mucosa stimulates afferent nerves of the intramural plexuses which, through a cholinergic reflex, in turn stimulate secretion of pepsinogen. The several modes of stimulation are synergistic. Infusion of secretin increases the rate of pepsinogen secretion 3 times when the contents of the stomach are neutral and 12 times when the contents are acid. Pepsinogen secretion is inhibited by atropine. In man or the dog, administration of histamine causes a transitory burst of pepsinogen secretion, which has been attributed to the washing-out of the precursor from the tubules of the gastric glands. As the dose of histamine is increased, pepsinogen secretion increases, along with acid secretion, until the maximal rate of acid secretion (about 10 times basal) is reached. The rate of pepsinogen secretion is then 3–4 times basal.

There are minor qualifications to the generalization that there is a correlation between acid and pepsin outputs. For example, secretin, which stimulates pepsinogen secretion, inhibits acid secretion. Unless otherwise qualified, the statement that one or another factor influences acid secretion can be taken as applying to pepsinogen as well.

Plasma Pepsinogen and Uropepsinogen

Blood plasma acidified to pH 2.0 has a small proteolytic activity, ranging between 1/500th and 1/5,000th that of gastric juice; this activity has been attributed to pepsin. Pepsinogen is present in lymph collected from small lymphatic vessels on the surface of the stomach, and it is doubtless present in gastric interstitial fluid. Plasma pepsinogen falls to low levels after gastrectomy, and it is reduced in atrophic gastritis. There are variations in the concentrations of the different pepsinogens in plasma, but the physiological and diagnostic significance of the variations is not understood. There is a rough correlation between the concentration of pepsinogen in gastric venous blood and the rate of gastric secretion, but the relation between systemic plasma concentration and gastric secretory activity is very poor. The plasma level depends not only on the rate of addition of the enzyme to blood but on the rate of removal as well.

Pepsinogen is excreted into the urine, and it is known as uropepsinogen. The particular kinds of pepsinogen found in urine are the same as those found in plasma.

The renal glomerular membrane is relatively permeable to proteins whose molecular weight is below 68,000, and so pepsinogen passes into the glomerular filtrate. In company with other filtered proteins, it is probably reabsorbed to some extent by the renal tubules. If this is the case, its excretion is affected not only by the blood pepsinogen concentration, which determines its rate of filtration, but by factors affecting reabsorption. Adrenal cortical hormones increase uropepsinogen excretion by increasing its renal clearance; this means that these hormones either increase glomerular permeability to pepsinogen or decrease its rate of tubular reabsorption. As a result, the rate of excretion of the enzyme in the urine does not reflect its rate of liberation into the blood from the gastric mucosa, and the latter does not accurately reflect the prevailing rate of gastric secretion. Therefore the rate of uropepsinogen excretion is not a measure of gastric secretion. This conclusion is confirmed by clinical studies, which have shown that, although the rate of uropepsinogen excretion is, on the average, higher in persons having a high rate of gastric secretion than in those having a low rate, only a small percentage of persons with gastric hypersecretion have urinary pepsinogen levels sufficiently elevated to place them beyond the normal range of variation. Only when the mucosa is sufficiently atrophic (as it is in pernicious anemia) to be unable to secrete pepsinogen is the reduced concentration of the enzyme in plasma and urine pathognomonic.

Gastric Rennin, Gelatinase and Tributyrase

The gastric juice of newborn calves contains a milk-clotting enzyme, rennin; but the enzyme is probably absent from human stomachs. Pepsin itself has powerful milk-clotting activity and performs rennin's function in the human stomach. Crude pepsin preparations contain a distinct and separable enzyme, gelatinase, which liquefies gelatin about 400 times faster than does pepsin. Its close association with pepsin suggests that it, too, comes from chief cells.

Gastric juice also contains a lipase that is stable at pH 2.0 and has a broad pH optimum in the range of pH 4.0 – 7.0. It catalyzes hydrolysis of ester bonds at the 1 and 1′ positions of triglycerides. It is much more active against a substrate containing short-chain or medium-chain fatty acids than against long-chain triglycerides. Because the only natural source of short-chain triglycerides is milk, the enzyme is more important for infants than for adults. A lipase is also secreted by pharyngeal glands.

Acid hydrolysis of triglycerides in the stomach is probably insignificant.

Gastric Hydrochloric Acid

When gastric juice is collected either from the human stomach or from a pouch of the oxyntic gland area of the dog stomach, its composition is characteristically found to be a function of the rate of secretion (Fig 8 – 1). As the rate of secretion increases, hydrogen ion concentration rises, sodium concentration falls and chloride concentration remains nearly constant. At highest rates of secretion, the juice is a nearly isotonic solution of almost pure hydrochloric acid mixed with a very small amount of potassium and sodium chlorides. The upper limit of acid concentration in man is 150 mN, and in the dog it is about 160 mN. If judged by the interest it arouses among physiologists and laymen alike, the most dramatic component of the juice is the acid, which provides optimally low pH for pepsin activity. This component has never been caught directly as it issues from the cells secreting it, nor has it been obtained absolutely uncontaminated by at least a small fraction of other gastric juices. Consequently, there is no totally unequivocal answer to the questions: Is the composition of the acid juice constant as it is secreted at various rates, or does its composition vary with the rate of secretion? Is the acid juice pure hydrochloric acid, or does it contain sodium and potassium salts?

The relation between rate of secretion and composition is not invariable; juice high in

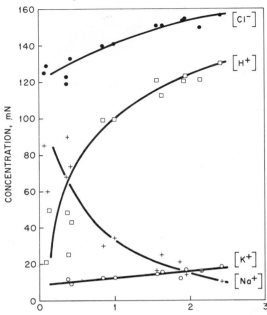

Fig 8–1.—Relation between concentration of electrolytes in the gastric juice of a normal young man and rate of secretion. Secretion was stimulated by intravenous histamine infusion at various constant rates, and the juice to be analyzed was collected after steady rate of secretion had been reached. (Adapted from Nordgren, B.: Acta Physiol. Scand. 58 [Suppl. 202]:1, 1963.)

acid can be obtained at low rates of secretion. When a single subcutaneous dose of histamine is given to a dog, the acidity of the juice collected from a pouch of the oxyntic gland area of the stomach rises to a maximum as the rate of secretion rises. However, as the rate of secretion falls in the fourth quarter-hour after histamine injection, the hydrogen ion concentration of the juice remains high and continues to be so throughout the whole period of declining rate of secretion. Other experiments show that acid concentration can be independent of the rate of secretion; when secretion is stimulated by continuous administration of histamine, injection of atropine profoundly reduces the volume, but not the acidity, of the juice secreted.

Three theories explain variations in the composition of gastric juice:

1. The acid juice as it issues from the secreting cells does vary in composition; at low rates of secretion it contains much sodium and little acid, but as the secretory rate rises, the concentration of sodium falls and that of hydrogen ion increases until at the maximum rate of secretion the juice is nearly pure, isotonic HCl.

2a. Fluid secreted by the oxyntic cells has a constant composition independent of its rate of secretion; it is nearly pure, isotonic HCl with perhaps a very small admixture of NaCl and KCl. On the surface of the mucosa it mixes with one or more components secreted by other cells of the mucosa. These components likewise have constant composition; they are similar to an ultrafiltrate of plasma, containing much sodium and some bicarbonate, and upon mixing with acid juice the bicarbonate reacts with the acid so that some hydrogen ions disappear and osmols are lost as carbon dioxide evolves from the mixture. When acid secretion is stimulated by histamine or gastrin, the rate of secretion of the additional components changes slightly, if at all. Consequently, the composition of pure oxyntic juice can be deduced by extrapolation of the analytic data to an infinite rate of secretion.

2b. A variant of this second theory is that the composition of the other components changes with their rate of secretion.

3. Gastric secretions are modified as they flow over the surface of the mucosa by exchanges with interstitial fluid. Hydrogen ions diffuse slowly or rapidly down a very steep concentration gradient from the lumen into the mucosa, and sodium ions diffuse along an electrochemical gradient from interstitial fluid into the lumen. The extent to which juice is modified depends on the length of time it is in contact with the mucosa and on the permeability of the mucosa.

Although gastroenterologists have supported one or another of these theories with the fervor of Big-enders attacking Little-enders, there are no experimental observations that establish one as correct and reject the others as error. On the contrary, a mixture of all three theories appears to be required to explain the facts.

At very high rates of secretion, gastric juice does approach in composition an isotonic solution of HCl, containing perhaps no more than $3-5$ mN Na and less than 1 mN K; but no amount of manipulation of the data by mathematical extrapolation can prove that this is the composition of acid juice secreted at low rates. A fluid whose composition is similar to that of an ultrafiltrate of plasma can be collected from the surface of the gastric mucosa under very special conditions, but there is no direct experimental evidence that the nonacid juice mixed with acid secretion has this composition. Finally, both sodium and hydrogen ions can cross the gastric mucosa; and when the permeability of the mucosa is high such exchanges grossly modify acid juice secreted by the mucosa. However, the part such exchanges play under physiological conditions has not yet been clearly demonstrated.

Sodium and Potassium Outputs in Acid Juice

Sodium in gastric juice comes from two sources: secretion in acid and nonacid juices from both pyloric and oxyntic gland areas and from diffusion across the mucosa. Under normal circumstances the latter contribution is small. As the rate of secretion of acid juice rises, both concentration (mN) and output (mEq/min) fall. If, however, the gastric mucosa is abnormally permeable, as after damage by acetylsalicylic acid in acid solution, sodium output rises with increasing volume output, because a large amount of sodium diffuses across the mucosa from interstitial fluid. Sodium output is also large in protein-losing gastropathy when the mucosa sheds plasma.

Potassium concentration in the basal secretion of the unstimulated human stomach is higher than in plasma, being $14-18$ mN. In stimulated secretion, the concentration of potassium is constant when the rate of secretion is constant. There is a transient increase of potassium concentration of a few milli-equivalents per liter, lasting 20 min or more, when secretion rate rises, and a corresponding transient fall in potassium concentration when secretion rate decreases. In man, the potassium concentration during steady-state secretion stimulated by intravenous administration of histamine is between 8 and 20 mN. It is lower in the dog, the average being 6 mN, with a range of 3 – 9. Because potassium concentration varies in a relatively narrow range, the output (concentration times rate of secretion) is positively correlated with the output of hydrogen ions.

Potassium in gastric juice is partly derived from intracellular potassium. If radioactive potassium (^{42}K) is injected intravenously and acid secretion is stimulated before the iso-tope has had a chance to equilibrate with intracellular potassium, the specific activity of potassium in the juice is only about half that of plasma potassium.

Loss of gastric juice through vomiting or drainage may severely deplete the body's stores of sodium and potassium.

Source of Hydrochloric Acid

Acid is secreted by large cells distinguished by having intracellular canaliculi lined with microvilli. The cells contain abundant tubular and vesicular components that undergo characteristic changes when the cells are stimulated to secrete (Fig 8 – 2). The cytoplasm of the cells is densely packed with mitochondria. Hydrogen ions are probably

Fig 8–2.—Electron micrographs of non-secreting and secreting oxyntic cells of the mouse gastric mucosa. **Left,** a nonsecreting cell. The cell's cytoplasm is filled with tubules and vesicles that do not communicate with the plasma membrane. A few intracellular canaliculi are seen as circular voids containing a few microvilli. **Right,** a secreting cell. A long intracellular canaliculus containing many interdigitating microvilli lies horizontally above the nucleus, and seven other canaliculi are seen in cross or tangential section below the nucleus. Mitochondria surround each canaliculus. (Courtesy of S. Ito.)

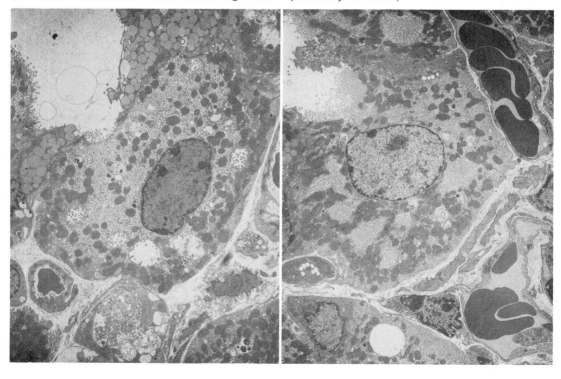

secreted at high concentration into the canaliculi by reactions occurring in the canalicular membrane.

In the mammalian stomach, acid-secreting cells are on the walls of the gastric tubules, and for this reason they are called *parietal* cells (Latin, *paries* = wall). However, the stomachs of many species of animals contain acid-secreting cells that are not on the walls of glands. For this reason, a general, functional designation is better than a morphologic one, and the acid-secreting cells will be called *oxyntic* cells (Greek, *oxys* = sharp). That part of the gastric mucosa containing a dense population of oxyntic cells is called the *oxyntic glandular mucosa.* However, the mucosa lining the gastric antrum, the *pyloric glandular mucosa,* also contains oxyntic cells.

Measurements and Standards of Acidity

In studying gastric acidity, we must know two quantities: the acidity of a particular sample and the amount of acid secreted in a given time.

The acidity of a solution is the *activity* of hydrogen ions in it, not their concentration. Activity is a dimensionless quantity whose symbol is written a_{H^+}. Although physical chemists are not in complete agreement on the subject, the most commonly accepted definition of pH is the negative logarithm to the base 10 of the activity of hydrogen ions in a solution.

$$pH = -\log_{10} a_{H^+} \qquad (8.1)$$

In an infinitely dilute solution the activity of hydrogen ions is exactly equal to their concentration $[H^+]$. In more concentrated solutions, activity is usually less than concentration, and this fact is expressed by the equation

$$a_{H^+} = f[H^+] \qquad (8.2)$$

Here f is the activity coefficient whose numerical value is less than 1 and which must be experimentally determined.

The concentration of hydrogen ions in blood and body fluids except gastric juice is so small that their activity coefficient is probably close to 1, and there is no reason to think that the coefficient varies under most physiological circumstances. Therefore, for these body fluids only, substitution of 1 for f in equation (8.1) gives the familiar equation

$$pH = -\log_{10} [H^+] \qquad (8.3)$$

Gastric juice is not a dilute solution of hydrogen ions, and the activity coefficient of it is less than 1. Its value depends upon both the concentration of hydrogen ions and the sum of the concentrations of sodium and potassium ions. In a solution whose pH, accurately measured, is found to be 1.00 and in which $[Na^+] + [K^+] = 50$ mEq/L, the activity coefficient is 0.810. Thus, the activity of hydrogen ions is 0.100, but the concentration of hydrogen ions is 0.100/0.810, or 0.124N.*

The amount of acid secreted is estimated by titration with a standard base. A sample of gastric juice properly collected in a given interval may not contain all the acid secreted in the collection period, for acid may be lost by the flow of gastric contents through the pylorus, by back-diffusion of acid into the mucosa and by neutralization by bicarbonate contained in nonacid secretions. Loss through the pylorus can be minimized by good collection technique or estimated by indicator dilution methods, but acid lost through back-diffusion or by bicarbonate neutralization can be estimated only by calculations of doubtful validity. Acid in gastric contents may also be buffered by proteins and by other buffers contained in nonacid secretions. This moiety of acid can be recovered if the titration with a standard base is carried to the correct end-point. In most instances the correct end-point lies between

---

*See Moore, E. W., and Scarlata, R. W.: The determination of gastric acidity by the glass electrode, Gastroenterology 49:178, 1965; and Moore, E. W.: Determination of pH by the glass electrode: pH meter calibration for gastric analysis, Gastroenterology 54:501, 1968. The latter paper contains a table of activity coefficients of hydrogen ions as a function of the concentrations of sodium and potassium in gastric juice.

the pH of plasma, 7.4, and about 8.3.[†] Since the latter pH is the one attained in the routine titration of gastric contents when phenolphthalein is used as the indicator, the usual estimation of total acidity of a sample of gastric contents is reasonably accurate.

In practice, a sample of gastric contents is titrated with standard NaOH, first to pH 3.0, as judged by the color change of diaminoazobenzene (Topfer's reagent); and the acid found is called "free acid." Then the titration is continued to pH 8.0, as judged by the color change of phenolphthalein, and the difference between total and free acid is called "combined acid." The free and combined acids are often reported in terms of clinical units, which are the number of milliliters of 0.1 N NaOH required to neutralize 100 ml of gastric juice. Clinical units are identical with milliequivalents per liter, but one suspects that if the person who coined the definition had realized it would give such a sensible result, he would have changed his definition. The acid of gastric juice, determined by titration, should always be expressed in milliequivalents secreted in a given time or as milliequivalents per liter (mN), and the term "clinical unit" or its even more ridiculous alternative, "milligrams of HCl," should be avoided.

The determinants of the acidity of any particular sample of gastric juice are the rate of secretion and the rate of neutralization or dilution. Acidity cannot be higher than that of the undiluted, unneutralized acid secretion, in man about 150 mN,[‡] and consequently the only sense in which hyperacidity can exist is that the sample of gastric juice is more acid than some supposedly normal sample. A better term is hypersecretion, in which the volume secreted in a given time is above normal. Then, if diluting or neutralizing processes are no more than normal, hypersecretion will result in high acid concentration.

Three data are important in assessing gastric secretion: whether the stomach can secrete acid at all, how much it secretes under standard conditions and how much it can secrete under the strongest stimulation. Total inability to secrete acid, or achlorhydria (rarely encountered except in subjects with pernicious anemia), is diagnosed when the pH of juice aspirated from the stomach fails to fall below 6.0 after adequate stimulation. The usual stimulus is the "standard dose" of histamine, [‖] 0.01 mg of histamine acid phosphate (or 0.0036 mg of histamine base) per kg of body weight given subcutaneously. However, this may not be enough. In one large series, 58 subjects were encountered who secreted no acid in response to this dose, but half of them secreted some acid when given 4 times the dose.

In order to distinguish only between presence or absence of acid, tubeless gastric analysis is sometimes used. The subject is fed an acidic ion exchange resin that has been combined with the base quinine or the basic blue dye azure A. At pH below 3.5, hydrogen ions displace the base from the resin, and the free base is absorbed and excreted in the urine, where its presence indicates acid in the stomach. Judged by comparison with results of maximal histamine stimulation and sampling of gastric contents by intubation, the diagnosis of achlorhydria made by the tubeless method is incorrect in 4 of 10 instances. On the other hand, the tubeless test indicates secretion of acid by persons who are otherwise found to be achlorhydric in 2 of 100 cases. On account of its un-

---

[†]See Makhlouf, G. M.; Blum, A. L., and Moore, E. W.: Undissociated acidity of human gastric juice: Measurement and relationship to protein buffers, Gastroenterology 58:345, 1970.

[‡]An occasionally encountered exception is the acidity of stomach contents of a person who has tried to kill himself by drinking strong acid.

[‖]Histamine is a base, and it is usually dispensed as histamine acid phosphate. One milligram of histamine base is contained in 2.75 mg of histamine acid phosphate, and 1 mg of histamine is contained in 1.7 mg of the more rarely used histamine dihydrochloride. When the dose is reported, it must be made clear how much of which form of the compound was used.

reliability, the tubeless method is suitable only for mass screening, not for individual diagnosis.

Basal Secretion

In the absence of extraneous stimuli, the human stomach secretes acid at a low rate. The magnitude of this basal secretion is defined by one study of 615 men and 634 women of all ages with gastroenterologic problems but without disease of the stomach. After the subject had fasted overnight, his stomach was intubated and its residual contents removed. Then secretion was collected by continuous suction for 60 min, and its acid was titrated to the end-point of pH 3.5. Since this titration does not measure acid buffered within the stomach, it underestimates total acid secretion by a small amount. For the men, the range of secretion was 0–17 mEq of acid per hour, with a mean and standard deviation of 2.4 ± 2.85. When the distribution is not a normal one, the mean plus or minus 2 times the standard deviation may not actually include 95% of the control group. Therefore the upper limit of normal was set so that there was 0.975 probability that no more than 5% of a similar control population would lie above it. This upper limit of normal was 6.6 mEq of acid per hour. For the women, the range was 0–15 mEq per hour, with a mean and standard deviation of 1.3 ± 2.0 and an upper limit of normal of 4.1.

Maximal Acid Output

The maximal ability of the stomach to secrete acid is measured by continuously collecting gastric secretions after the subject has been given what is thought to be a sufficiently strong stimulus. The rate of acid secretion in the quarter hour in which secretion is greatest is multiplied by 4, and this is the *peak acid output* or *maximal acid output* in milliequivalents per hour.

For many years histamine, or its synthetic congener betazole, was the stimulant used, and a vast body of data on the response to histamine has been collected. Although histamine is being replaced by pentagastrin as a stimulus, the experience gained with histamine still provides a valuable standard.

When histamine is given by intravenous injection, a constant blood level of the drug is quickly established, and the rate of secretion reaches a steady value. To avoid the inconvenience of intravenous injection, a single subcutaneous dose of histamine is frequently used. In that case, acid secretion rises in the first 15 min, reaches a peak between 15 and 45 min and then declines. An

TABLE 8–2.—RESULTS OF AUGMENTED HISTAMINE TEST ON NORMAL ADULTS IN 6 SEPARATE STUDIES (ACID SECRETED IN FIRST 60 MIN AFTER SUBCUTANEOUS INJECTION OF 0.04 MG HISTAMINE ACID PHOSPHATE/KG, EXPRESSED AS mEq/HR)*

	MALES			FEMALES		
STUDY	NO. OF SUBJECTS	MEAN	RANGE	NO. OF SUBJECTS	MEAN	RANGE
1	27	22.2				
2	14	22.4	10–35	18	14.6	0.1–31
3	31	23.2		15	15.0	
4	29	22.7	10–42	28	17.2	6–35
5	30	23.3		12	17.7	
6	15	28.8	11–17	21	17.1	3–47

*From Bock, O. A. A., *et al.*: Gut 4:112, 1963; references for the 6 studies are contained in this paper.

TABLE 8–3.—BASAL ACID OUTPUTS, AND
PEAK ACID OUTPUTS AFTER HISTAMINE
STIMULATION AND DURING A MEAL IN
6 NORMAL SUBJECTS AND 7 PATIENTS WITH
DUODENAL ULCER, mEq PER HOUR ± S.E.*

	BASAL	PEAK HISTAMINE	PEAK MEAL
Normal subjects	1.4 ± 0.7	34.5 ± 2.8	30 ± 4
Duodenal ulcer patients	7.5 ± 2.5	58.2 ± 7.2	64 ± 7

*Adapted from Fordtran, J. S., and Walsh, J. H.: J. Clin. Invest. 52:645, 1973, with additional data from J. S. Fordtran.

antihistaminic drug that blocks all the effects of histamine except stimulation of acid secretion is given, and histamine is injected at 0.04 mg of the acid phosphate per kg. This is the *augmented histamine test*. The data shown in Table 8–2 show that the results are remarkably uniform.

The response to the augmented histamine test is reduced to about half the normal value in superficial gastritis and almost to zero in gastric atrophy. On the other hand, it is raised in patients with duodenal ulcer (Table 8–3) and one unfortunate person with a gastrinoma holds the world's record for secreting 111 mEq of acid in 60 min.

Pentagastrin, which does not require medication to cover side-effects, is now substituting for histamine in tests of secretory capacity, and the usual dose is 6 μG per kg given subcutaneously or intramuscularly. In one large study, patients chosen to give a wide range of secretory responses were given pentagastrin, and their peak acid outputs were measured. On another occasion, some of the subjects were given histamine, some were given betazole and some were given a repeat dose of pentagastrin. The results given in Table 8–4 show that the peak acid output following pentagastrin matches that obtained by histamine.

Many similar tests have shown that, on the average, men secrete more than women, that the rate of secretion falls off after the age of 50, that patients with duodenal ulcer secrete more than normal subjects, and that patients with gastric ulcer secrete less and patients with gastric cancer far less than control subjects. These data define the average, but individuals may not conform to the mean of their class. For one large study of stimulated secretion, the upper limit of normal was set so that there was a 0.975 probability that no more than 5% of the normal population would be above the limit. Nine percent, or one of 11, of those with gastric ulcer, a class secreting on the average less than normal, actually secreted above the upper limit of normal. Although the mean rate of secretion in patients with duodenal ulcer was twice as high as in control subjects, more than half of these patients secreted amounts of acid that fell within the normal range.

TABLE 8–4.—INTRAMUSCULAR PENTAGASTRIN COMPARED WITH OTHER
STIMULI AS TESTS OF GASTRIC SECRETION. PEAK ACID OUTPUT IN
mEq ± S.D./HR.*

GROUP	N	PENTAGASTRIN, IM 6 μG/KG	COMPARISON TREATMENT OF SAME SUBJECTS	COMPARISON RESULTS
1	26	21.86 ± 3.32	Histamine, IV, 0.04 μg/kg·hr	24.18 ± 3.43
2	93	29.68 ± 1.53	Histamine, SQ, 0.04 μg/kg	33.12 ± 1.62
3	30	43.61 ± 2.58	Betazole, SQ, 2 mg/kg	46.83 ± 2.53
4	34	21.28 ± 2.70	Pentagastrin, SQ, 6 μg/kg	21.44 ± 2.14

*Multicentre Study: Lancet 1:341, 1969.

Parietal or Oxyntic Cell Mass

The secretory response to stimulation depends in part upon the total number of oxyntic cells in the gastric mucosa. The parietal or oxyntic cell mass has been estimated by counting the cells in stained sections of the mucosa, chosen to give an adequate sample of the total acid secreting volume. In a group of patients scheduled to undergo gastrectomy, the augmented histamine test gave rates of secretion from 0.6–82.7 mEq HCl per hour. After a portion of the stomach had been removed, the test was repeated, and the number of oxyntic cells in the resected portion was estimated. The difference between the rates of acid secretion in the two tests was attributed to the oxyntic cells contained in the part of the stomach removed. The correlation between the decrease in maximal acid output and the number of oxyntic cells removed was almost linear over a very wide range.

Gastrin is a trophic hormone for the secretory cells of the digestive tract (see Chapter 12). In rats, increasing or decreasing the circulating concentration of gastrin increases or decreases the maximal secretory capacity. Although there are no comparable data on human subjects, it is probable that the continuously high concentration of gastrin in the plasma of patients with gastrinoma is responsible for their increased oxyntic cell mass and their maximal secretory capacity. It is likely that the exaggerated output of gastrin following a meal in duodenal ulcer patients contributes to their high maximal secretory capacity.

Secretion During a Meal

Food is the natural stimulus for secretion, but it is difficult to measure gastrointestinal functions during a 10-course dinner at the Duchess of Guermantes'. The nearest approach is by the method of intragastric titration.

After a subject has fasted, a double-lumen tube is passed into his stomach, and the resid-ual contents are aspirated. The pH of the stomach is raised to 5.5 by infusion of a standard bicarbonate solution. Then the subject eats a meal. During eating and until the stomach is empty again, the contents of the stomach are frequently sampled through one lumen, and the pH of the sample is measured. The sample is then returned to the stomach. Infusion of the standard bicarbonate solution through the other lumen is adjusted so that the pH of the contents of the stomach remains at pH 5.5. The rate of infusion of bicarbonate measures the rate of acid secretion, and the total amount of bicarbonate infused measures the cumulative secretion of acid.

In the example cited here, the meal was 5 oz of ground sirloin steak cooked and seasoned with salt and pepper, two pieces of toast and 360 ml of water. The meal contained 39 gm protein, 30 gm fat, 30 gm carbohydrate and 546 kilocalories. Its total volume was 500 ml and, when homogenized, its pH was 5.5. The subject ate the meal over a period of 30 min, chewing it thoroughly before swallowing it.

The rate of secretion of acid by 6 normal subjects and by 7 patients with duodenal ulcer is given in Figure 8–3, left, and the cumulative acid secreted is given in Figure 8–3, right. Basal secretion, peak rate of secretion during the augmented histamine test and peak rate of secretion in response to the meal are given in Table 8–3. In each group, the natural stimulus of food caused a peak output equal to that achieved with histamine stimulation.

If acid is not neutralized by intragastric titration after a similar meal is eaten, the rate of gastric secretion is greatest during the first hour. The volume of gastric contents remains constant despite rapid emptying, for the rate of secretion equals the rate of emptying. During the first hour, the pH of gastric contents is greater than 3.0. Gastrin release is stimulated by food but is not inhibited by acid, and plasma gastrin concentration rises to a peak. During the second hour, the pH of gastric contents falls below

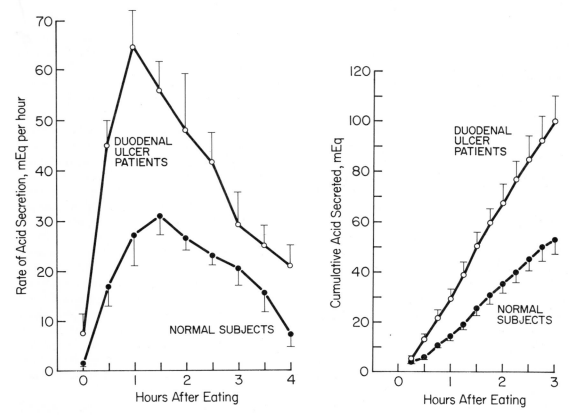

Fig 8–3.—Rate of acid secretion and cumulative acid secreted by 6 normal subjects and 7 duodenal ulcer patients in response to a meal. **Left,** rate of acid secretion. **Right,** cumulative acid secreted. (Adapted from Fordtran, J. S., and Walsh, J. H.: J. Clin. Invest. 52:645, 1973, with additional data supplied by J. S. Fordtran.)

3.0 and becomes 2.0 by the end of the hour. Gastrin release is now inhibited, and plasma gastrin concentration falls. Gastric secretion almost stops by the end of the second hour, although some food is still in the stomach.

Gastric Blood Flow in Relation to Secretion

Arteries supplying the stomach pierce the muscular coat and form an arterial plexus, which gives small branches to the muscular layer and many branches to the mucosa. The transition from arterioles to capillaries occurs in the mucous membrane, and there are many anastomoses of arteries within the mucosa. The capillaries unite to form a superficial venous plexus around the orifices of the glands. The veins then form a more deeply placed venous plexus before going out through the muscularis. In addition, arteriovenous shunts arise from the mucosal arteries just after they pierce the submucosa and drain into veins of the submucosal venous plexus. The shunts allow blood to bypass the capillary bed, and consequently the actual amount of nutrient blood reaching the capillaries is a function not only of the total blood flow but also of the patency of the shunts. There may be a high blood flow with low oxygen extraction if blood passes predominantly through the shunts, or high blood flow with high oxygen extraction if the shunts are closed and blood flows chiefly through the capillary network.

Because of the frequent arterial anastomoses, it is possible to ligate arteries to control hemorrhage without cutting off completely the blood supply to the mucosa. In fact, all major arteries to the stomach must be simultaneously ligated before the whole stomach is deprived of its blood supply.

Effective blood flow through the gastric mucosa is measured by a clearance method. In the dog, aminopyrine is a base that is both water and fat soluble. Its acid dissociation constant (pKa) is about 5, so that at the pH of blood it exists chiefly in the basic form. Because it is fat soluble, it diffuses rapidly through cell membranes from blood into tissues, and its concentration in cell water is, at equilibrium, equal to its concentration in water of the blood. In acid solution, aminopyrine picks up a proton; therefore, the concentration of its basic form is vanishingly small in acid solution, and a steep concentration gradient for the basic form is established between blood and the acid solution. Consequently, aminopyrine diffuses from blood into the acid solution, and this rate of diffusion is so fast that all or almost all of the aminopyrine contained in blood flowing through the mucosa is trapped in the acid solution. The acid solution may be either acid secreted by the mucosa or an exogenous solution placed in contact with mucosa.

The amount of aminopyrine brought to the mucosa per minute is equal to the blood flow through the mucosa per minute multiplied by the concentration of aminopyrine in the blood. If all the aminopyrine brought to the mucosa by the blood is cleared from the blood by being trapped in the acid solution, the amount trapped is equal to the amount brought to the mucosa by the blood. Since the amount trapped per minute and the concentration of aminopyrine in the blood can be measured, the blood flow can be calculated by dividing the amount trapped per minute by the blood concentration. In principle, this method of measuring effective mucosal blood flow is similar to the clearance method used to measure renal blood flow, and it is subject to the same theoretical and practical limitations.

Aminopyrine cannot be used in man on account of its toxicity, but the dye, neutral red, can be substituted.

Under resting conditions, effective mucosal blood flow measured by aminopyrine clearance is approximately equal to 50% of the total blood flow through mucosa and muscle measured by collection of venous effluent blood. Flow through mucosa and muscle are poorly correlated, for flow through the mucosa varies under many conditions without a corresponding change in flow through the muscle. In the nonsecreting stomach, effective mucosal blood flow is reduced by infusion of vasopressin or norepinephrine, and it is increased by infusion of epinephrine. When acid secretion is stimulated by infusion of histamine, the mucosal blood flow increases parallel to the secretion rate. Infusion of gastrin also increases mucosal blood flow, but the increase per unit increment of secretion is less than that caused by histamine infusion. Secretin reduces blood flow as well as gastrin-stimulated secretion of acid. Blood flow is increased by cholinergic stimulation.

Increased blood flow in itself does not stimulate secretion, but blood flow is permissive in the sense that adequate flow is required if the stomach is to secrete. Profound vasoconstriction caused by sympathetic stimulation or by vasoconstrictor drugs is followed by depression of secretion. When the blood flow of a stomach is being regulated by perfusion and the stomach is stimulated by histamine or through the vagus, increase in blood flow may be followed by increased secretion; but when the stomach is not stimulated, an increase in blood flow does not initiate secretion. Blood vessels of the stomach do not show autoregulation, for an increase in perfusion pressure is followed by a decrease in vascular resistance.

When blood flow to the stomach increases, the mucosa becomes wet and swollen, and the rugae are full and smooth. There is no change in blood flow with small contractions, but the mucosa becomes engorged and dark with each strong contraction. This is probably caused by temporary constriction of venous outflow by muscular contraction.

Mechanism of Acid Secretion

The primary process of acid secretion is the one that raises hydrogen ion concentration from 0.000,05 mN in the plasma to 150–170 mN in gastric juice. This requires a minimum of 1,532 calories per liter of juice secreted. Cells secreting acid use oxygen and any number of substrates, both carbohydrate and lipid. They secrete hydrogen ions in the direction of the lumen of the stomach and an equal number of bicarbonate ions in the direction of the blood. Bicarbonate exchanges for chloride at the nutrient surface of the cells, and chloride accompanies hydrogen ions into the secreted fluid. The cells also secrete potassium, and perhaps sodium, ions, but the role of these two ions is utterly obscure. Water follows passively, and the secretion is isotonic with the secreting cells.

Despite a large amount of work, the intimate mechanism of acid secretion is still unknown. There are two general types of theories explaining secretion: (1) a redox scheme and (2) an ATPase scheme.

According to the redox scheme, there are four steps in the secretion of acid:

1. A high level of potential energy is generated within the cell. Oxyntic cells contain many mitochondria capable of producing ATP, and ATP is the immediate source of energy. Mitochondria may also generate reducing equivalents, which are transported to the secretory membrane of the intracellular canaliculi.

2. At the secretory membrane, a cycle of oxidation and reduction separates protons and electrons. A reduced substrate is oxidized; the proton is delivered to the secreted fluid; and the electron is delivered to the respiratory chain. After passing along the chain, the electron is eventually delivered to oxygen. There may actually be a cascade of successive oxidations and reductions, each separating a proton and an electron. If the same electron is passed through many cycles from substrate to oxygen, there is no need for a fixed relation between the number of protons separated and the number of electrons eventually delivered to oxygen. The only limit on proton generation is the energy difference between the beginning and the end of the series of reactions.

3. Electrons carried along the respiratory chain are eventually given to oxygen. It is possible that the links in the respiratory chain are identical with those separating the secreted proton and the electron.

4. At some place in the series of reactions, a hydroxyl ion is generated for every proton secreted. The hydroxyl ion reacts with carbon dioxide coming from metabolism or from the blood to form bicarbonate, and bicarbonate is exchanged for chloride. The reaction between hydroxyl ions and carbon dioxide is catalyzed by carbonic anhydrase, which is abundantly present in the oxyntic cells, probably bound to the secretory membrane (Fig 8–4).

The ATPase hypothesis rests on the fact that an ATPase is contained in the membrane of the tubovesicles and canaliculi of the oxyntic cells. This ATPase is activated by bicarbonate ions, not by sodium and potassium. Furthermore, it is inhibited by thiocyanate, which is known to inhibit acid secretion.

Membrane-bound vesicles can be prepared from oxyntic cells, and these vesicles contain bicarbonate-activated ATPase. The vesicles secrete acid into their contained fluid when they are supplied with ATP. Mitochondria generate ATP when the cells are stimulated to secrete. ATP is translocated to the secretory membrane, where it is hydrolyzed, and its energy is used to transport a proton into the gastric juice. At the same time, a bicarbonate ion is transferred to the cells' cytosol and eventually exchanged for chloride at the nutrient surface. Details of the reactions occurring at the secretory

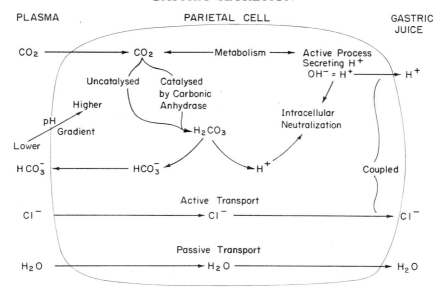

PLASMA   PARIETAL CELL   GASTRIC JUICE

Fig 8–4.—Role of CO_2 in intracellular neutralization during secretion of acid by the gastric mucosa.

membrane are unknown. It is possible that they contain oxidation-reduction steps, and consequently the two schemes are not mutually exclusive.

An almost infinite number of versions of these two general schemes can be developed by anyone with enough imagination and scratch paper, and almost that many have, in fact, been proposed. There is experimental evidence that can be made to support or to invalidate any step in each scheme. That evidence cannot be summarized here, and the curious reader can find in the references at the end of this chapter reports of the gladiatorial combats over the mechanism of acid secretion. The author of this book, having been carried from the arena more dead than alive over 20 years ago, can give him no help.

Gastrin, histamine, acetylcholine and other compounds stimulating acid secretion may act through cyclic AMP, but there is no agreement on the subject. Exogenous cyclic AMP stimulates acid secretion by isolated frog gastric mucosa, and secretion stops when tissue cyclic AMP is exhausted. Theophylline, a methyl xanthine, inhibits the phosphodiesterase, which destroys cyclic AMP, causes an increase in tissue content of cyclic AMP and stimulates acid secretion.

Several prostaglandins inhibit acid secretion, but the means by which they do so is unknown. Because carbonic anhydrase is present in excess, inhibitors of the enzyme are effective only in very large doses.

When the gastric mucosa secretes acid, it also removes acid from the blood in the form of the potentially acid carbon dioxide molecule, and it returns an equal number of bicarbonate ions. The pH of the plasma is described by the equation

$$pH = 6.10 + \log \frac{[HCO_3^-]}{[CO_2]} \qquad (8.4)$$

Because the numerator of the fraction rises and the denominator falls as blood flows through a secreting stomach, gastric venous blood has a higher pH than has arterial blood. This alkaline tide that accompanies acid secretion can sometimes be detected by a rise in the pH of the urine and a compensatory rise in the P_{CO_2} of alveolar air. This process is not, however, the one responsible for the alkaline tide following wakening; that tide is, instead, the result of the quick fall in alveolar and blood P_{CO_2} subsequent to the relief of sleep-induced depression of the respiratory center. The latter kind of alkaline tide occurs in achlorhydric, as well as in normal, subjects.

Secretion of Chloride and the Origin of Potentials

Chloride moves from a plasma concentration of about 108 mN to about 170 in the gastric juice. The mucosal surface of the oxyntic gland area is electrically negative to the serosal side, the potential difference being about 60–80 mV in the resting stomach and a little lower in the secreting one. Therefore chloride is transported against an electrochemical gradient; it must be secreted by an active, energy-consuming process or a chloride pump. The pump is electrogenic in the sense that it can produce a net transport of electric charge. It can operate without an exchangeable anion on the mucosal surface of the stomach, and it causes a net flow of chloride ions from the cells to the lumen. When the gastric mucosa is not secreting acid and when its mucosal surface is bathed by a neutral, sodium-containing solution, some mucosal cells pump sodium from the mucosal to the serosal surface. The direction of pumping, movement of a cation from mucosa to serosa, tends to make the mucosal surface negative with respect to the serosal surface. The combined action of the two pumps establishes a potential difference of 60–80 mV across the nonsecreting oxyntic glandular mucosa. Bathing the mucosal surface with acid abolishes sodium pumping and reduces the potential difference.

Cells of the mucosa are impermeable to sulfate or 2-hydroxyethylsulfonate ions; and when excised but viable sheets of the mucosa are placed in salt solutions whose chloride ions are replaced by either of the other anions, the chloride potentials are abolished. Then, if the tissue is stimulated to secrete acid, an electrogenic proton or hydrogen ion pump is revealed. This pump transports protons across the mucosal surface of the cells,

making that surface positive in the external circuit. The pump carries positive charges to the mucosal surface so that, in the absence of chloride pumping, the mucosal surface becomes positive with respect to the serosal surface; the direction of the potential difference is reversed. Artificial increase of the current flow through the cells by placing a cathode on the mucosal surface and an anode on the serosal surface enhances acid secretion. On the other hand, if the proton pump is bucked by placing an anode on the mucosal surface and a cathode on the serosal surface, acid secretion is reduced. The potential difference between the two electrodes must be, on the average, 140 mV before acid secretion is stopped.

Some persons believe that an electrogenic sodium pump that transports sodium ions from the gastric lumen to the blood contributes to the potential difference.

When the mucosa is stimulated to secrete acid, both the chloride and the proton pumps increase the rate at which they transport their ions. The two pumps are closely coupled by some unknown means that keeps them nearly in step. Because they transport equal, or very nearly equal, quantities of anions and cations, the tendency of the chloride pump to make the mucosal surface negative is bucked by the tendency of the proton pump to make it positive; the potential difference across the mucosa falls to 30–50 mV, the mucosal surface being negative in the external circuit with respect to the serosal surface.

An odd aspect of the chloride pump is that, if given a chance, it prefers to transport bromide rather than chloride. If there is bromide in the plasma, as, for example, when bromide salts are ingested, hydrobromic acid is secreted, and the ratio of bromide to chloride in the gastric juice exceeds that in the plasma by a factor of about 1.5. The gastric mucosa also actively secretes iodide ions into the gastric juice. When the plasma iodide concentration is low, the ratio of concentration of iodide in the gastric juice to that in plasma is very high (about 15:1), but the ratio falls to

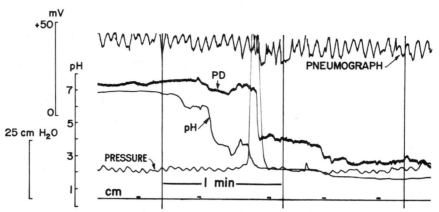

Fig 8–5.—Recording of the pH, potential difference *(PD),* and pressure from the gastroduodenal junction in man. The tip of the probe by which pH, potential difference and pressure were measured was withdrawn 1 cm at each signal mark from the left (the duodenum) to the right (the stomach). The pylorus is recognized by the sharp change of potential difference and the simultaneous peak in the pressure recording. The pH started to decrease distal to the pylorous. (From Andersson, S., and Grossman, M. I.: Gastroenterology 49:364, 1965.)

Fig 8–6.—Effect of solutions of aspirin and of isotonic saline upon the potential difference across the human gastric mucosa. Points represent the mean ±S. E. of observations on 10 normal subjects. The mucosal surface is negative with respect to the blood. (From Geall, M. G.; Phillips, S. F., and Summerskill, W. H. J.: Gastroenterology 58:437, 1970.)

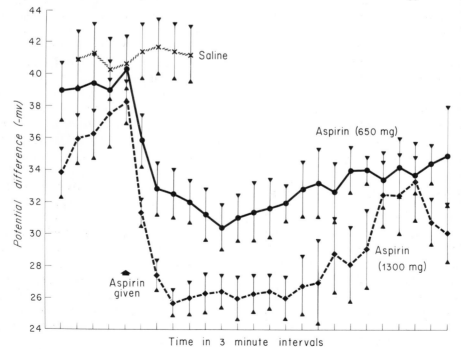

unity as plasma iodide rises to 3 mN. The mucosa's ability to concentrate iodide is depressed by thiocyanate and perchlorate, agents which also depress thyroidal accumulation of iodide.

The proton and chloride pumps responsible for the large potentials of the oxyntic gland area are probably located within the oxyntic cells. The contributions of the surface epithelial cells to the electric properties of the stomach are unknown. The potential difference across the pyloric glandular mucosa, which contains few oxyntic cells, is less than 10 mV, the mucosa being negative with respect to the serosa.

The potential difference across the duodenal mucosa is oriented in the opposite direction, the mucosal surface being positive with respect to the serosal. If a probe bearing an electrode is passed into the duodenum, the potential difference between the mucosal surface and the skin or blood can be measured. When the probe is pulled slowly toward the stomach, an abrupt reversal of potential difference occurs as the probe passes through the pyloric sphincter to come into contact with the pyloric glandular mucosa (Fig 8-5). This reversal of potential difference can be used to identify the location of the pyloric sphincter.

The potential difference may also be used to assess the integrity of the mucosa. Agents that damage the gastric mucosal barrier to acid lower the potential difference. The effect of aspirin is shown in Figure 8-6. Drinking 25 or 50 ml of 80-proof whiskey has the same effect.

Osmotic Pressure of Gastric Juice

The major components contributing to the osmotic pressure of gastric juice are H^+, Na^+, K^+ and Cl^-. Other constituents (Ca^{2+}, Mg^{2+}, PO_4, glucose, urea and proteins) together make up less than 10 mOsm.

If hydrogen and chloride ions, and perhaps potassium ions, are actively secreted and if the secreting cells are permeable to water, then, at the same time that the ions are se-creted, water should move from blood to gastric juice, merely to maintain osmotic equilibrium. Were this the sole process involved in water movement, gastric acid secretion should always be isotonic with plasma. Nevertheless, in samples of gastric juice having low acidity, there is systematic deviation from isotonicity in the direction of hypotonicity of the juice. The osmotic pressure of human basal secretion collected by stomach tube is 171-276 mOsm. For the dog whose juice can be collected from a pouch of the oxyntic gland area without contamination by other secretions, the maximal deviation from isotonicity is about 40 mOsm of a total of 310 mOsm. Part of this can be explained on the theory that actual samples of gastric juice are a mixture of two or more components. Low acidity of the sample means that it has been reduced by admixture of an originally isotonic acid juice with an alkaline juice that is also isotonic. Since the alkaline juice contains bicarbonate, mixing of it with acid produces volatile carbon dioxide, which disappears from the osmotic balance, leaving a fluid of low osmotic pressure.

When acid secretion is stimulated by histamine, the osmotic pressure of gastric juice rises as the rate of secretion increases. In 4 of 7 human subjects, the ratio of the osmotic pressure of gastric juice to the osmotic pressure of plasma rose to between 0.97 and 1.04; the juice became isotonic. In three other subjects the juice was always hypotonic, the ratio ranging from 0.76-0.88. Juice collected from the oxyntic glandular mucosa of the dog stomach has a similar range of osmotic pressure. In two series of experiments the ratio of the osmotic pressure of juice secreted under maximal histamine stimulation to that of plasma was 1.05-1.10; but in a third series of apparently identical experiments, juice was hypotonic to gastric venous plasma in 20 of 30 dogs and isotonic in the rest. If the osmotic pressure of plasma is varied in either direction by infusion of hypo- or hypertonic solutions, there are corresponding changes in the osmotic pressure of acid gastric juice.

When acid juice is secreted, two osmotically active ions, H^+ and Cl^-, are removed from plasma, together with enough water to keep the juice nearly isotonic with plasma. However, the hydrogen ion is manufactured de novo by the mucosa, and the mucosa exchanges a new bicarbonate ion for a chloride ion. The result is that the water of the gastric juice is extracted from plasma, but the total mass of osmotically active particles in plasma is unchanged. In effect, water, not osmols, is removed, and the total osmotic pressure of plasma and other body compartments rises slightly. When acid juice is neutralized and the products are reabsorbed in the intestine, the original osmotic balance is restored. If acid juice is lost by vomiting or gastric drainage, dehydration, consisting of both loss of extracellular fluid and increase in its osmotic pressure, as well as metabolic alkalosis, occurs.

Gastric Urease

The enzyme urease occurs in the gastric mucosa of man and many other mammals. Much of it is contained within bacteria contaminating the mucosa, but some urease not of bacterial origin may be contained in the mucosal cells. There is evidence that fetal gastric mucosa, presumed to be uncontaminated with bacteria, contains urease; that sterile samples of the cytoplasm of the surface epithelial cells of mouse stomach have urease activity; and that more enzyme is in the mucosa than can be accounted for by its bacterial population. Bacteria containing urease are also present in the nose and pharynx, and their number, together with the urease activity of the tissues, can be temporarily reduced by heavy antibiotic medication. Urea available for hydrolysis comes in contact with the enzyme as water containing it flows from the plasma through the mucosa during gastric secretion. The result is that, during acid secretion, a small amount of urea is hydrolyzed in the mucosa.

The products of urea hydrolysis are alkaline: 2 M of ammonia to 1 M of carbon dioxide. Because the amount of ammonia formed is only 1/500th the amount of acid secreted, hydrolysis of urea does not contribute to neutralization of acid in normal subjects. When large amounts of urea (15 – 25 gm) are ingested, its hydrolysis may neutralize up to half the acid secreted, and the gas eructed actually tastes faintly of ammonia. The histamine-stimulated juice of uremic patients whose blood urea concentration is about 6 times normal contains 60% as much acid as normal controls, and there is ammonia in the juice. In such a patient, ammonia released in the stomach may contribute to hyperammonemia and encephalopathy. A fall in gastric juice ammonia, either spontaneous or induced by antibiotic medication, is accompanied by an equivalent increase in gastric acidity.

Exchange of Water and Electrolytes across the Resting Gastric Mucosa

Water molecules cross the gastric mucosa rapidly in both directions. The net rate of flow in either direction is the difference between the two rates:

Flux rate$_{lumen\ to\ blood}$ − Flux rate $_{blood\ to\ lumen}$

$$= \text{Net flow*}$$

The flux rate from lumen to blood can be estimated by placing water labeled with deuterium oxide or tritiated water (D_2O or THO) in the stomach and measuring the specific activity of the isotope in the lumen at frequent intervals. The results obtained when 50 ml of a solution containing D_2O was

---

*The direction of one-way fluxes and of net flow should always be specified in such a way as to make them instantly recognizable. Here the anatomical description, lumen to blood or mucosal surface to serosal one, will be used. Another system having wide, but not universal, currency is:

One-way flux, lumen to blood	= Insorption
One-way flux, blood to lumen	= Exsorption
Net flow, lumen to blood	= Absorption
(insorption greater than exsorption)	
Net flow, blood to lumen	= Enterosorption
(exsorption greater than insorption)	

See Code, C. F.: The semantics of the process of absorption, Perspect. Biol. Med. 3:560, 1960.

placed in contact with the oxyntic glandular mucosa of a dog are shown in Figure 8–7. Fifty percent of the isotopically labeled water was absorbed in less than 40 min; but because there was only a small net movement of water, an almost equal number of water molecules must have entered the lumen as the labeled molecules left. These large fluxes of water in each direction across the gastric mucosa are slightly, or not at all, affected by the osmotic pressure of the contents of the lumen.

In the middle of the experiment described in Figure 8–7, an infusion of histamine was started, and copious secretion began. Tritiated water was placed in the lumen, and measurement of the specific activity of tritium showed that, despite a large net output of water from the mucosa in its secretions, flux of water from lumen to blood remained high.

Flux of sodium in either direction across the resting gastric mucosa is small. There is usually an electrochemical gradient favoring net movement of sodium into the lumen, for interstitial sodium concentration is about 145 mN, and the surface of the mucosa is electrically negative with respect to the interstitial fluid. Nevertheless, net output of sodium is only a few microequivalents per cm^2 an hour. The unidirectional flux from lumen to blood is about one fifth the net output. Flux of sodium is reduced when the contents of the lumen are acidified, either by secretion by the mucosa itself (see Fig 8–7) or by exogenous HCl. The fluxes of potassium across the resting mucosa are still smaller than those of sodium, but they are unaffected by the acidity of the luminal contents.

When acid is present in the lumen, hydrogen ions diffuse slowly from the lumen into the mucosa; the rate is less than that at which sodium leaves the mucosa. If 250 ml of 100 mN HCl is placed in the stomach of a normal human subject, the concentration of acid

Fig 8–7.—Fluxes across the oxyntic glandular mucosa of a dog's stomach. At the beginning of the experiment a solution containing D_2O, sodium labeled with ^{22}Na and potassium labeled with ^{42}K was placed in contact with the mucosa, and the rates of disappearance of the isotopes were measured. Then an infusion of histamine was started, and tritiated water was added to the solution. When acid secretion began, the rate of disappearance of ^{22}Na fell to zero, but fluxes of water and potassium into the mucosa were unaltered. (From Code, C. F., et al.: J. Physiol. 166:110, 1963.)

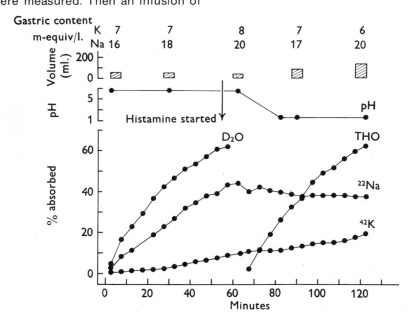

recovered 15 minutes later may be 90–95 mN. Secretion of neutral fluid containing bicarbonate is responsible for part of the diminution in the concentration of acid, and back-diffusion of acid into the mucosa is responsible for the rest. In the dog, back-diffusion of acid into the pyloric glandular mucosa is faster than into the oxyntic glandular mucosa.

The Gastric Mucosal Barrier*

The property of the gastric mucosa that allows it to contain an acid solution and that prevents rapid penetration of itself by hydrogen ions is the *gastric mucosal barrier*. The ability of the mucosa to prevent rapid diffusion of sodium ions from its interstitial space into the lumen is likewise a property of the gastric mucosal barrier.

In practice, measurement of sodium exchanges across the mucosa gives a more reliable estimate of the integrity of the barrier than does measurement of disappearance of acid. The reason is that the actual flux of hydrogen ions into the mucosa cannot be measured if, on one hand, an unknown amount of alkaline fluid that neutralizes acid is being shed by the mucosa or if, on the other hand, acid is being secreted by the mucosa. In the first instance, disappearance of acid is greater than its rate of diffusion into the mucosa; and in the second, the apparent rate of disappearance of acid is much smaller than the rate of back-diffusion.

Damage to the Gastric Mucosal Barrier

The gastric mucosal barrier is broken by many compounds; among them are aliphatic acids such as acetic, propionic and butyric acids, which, because they are un-ionized and fat soluble, diffuse through cell membranes into the gastric mucosa. Salicylic acid and acetylsalicylic acid in acid solution are likewise un-ionized and fat soluble, and they are rapidly absorbed from acid gastric contents into the mucosa. Once absorbed, they kill the surface epithelial cells, causing desquamation and barrier breaking. The salicylates in neutral solution are ionized and water soluble, and they are absorbed only slowly into the gastric mucosa from neutral gastric contents, causing no damage. Other compounds breaking the barrier are those attacking the lipid components of the cell membranes: detergents, natural or synthetic. Among the natural detergents the most important are bile acids and lysolecithin, both present in normal duodenal contents. Patients with gastric ulcers have abnormally great regurgitation of duodenal contents, and many surgeons believe that gastric ulceration is caused by damage inflicted upon the barrier by regurgitated bile acids and lysolecithin. Ethanol is absorbed equally rapidly from neutral and acid solutions, and it also breaks the barrier.

Although it is reasonable to suppose that ischemia damages the gastric mucosa, there is no evidence to support the supposition. Prolonged ischemia resulting either from sympathetic discharge or the administration of large doses of vasoconstrictors does not break the barrier in the dog or in primates. Gastric blood vessels eventually escape from vasoconstrictor influences, and blood flow through the mucosa may return to normal in the face of continued sympathetic discharge or exogenous vasoconstrictors. When the mucosa is stimulated to secrete, oxygen demand resulting from increased metabolism overcomes vasoconstriction and results in active hyperemia. However, the combination of ischemia and attack by bile acids has been found to break the barrier in experimental animals, and the combination may be equally deleterious in man. Other combinations are more damaging than their individual components: that of acid, aspirin and alcohol is particularly effective in breaking the barrier and causing bleeding.

---

*The *gastric mucosal barrier* is not to be confused with the *gastric barrier*. The latter is the ability of acid in the stomach to kill microorganisms, including ingested pathogens, so that chyme delivered to the small intestine is nearly sterile. The gastric barrier is particularly important in ruminants whose stomachs are a vast fermentation vat.

Consequences of damage to the mucosa are depicted in Figure 8–8. Acid, which diffuses slowly through the normal mucosa, rapidly enters one whose barrier has been broken. Acid destroys mucosal cells and liberates histamine. Histamine stimulates acid secretion, causes vasodilatation and increases capillary permeability to proteins. The mucosa becomes edematous, and fluid derived from interstitial fluid is forced through the mucosa. This fluid may contain a large amount of plasma proteins. Acid stimulates the intramural plexuses, and motility of the stomach is increased. It also stimulates pepsin secretion, perhaps by way of histamine liberation and perhaps through its effect on the plexuses. Mucosal capillaries may be destroyed by acid so that interstitial hemorrhage and frank bleeding occur. Bleeding is more frequent and copious when there is concurrent cholinergic stimulation, probably because contraction of gastric muscle increases venous and capillary pressures.

The rate at which plasma proteins normally leak through the gastric mucosa is of the order of a gram a day, but in some circumstances the gastric mucosa may shed such large amounts of plasma proteins that hypoalbuminemia and anasarca result. Plasma proteins are shed in protein-losing gastropathy, a condition that sometimes may be caused by allergic response of the gastric mucosa to foreign proteins. It is also caused by irrigation of the mucosa with a solution of ethanol or of compounds attacking the protein components of the mucosal membrane. Plasma proteins then leak into the lumen through incompetent tight junctions between epithelial cells. The precipitated, acidified hemoglobin, the "coffee grounds" found in a bleeding stomach, may conceal leakage of plasma proteins. When frank bleeding occurs in hemorrhagic gastritis caused by aspirin or alcohol injury, the hematocrit of the shed blood may be as low as 10%, indicating that a disproportionately greater amount of plasma has been lost. Plasma shedding also occurs in acute gastric focal necrosis.

Antacids

Back-diffusion of acid through a broken gastric mucosal barrier or the effects of acid upon a duodenal ulcer can be prevented by neutralizing gastric contents with an antacid.

Antacids are bases, but antacid preparations differ greatly in their ability to neutralize acid in vitro and in vivo on account of

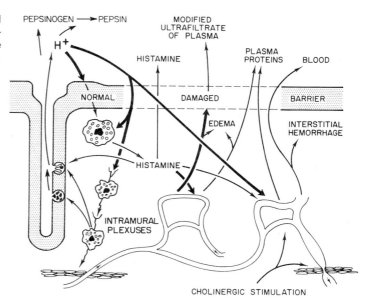

Fig 8–8.—Pathophysiological consequences of the back-diffusion of acid through the broken mucosal barrier.

their differing chemical composition, physical state and reactivity. When 1 ml of a commercial preparation of aluminum phosphate was stirred with 100 ml of water, addition of only 4 ml of 0.1 N HCl brought the pH to 3.0. In contrast, 35 ml of the same acid was required to bring 1 ml of a commercial mixture of aluminum hydroxide, calcium carbonate and magnesium hydroxide to pH 3.0. For in vivo tests, patients with duodenal ulcers were given a standard meal of 150 gm of cooked and seasoned ground beef, two pieces of toast with butter and 180 ml of water. One hour after the meal, the subjects were given either 60 ml of water or 60 ml of the same antacids cited above. The acidity of their gastric contents was measured an hour later. With water, the acidity averaged 69 mN H⁺. With aluminum phosphate, the acidity was 55 mN H⁺, but with the aluminum-calcium-magnesium mixture it was 4 mN H⁺. Results of in vitro and in vivo tests of many antacid preparations are available to the physician who wants to be sure that his prescribed therapy actually works.*

Growth and Regeneration of the Gastric Mucosa

Epithelial cells of the gastric mucosa are in a dynamic state of growth, migration and desquamation. Cells are shed by the normal human gastric mucosa at the rate of a half million cells a minute. The mucosa responds to mild injury by desquamation of its surface layer, followed by regeneration. The actual state of the mucosa, healthy or not, at any one instant or place is the result of a balance between protective and restorative forces and destructive forces.

The turnover time of cells in the epithelium of the lip, buccal mucosa, tongue and esophagus, estimated by counting the fraction of nuclei of an excised sample that are in mitosis, ranges from 4–15 days. The human pyloric gland area renews itself much more

rapidly, in from 1.3–1.9 days, by division of cells in the basal parts of the glands followed by migration to the surface. In the oxyntic gland area, mitosis occurs in poorly differentiated surface epithelial cells at the base of the pits. The new cells, whose mucus content increases as they migrate, move up the sides of the pits to replace the old cells that have been lost from the surface. In normal circumstances, time required for renewal of surface epithelial cells is 1–3 days. The other cells of the glands are derived from the neck chief cells. In the embryo, all gastric glands are first lined with mucoid cells, some of which turn into oxyntic and chief cells; regeneration occurs in a similar order. If a piece of mucosa is excised down to the submucosa, an epithelial ledge composed of mucus-containing cells and derived from the surface epithelium appears at the rim of the lesion. It grows over the denuded submucosa, and from it cellular projections extend into the underlying granulation tissue. These ingrowths develop lumina, and mucus soon appears in the constituent cells, which become transformed through intermediate types, gradually losing mucus, into chief and oxyntic cells. Argentaffin cells also appear in the healing area, but transitional forms have not been seen. Mucous cells that remain in the neck are extruded within a week.

Cell division in the gastric mucosa has a circadian periodicity; in the rat, the highest rate of cell division occurs at 10 A.M. The cells are partially under the control of pituitary hormones, for the mucosa undergoes profound involution after hypophysectomy or adrenalectomy. The oxyntic cells are reduced in size, and their content of cytochrome c oxidase, DPN diaphorase and other dehydrogenases, as measured by histochemical tests, declines. The mucosa's ability to secrete pepsinogen is reduced as much as 80%. Treatment of hypophysectomized animals with a mixture of somatotrophin, thyroid hormone and cortisone enlarges the mucosal cells but does not return them to normal size or function.

Adrenal cortical hormones also affect the

*Fordtran, J. S., Morawski, S. G., and Richardson, C. T.: *In vivo* and *in vitro* evaluation of liquid antacids, N. Engl. J. Med. 288:923, 1973.

stomach by their influence on the response of the tissues to damage. Growth of the epithelium following chemically induced desquamation is not influenced by administration of ACTH or cortisone, but the healing of excision ulcers is delayed. These hormones also cause degranulation of the mast cells in the gastric mucosa.

In man, the layer of mucus covering the gastric mucosa is 1 – 1.5 mm thick. The layer is neutral or alkaline, except where it is in direct contact with pools of acid secretion, for its aqueous menstruum is the alkaline secretion of the surface cells. Mucus over the resting stomach is transparent; but during active secretion, it becomes opaque and rolls up into whitish flakes. Mucus clings

tenaciously to the mucosa, provides lubrication and holds the alkaline secretion against the surface. Mucus is not in itself a barrier to diffusion; small ions such as hydrogen and chloride diffuse through such a gel almost as rapidly as they diffuse through water alone.

Gentle stroking or other mild physical stimulation of the mucosa causes the secretion of a thick layer of mucus. Secretion of mucus increases from 5 to 6 times following the rubbing of NaCl crystals over the mucosa. When the stomach is strongly stimulated, its mucosa is engorged; the folds are thick, red and succulent. Then the slightest injury brings out small hemorrhagic spots and erosions. The supranuclear parts of the surface epithelial cells are shed; and with

Fig 8–9.—An electron micrograph of a section of the tip of an unflattened rugal fold of a mouse stomach. The stomach has been filled for 8 min by oral instillation of a 20 mM solution of aspirin in 1 mN HCl and 145 mN NaCl. The tissue was then fixed and prepared for electron microscopy. The surface epithelial cells have been severely damaged, and their destruction has exposed a swollen, disintegrating capillary within whose lumen the red cells have hemolyzed. Underlying gland cells and deeper blood vessels (not seen in this portion) show no damage. (From Hingson, D. J.: Ph.D. Thesis, Harvard University, 1970; courtesy of D. J. Hingson and S. Ito.)

more severe injury, the cells themselves are desquamated (Fig 8–9). Gastric digestion of a meal is a mild assault on the mucosa. Mucous secretion is stimulated by mechanical contact with food; and during digestion, the mucous layer varies in depth from one area to another as does its content of desquamated cells. The crypts widen, and irregular gaps appear in the surface epithelium. With more serious injury, such as that caused by exposure of the gastric mucosa to 5% aqueous eugenol, 60% ethanol or 160 mN acetic acid, there is vigorous mucous secretion, followed by desquamation of the surface cells. Repeated insult exhausts the ability to secrete mucus; the layer of surface cells is stripped off, and only a bloody transudate flows from the mucosa.

Rapidity of repair depends on the extent of injury. When only restricted desquamation has occurred, healing is complete in 48 hours. Mucus first bridges the gap; and then new surface cells, migrating from the sides, close it. If injury has sliced off the surface layer, with almost complete destruction of foveolae, round or low cylindrical cells form a new surface within 1 hour; and by 36 hours the mucosa has partially recovered its barrier function. Columnar cells, already differentiated, migrate from the crypts; and if the crypts themselves have been destroyed, they are reformed. If injury has been severe, stripping off all but the deepest parts of the glandular tubules, complete regeneration and reorganization of the mucosa, with restoration of unimpaired function, may require 3–5 months.

REFERENCES

Babkin, B. P.: *Secretory Mechanism of the Digestive Glands* (2d ed.; New York: Paul B. Hoeber, Inc., 1950).

Baron, J. H.: Acid secretion in duodenal ulcer disease before and after surgery, Mt. Sinai J. Med. 41:732, 1974.

Brassinne, A.: Gastric clearance of serum albumin in normal man and in certain gastroduodenal disorders, Gut 15:194, 1974.

Code, C. F.: Histamine and gastric secretion: A later look, 1955–1965, Fed. Proc. 24:1311, 1965.

Davenport, H. W.: Physiological structure of the gastric mucosa, in Code, C. F. (ed.): *Handbook of Physiology:* Sec. 6. *Alimentary Canal,* Vol. II (Washington, D.C.: American Physiological Society, 1967), pp. 759–780.

Durbin, R. P.: Electrical potential difference of the gastric mucosa, in Code, C. F. (ed.): *Handbook of Physiology:* Sec. 6. *Alimentary Canal,* Vol. II (Washington, D.C.: American Physiological Society, 1967), pp. 879–888.

Engle, G. L.: Memorial lecture: The psychosomatic approach to individual susceptibility to disease, Gastroenterology 67:1085, 1974.

Fordtran, J. S., and Walsh, J. H.: Gastric acid secretion rate and buffer content of the stomach after eating, J. Clin. Invest. 52:645, 1973.

Gregory, R. A.: *Secretory Mechanisms of the Gastrointestinal Tract* (London: Edward Arnold & Co., 1962).

Harris, J. B., Alonso, D., Park, O. H., Cornfield, D., and Chacin, J.: Lipoate effect on carbohydrate and lipid metabolism and gastric H⁺ secretion, Am. J. Physiol. 228:964, 1975.

Hirschowitz, B. I.: Secretion of pepsinogen, in Code, C. F. (ed.): *Handbook of Physiology:* Sec. 6. *Alimentary Canal,* Vol. II (Washington, D.C.: American Physiological Society, 1967), pp. 889–918.

Hunt, J. N., and Wan, B.: Electrolytes of mammalian gastric juice, in Code, C. F. (ed.): *Handbook of Physiology:* Sec. 6. *Alimentary Canal,* Vol. II (Washington, D.C.: American Physiological Society, 1967), pp. 781–804.

Ivey, K. J.: Gastric mucosal barrier, Gastroenterology 61:247, 1971.

Jacobson, E. D. (ed.): Symposium on the gastrointestinal circulation, Gastroenterology 52:328, 1967.

Jacobson, E. D.: Secretion and blood flow in the gastrointestinal tract, in Code, C. F. (ed.): *Handbook of Physiology:* Sec. 6. *Alimentary Canal,* Vol. II (Washington, D.C.: American Physiological Society, 1967), pp. 1043–1062.

Jeffries, G. H.: Gastric secretion of intrinsic factor, in Code, C. F. (ed.): *Handbook of Physiology:* Sec. 6. *Alimentary Canal,* Vol. II (Washington, D.C.: American Physiological Society, 1967), pp. 919–924.

Johnson, L. R., Lichtenberger, L. M., Copeland, E. M., Dudrick, S. J., and Castro, G. A.: Action of gastrin on gastrointestinal structure and function, Gastroenterology 68:1184, 1975.

Kimberg, D. V.: Cyclic nucleotides and their role in gastrointestinal secretion, Gastroenterology 67:1023, 1974.

Ritchie, W. P., Jr.: Acute gastric mucosal damage induced by bile salts, acid and ischemia, Gastroenterology 68:699, 1975.

Sachs, G., Heinz, E., and Ullrich, K. J. (eds.): *Gastric Secretion* (New York: Academic Press, 1972).

Semb, L. S., and Myren, J. (eds.): *The Physiology of Gastric Secretion* (Baltimore: Williams & Wilkins Co., 1968).

Wolf, S.: *The Stomach* (New York: Oxford University Press, 1965).

Wormsley, K. G.: The pathophysiology of duodenal ulceration, Gut 15:59, 1974.

9

Pancreatic Secretion

THE PANCREAS SECRETES a juice having two major components, an alkaline fluid and enzymes, into the duodenum. The two components occur in variable proportions depending on the stimuli. The alkaline fluid component, ranging in volume from 200–800 ml per day, has a high concentration of bicarbonate, which neutralizes acid entering the duodenum and helps to regulate the pH of intestinal contents. The enzyme component contains the major enzymes necessary for the adequate digestion of carbohydrate, fat and protein. Pancreatic secretion, by its influence on duodenal contents, also affects gastric secretion and emptying.

Structure and Innervation of Pancreas

Secretory acini of the pancreas are composed of cells containing zymogen granules; and, because of the roughly spherical shape of the acinus, the cells tend to resemble truncated pyramids. The acini are connected with excretory ducts by long intercalary and intralobular ducts whose walls are formed by cells that do not contain zymogen granules. Cells of these ducts frequently project into the acinus itself, where they are known as the centroacinar cells. As the intercalary ducts merge into excretory ducts, the cells lining them become cuboidal epithelium.

Cells lining the largest ducts stain for mucin, and small mucous glands open into the main pancreatic ducts. The walls of the largest ducts contain elastic fibers and smooth-muscle cells; when the walls contract, flow of juice into the duodenum is prevented.

Two or more ducts, their number and arrangement depending on the embryologic history of the gland, enter the duodenum. In approximately half the human subjects examined at autopsy, the main pancreatic duct and the bile duct were joined to form an ampulla as they entered the duodenum. In the other half, the ducts entered separately or were divided by a septum running to within 2 mm of their common orifice.

The pancreas contains islets of α and β cells, which secrete glucagon and insulin internally. These islets and the acini, which form the external secretion, are served by separate arterioles. Those to the islets break up into large sinusoids and drain into smaller capillaries, which surround adjacent acini. Other acini receive blood from capillaries arising directly from arterioles. The significance of the first form of circulation to the acini is unknown, but glucagon and insulin are not constituents of the external secretion.

Preganglionic parasympathetic fibers reach the pancreas from the vagus nerves. They synapse within the pancreas with postganglionic fibers whose unmyelinated fibers innervate acinar cells, islet cells and smooth-muscle cells of the ducts. There is a nervous connection between the pancreatic ganglia and the intrinsic ganglia of the intestinal wall, but the functional significance of the connection is not known. Preganglionic sympathetic fibers synapse with postganglionic cells in the celiac and associated ganglia, and the postganglionic fibers are distributed to the pancreatic blood vessels. In addition, sympathetic secretory fibers that do not synapse in the celiac ganglion occur in the splanchnic nerves. Myelinated afferent fibers from the pancreas ascend through the same nerves. Some of these derive from receptors analogous to those in the carotid sinus. Reduction of blood pressure within the pancreas reflexly evokes sympathetically mediated vasoconstriction and cardiac acceleration, resulting in elevation of systemic blood pressure. Afferent fibers from the pancreas also mediate the sensation of pain; that from the head of the pancreas is localized in the mid-epigastrium and that from the tail, in the left upper quadrant.

Methods of Collecting Pancreatic Juice

The best method of collecting pancreatic juice from experimental animals is to form a permanent duodenal fistula by inserting a metal and plastic tube through the body wall and into the duodenum, opposite the pancreatic duct. When not in use, the tube is closed, so that pancreatic juice and intestinal contents flow naturally through the intestine. Digestive functions and the acid-base balance remain normal. For collection of the juice, the cap is removed from the fistula, and an appropriately shaped glass cannula is passed through the fistula into the duct. At the same time, substances can be placed in the duodenum to test their effect on secretion. For complete nervous isolation of the pancreas, a portion of it can be transplanted to the mammary gland of a lactating bitch, so that the duct drains through the nipple. In man, the standard method of collecting pan-

creatic juice is to use a double-lumen tube; one lumen drains the stomach, the other the duodenum. Occasionally, a multilumen tube fitted with two balloons is used. One balloon is placed above and the other below the entrance of the duct into the duodenum; juice is aspirated through a lumen whose orifice is between the two balloons.

Composition of the Aqueous Component

The distinguishing characteristic of the aqueous component is its high bicarbonate content. At the lowest rates of secretion, the concentration of bicarbonate is 20 mN or more. Its concentration rises as the rate of secretion increases and approaches a maximum characteristic of the particular man or animal. Curves illustrating the relationship in man are given in Figure 9–1. For man, the maximum bicarbonate concentration averages 140 mM. When plasma bicarbonate rises in metabolic alkalosis or falls in metabolic acidosis, pancreatic juice bicarbonate shows corresponding changes. Dissociation between rate of secretion and bicarbonate concentration occurs in chronic pancreatitis, in which the bicarbonate concentration of pancreatic juice is low at high rates of secretion, and during inhibition of secretion by anticholinergic drugs, in which bicarbonate concentration of the juice is high at low rates of secretion. In man, but not in the perfused cat pancreas, bicarbonate concentration tends to fall upon prolonged stimulation, although volume rate of secretion remains constant.

The Pco_2 of pancreatic juice is not far from 40 mm Hg, and consequently its pH ranges from 7.6–8.2.

The concentrations of sodium and potassium in pancreatic juice are independent of the rate of secretion and very nearly equal to their concentrations in plasma water. If plasma composition remains constant, both the sum of the cations and the sum of the anions in pancreatic juice are independent of the rate of secretion. Since chloride is, in addition to bicarbonate, the second major anion, chloride concentration falls as bicarbonate

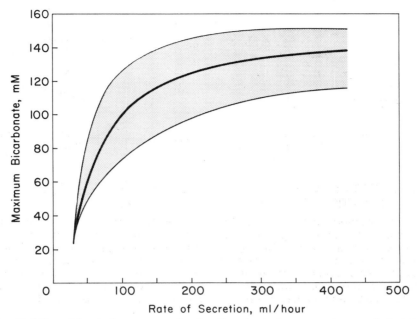

Fig 9–1.—Relationship between maximal bicarbonate concentration of pancreatic juice and rate of secretion in 65 normal human subjects. The heavy line is the line of best fit, and the shaded area encloses all the observations. (Adapted from Dreiling, D. A., and Janowitz, H. D.: Am. J. Dig. Dis. 4:137, 1959.)

concentration rises. The osmotic pressure of pancreatic juice is identical with that of plasma. Experimental variations in the osmotic pressure of fluid perfusing the gland are promptly followed by corresponding changes in that of pancreatic juice.

Secretion of the Aqueous Component

Cells of the intralobular and immediate extralobular ducts secrete an isotonic fluid containing a high concentration of bicarbonate. Secretion by the cells of the intralobular ducts is apparently spontaneous, but secretion by the cells of the extralobular ducts is stimulated by the hormone secretin. Fluid collected by micropuncture from the ducts of the isolated rabbit pancreas has the composition $[Na^+]$ 157 mN, $[K^+]$ 7 mN, $[Cl^-]$ 50 mN, and $[HCO_3^-]$ 110 mN (Fig 9–2, left). As this fluid flows down the collecting ducts, some of its bicarbonate exchanges for chloride of the plasma. There is no net exchange of sodium or potassium. The bicarbonate

concentration of the fluid leaving the main collecting duct depends upon the rate of flow of the juice. When the rate of flow is fast, there is little time for bicarbonate to exchange with chloride, and the bicarbonate concentration of the juice approaches that of the fluid secreted in the intralobular and extralobular ducts (Fig 9–2, right).

There are species differences in the composition of the juice secreted by the intralobular ducts. In the rat, this fluid has a relatively low concentration of bicarbonate and a high concentration of chloride. However, the cells of the rat's collecting ducts secrete bicarbonate into the fluid passing through them and remove chloride. Therefore, in the rat, the concentration of bicarbonate in the fluid leaving the main collecting duct is highest when the rate of flow is lowest, and it falls as the rate of flow increases. The situation in the human pancreas has not been examined by micropuncture, but because the relation between flow rate and bicarbonate concentration in human pancreatic juice resembles

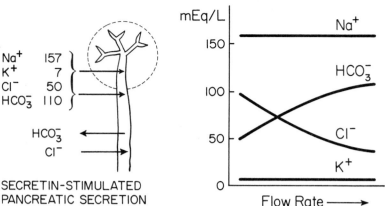

SECRETIN-STIMULATED
PANCREATIC SECRETION

Fig 9–2.—The composition of secretin-stimulated bicarbonate-containing fluid secreted by the extralobular ducts of the rabbit pancreas, and the effect of the rate of flow of the fluid through the collecting ducts upon the final composition of the juice. **Left,** the composition of the secretin-stimulated secretion of the extralobular ducts. The intralobular ducts spontaneously secrete a fluid of similar composition. **Right,** the effect of rate of flow of fluid through the collecting ducts upon composition of the juice delivered to the duodenum. (Adapted from Swanson, C. H., and Solomon, A. K.: J. Gen. Physiol. 62:407, 1973.)

that of the rabbit, it is likely that the primary fluid secreted in the human pancreas also has a high bicarbonate concentration. Pancreatic secretion in the cat and dog is similar to that in the rabbit.

The cells secreting bicarbonate-containing fluid actively transport sodium ions from the plasma into the luminal fluid, and they deliver an equal number of hydrogen ions to the plasma (Fig 9 – 3). As hydrogen ions enter the plasma, they react with bicarbonate ions, and carbon dioxide is liberated. Consequently, the partial pressure of carbon dioxide rises in pancreatic capillaries. Carbon dioxide, diffusing into the cells, is the chief source of the carbon dioxide that eventually becomes the bicarbonate of pancreatic juice.

As hydrogen ions are removed from the cells, an equal number of hydroxyl ions, derived from water, are left behind. The hydroxyl ions immediately combine with carbon dioxide, 95% of which comes from the plasma, and bicarbonate ions are formed. Bicarbonate ions accompany sodium into the juice. The reaction of hydroxyl ions with carbon dioxide is catalyzed by carbonic anhydrase, and the secretion of bicarbonate-

containing juice is partially inhibited by acetazolamide, the carbonic anhydrase inhibitor.

Secretion of bicarbonate is not the primary chemical process. Electric neutrality of solutions can be maintained only if anions accompany sodium pumped from plasma into pancreatic juice. Under normal circumstances, bicarbonate, and to a lesser extent chloride, accompanies sodium ions. However, if the pancreas is perfused or bathed with a physiological salt solution in which bicarbonate is replaced by acetate, the pancreas secretes an acetate-containing solution in which the concentration of acetate is about 3 times as great as it is in the perfusing fluid. Acetate can penetrate cells more readily than can chloride, and it accumulates in the secretion. Several other cell-penetrating anions, including sulfamerazine, can substitute for bicarbonate and be secreted into pancreatic juice.

The concentration of sodium within the pancreatic cells is low, and if sodium is to move from the cells into the juice, it must be pumped against an electrochemical gradient (see Fig 9 – 3). The pump that accomplishes this is probably on the luminal border of the

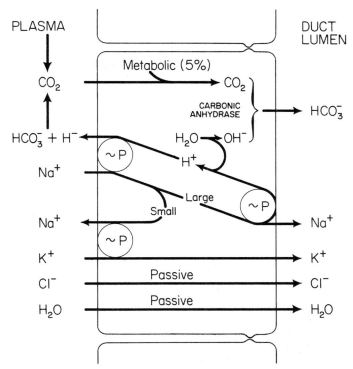

Fig 9–3.—The secretion of bicarbonate-containing fluid by the intralobular and extralobular ducts of the rabbit pancreas. (Adapted from Swanson, C. H., and Solomon, A. K.: J. Gen. Physiol. 62:407, 1973.)

cells, and it is coupled with the mechanism separating hydrogen and hydroxyl ions from water. The sodium pump uses energy derived from ATP, and it can be stopped by ouabain and other inhibitors of Na^+-K^+-activated ATPase. Because sodium moves down its electrochemical gradient from plasma to cell interior, there need not be a sodium pump on the contraluminal border of the cells. Extrusion of hydrogen ions requires energy, and extrusion may be coupled with transport of sodium into the cells.

Potassium must be accumulated within the cells by an active process, and, as in many other kinds of cells, potassium accumulation may be a 1:1 exchange for sodium. Chloride can move down its electrochemical gradient from plasma to juice, and there is no evidence that it is pumped. Because pancreatic juice is always iso-osmolar with the fluid perfusing the pancreas, water must move passively along an osmotic gradient established by mechanisms moving sodium and bicarbonate ions into the juice.

Secretion of the aqueous component is controlled by cyclic AMP. The concentration of cyclic AMP in the pancreas rises within 30 sec after the gland is stimulated by secretin, and it remains high as long as secretin is administered. The nonsecreting, perfused cat pancreas is stimulated to secrete by addition of cyclic AMP, and secretion is enhanced by addition of theophylline, the methyl xanthine that inhibits the enzyme destroying cyclic AMP.

Enzyme Component

The enzyme component of pancreatic juice is mixed in various proportions with the aqueous component. Canine pancreatic juice contains protein ranging from 0.1 to 10%, and human pancreatic juice has the same range of protein content. Secretion of enzymes adds very little volume to pancreatic juice, but most of the calcium in the juice appears to accompany the enzymes.

Pancreatic juice contains three major enzyme groups: amylytic, lipolytic and proteolytic. In addition, there are many other enzymes. Details of enzyme structure and molecular weight, activation and inhibition, sites of action on substrates and specificities can be found in standard textbooks of biochemistry.

Crystalline amylase isolated from human pancreas is an α-amylase, which splits the α-1,4-glucosidic bond as does salivary amylase. The end-products of exhaustive digestion of unbranched starch are maltose and maltotriose. The products of digestion of a branched starch are maltose, maltotriose and a mixture of dextrins containing α-1,6-branches. The enzyme, unlike salivary amylase, attacks raw, as well as cooked, starch. It is stable in the pH range $4-11$, and its pH optimum is 6.9. Amylase is secreted in the juice in an active state.

Pancreatic juice contains at least three lipolytic enzymes. The first, usually called lipase, is a glycerol-ester hydrolase that hydrolyzes a wide variety of insoluble esters of glycerol at an oil-water interface; it requires cooperation of surface-active agents such as bile salts. Its pH optimum depends on the substrate and ranges from $7.0-9.0$. The second enzyme hydrolyzes esters of secondary and other alcohols, such as those of cholesterol at an optimum pH of 8.0, and it requires bile salts. The third enzyme hydrolyzes water-soluble esters. Pancreatic juice also contains phospholipase A, which hydrolyzes lecithin to lysolecithin.

Fresh, uncontaminated pancreatic juice and extracts of pancreas have no proteolytic activity; but when suitably activated, they form several proteolytic enzymes. These are trypsin and a number of chymotrypsins. Their precursors, or zymogens, are trypsino-

gen and chymotrypsinogens. Trypsinogen is probably converted to trypsin in two ways: by action of trypsin itself and by action of enterokinase. At pH $7.0-9.0$, activation of pure trypsinogen occurs autocatalytically, the activated enzyme converting more zymogen into the active enzyme. Enterokinase is an enzyme that is a component of the brush border of epithelial cells of the duodenum in some species such as the hog, but it is apparently secreted into intestinal contents in other species such as man. Enterokinase contains a large amount of polysaccharide and it resists digestion by secreted proteolytic enzymes, but it is destroyed by bacteria in the colon. In the pH range from $6.0-9.0$, it catalyzes conversion of trypsinogen to trypsin. It is highly specific in that it activates only trypsin. Chymotrypsinogen is converted into the active enzyme only by trypsin. All these proteolytic enzymes attack the interior of peptide chains, and the end-products of action of pancreatic juice are a mixture of small polypeptides and amino acids. Previous action by pepsin is not necessary, and therefore protein digestion is unimpaired by absence of gastric juice.

Trypsin inhibitor, a polypeptide whose molecular weight is $5,000-6,000$, is present in both pancreatic juice and extracts of pancreas. It combines with trypsin at pH $3.0-7.0$ in the ratio of 1 molecule of trypsin to 1 of inhibitor; the product is enzymatically inactive. It also inhibits chymotrypsin, but less completely. Its concentration in human pancreatic juice is such that 1 ml of juice inhibits about 0.08 mg of trypsin. Its presence in the pancreas may protect the gland against autodigestion by small amounts of trypsin that may become active within it; but because its concentration in the juice or gland is much lower than that of trypsinogen, it does not prevent proteolytic activity by fully activated juice.

Pancreatic juice and extracts contain procarboxypeptidase, the zymogen precursor of carboxypeptidase. This enzyme attacks peptide chains at the end, liberating the amino acid with the free carboxyl group. Crude carboxypeptidase is activated by enterokinase; but this may be the result of contamination with trypsinogen, which, after activation by enterokinase, activates procarboxypeptidase. An elastase distinct from trypsin is also present, as are two nucleolytic enzymes, ribonuclease and deoxyribonuclease.

Question of Adaptation of Enzymes to Diet

The relative amounts of fat, protein and carbohydrate in the diet vary greatly. In the short run (a matter of a day or so), the relative proportions of lipolytic, proteolytic and amylytic enzymes in pancreatic juice do not vary as the diet is altered. Although there are minor deviations in the ratio of one kind of enzyme to another, the enzymes are secreted roughly in parallel. In long-term experiments, lasting weeks or months, manipulation of the proportions of foodstuffs in the diet does affect the enzyme content of the pancreas and pancreatic juice. For example, pancreatic glands of rats fed 18% casein contained twice as much proteolytic enzymes as glands of rats fed 6% casein. The glands of rats fed a high-starch diet synthesized amylase 3–4 times as rapidly as those fed a high protein diet. The lipase activity per mg of protein in the dog's pancreatic juice rose 11% when the animal was fed a high fat diet for 3 weeks, and the protease activity per milligram of protein rose 20% following a similar period on a high protein diet.

There is no evidence for either short-term or long-term adaptation in man.

Pancreatic Metabolism in Relation to Secretion

When the pancreas is stimulated by intravenously administered secretin, its oxygen consumption abruptly increases 2–3 times. Consumption declines again as secretion stops, but does not reach resting level until 30 min later.

Pancreatic metabolism has also been studied in vitro, using either slices of large glands or whole glands of the mouse that are thin enough to allow adequate diffusion of oxygen into them. These preparations, when stimulated by secretin, by cholecystokinin or by parasympathomimetic drugs, also increase their oxygen consumption by 15–50%. Two separate processes can be distinguished: (1) synthesis of new enzymes, and (2) the secretory process whereby enzymes are extruded from the gland.

Slices of pigeon pancreas taken from fasted birds previously given injections of pilocarpine to deplete their enzyme content aerobically synthesize α-amylase when supplied with 10 of the 16 amino acids constituting the enzyme; the ones required are the L-forms of tryptophan, arginine, threonine, valine, tyrosine, lysine, leucine, histidine, isovaline and phenylalanine. The addition in vitro of the same amino acids to the pancreas of fasted mice likewise increases α-amylase synthesis. The rate of synthesis is of the order of 0.002–0.004 mg of enzyme per mg of dry weight of pancreas per hour. This is far below the in vivo rate, for rat pancreas can secrete and synthesize 0.08 mg of enzyme per mg of gland protein per hour. This is 0.4 gm of protein enzyme secreted in 24 hours by a gland containing no more than 0.2 gm of protein.

Pancreatic acinar cells contain zymogen granules whose variation in size, number and distribution closely follows the cell's zymogen content. The granules are discharged by vagal stimulation and by agents releasing pancreozymin, and they have been seen to pass from the cells into the ducts. During fasting, when enzyme secretion is small, they increase in size and number. Zymogen granules have been isolated in pure state. They have been found to be stable at pH 5.5 and to dissolve at pH 7.2; this is evidence that intracellular pH in the neighborhood of the granules is near 5.5. Isolated bovine granules are composed of 95% protein: multiple forms of ribonuclease, three kinds of procarboxypeptidase, trypsinogen, chymotrypsinogen A and B, amylase and some unidentified proteins. There is exact correspondence between the proteins of lysed granules and those of bovine pancreatic juice. Such an exact correspondence does not occur in all species; there are large differences between the enzyme content of zymogen granules and the secretions of rabbit and rat pancreases stimulated by pancreozymin or a cholinergic drug. This means that, if granules are the only route of protein secretion, there must be selective extrusion of granules of different composition, or it means that not all secreted enzymes are contained in granules. Trypsin inhibitor is contained in the "cell sap," not the granules.

Synthesis and Secretion of Enzymes

Pancreatic enzymes are synthesized by the ribosomes of the rough endoplasmic reticulum in the acinar cells. There are at least two ways in which the enzymes reach pancreatic juice.

1. As the enzymes are synthesized, they cross the membrane of the rough reticulum into the cysternae. Within the cysternae they are confined in small, smooth-surfaced vesicles. These vesicles move to the Golgi complex where they are formed into zymogen granules. When exocrine stimulation is not occurring, zymogen granules accumulate in large numbers in the apical region of the cells. Following secretory stimulation, the membrane of the granules fuses with the plasma membrane on the apical border of the cells, and the contents of the granule are discharged into the lumen of the terminal pancreatic ducts. Each zymogen granule contains the whole array of digestive enzymes, and the enzymes are secreted in parallel when the contents of granules are discharged from the acinar cells.

2. Strong and continuous stimulation depletes the cell's content of zymogen granules, and during subsequent rest they are reformed. However, when the granule content of the cell is reduced to zero during prolonged stimulation, synthesis and secretion still occur without segregation of the enzymes into granules. It is probable that even when enzymes are being packaged and transported in granules, enzymes are also traveling from their site of synthesis on ribosomes through the cytoplasm to the terminal pancreatic ducts without being sequestered in zymogen granules. In this case, the enzymes need not be secreted in parallel. Stimulation of the pancreas by a peptide extracted from the duodenal mucosa, the candidate hormone *chymodenin* (see Chapter 12), selectively enhances the secretion of chymotrypsinogen while hardly, if at all, affecting the secretion of the other pancreatic enzymes.

Secretion of pancreatic enzymes in response to nervous or hormonal stimulation is mediated by guanosine-3',5'-monophosphate (cyclic GMP), not by cyclic AMP.

Acinar cells of the isolated, perfused pancreas take up chymotrypsinogen contained in the perfusion fluid, and they secrete the absorbed enzyme into the juice. Normal intestinal epithelium absorbs many intact protein molecules (see Chapter 16), and it is possible that some of the chymotrypsinogen secreted by the pancreas is making a second trip.

Nervous Control of Secretion

Stimulation of the peripheral ends of the vagus nerve evokes pancreatic secretion. The effect is chiefly to increase the enzyme content of the juice, although in some species (dog, but not cat), vagal stimulation causes a small increase in volume as well. Acetylcholine is the mediator; and parasympathomimetic drugs, including pilocarpine, are also effective. The response is blocked by atropine.

Stimulation of the splanchnic nerves to the pancreas can be followed by either increased or decreased secretion. Since the increase is blocked by atropine, it is concluded that the nerves contain cholinergic secretory fibers. Inhibition is the result of adrenergic vasoconstriction.

The cephalic phase of pancreatic secretion controlled through the vagus nerve is described in Chapter 12.

Hormonal Control of Pancreatic Secretion

Hormonal control of pancreatic secretion is exerted by at least two hormones: secretin and cholecystokinin-pancreozymin, abbreviated as CCK-PZ or simply as CCK. (See Chapter 12 for their structure and their relation to other hormones.) The gastrins, because they share the same terminal amino acid sequence with cholecystokinin, also stimulate pancreatic secretion when given in pharmacologic doses; whether they do so in physiological situations is not certain.

Acid in the duodenum or jejunum elicits pancreatic secretion. In the dog, the threshold is close to pH 4.0; more acid solutions cause copious secretion, and those above pH 4.5 have no effect. This response is mediated by secretin.

The liberation and action of secretin are independent of extrinsic innervation of intestine or pancreas. When animals are cross-circulated, so that the only connection between them is by way of the blood, instillation of acid into the duodenum of one causes pancreatic secretion in both. When the pancreas is acutely or chronically transplanted, so that all extrinsic innervation is destroyed, acid in the duodenum still evokes secretion. The hormone can be extracted from the mucosa of the upper small intestine, the amounts being progressively smaller the farther the mucosa is from the pyloric sphincter. Secretin is contained in cells, unimaginatively called S-cells, which can be identified in the duodenal mucosa by immunofluorescent techniques. These cells belong to the APUD series (Amine Precursor Uptake and Decarboxylation) derived embryologically from the neuroectoderm.

Secretin has been extracted from the intestines of a large variety of mammals, birds, reptiles, amphibians and teleost and elasmobranch fish. When injected intravenously, such extracts cause secretion of the alkaline component of pancreatic juice. Secretin also stimulates secretion of the bile acid-independent fraction of bile.

Peptones and essential amino acids in the duodenum elicit secretion of enzymes by the pancreas, an effect mediated by cholecystokinin. This hormone has only a small effect upon volume flow of bicarbonate-containing juice, but it potentiates the action of secretin. It increases the enzyme content of already flowing juice.

Trypsin in intestinal contents suppresses release of cholecystokinin and thereby exerts feedback control on enzyme secretion. Bile acids also suppress release of cholecystokinin. These facts, together with the fact that cholecystokinin is a trophic hormone for the pancreas, explain in part the large increase in the size of the pancreas and in the rate of enzyme secretion in rats and chickens fed the trypsin inhibitors found in soybeans and egg whites. The inhibitors bind trypsin, and this prevents trypsin from suppressing the release of cholecystokinin. Continuously high circulating levels of cholecystokinin then stimulate the pancreas to grow and to secrete.

Pancreatic acinar cells have a resting membrane potential of about -40 mV. Acetylcholine and cholecystokinin both cause depolarization to the extent of 15 mV; sodium conductance of the membrane is increased, and calcium released within the cells triggers secretion of enzymes.

Stimulation in Man

Pancreatic secretion in man is measured by the secretin test. A double lumen tube is passed into the fasting subject; one lumen is placed at the end of the duodenum and the other at the pylorus. Both tubes are maintained under constant suction. After the duodenal aspirate has become clear, an intravenous infusion of secretin is begun, and aspiration is continued for 60–80 min. Peak output usually occurs within 30 min of the start of stimulation, and it is a function of the dose. The results of tests on 47 normal human subjects are given in Table 9–1. Many studies have produced essentially similar results.

The ability of the pancreas to secrete en-

TABLE 9–1.—PANCREATIC SECRETION
IN 47 NORMAL HUMAN SUBJECTS GIVEN
2 UNITS OF SECRETIN PER KG OF
BODY WEIGHT*

Volume, ml/kg·hr	2.68 ± 0.24†
Volume, ml/hr	176 ± 20
Total bicarbonate output, μEq/kg·hr	199 ± 22
Total bicarbonate output, mEq/hr	13.5 ± 1.8
Peak volume, ml/10 min	50 ± 15‡
Peak bicarbonate output, μEq/kg·10 min	60 ± 6
Peak bicarbonate concentration, mEq/L	99 ± 2

*From Hartley, R. C.; Gambill, E. E., and Summerskill, W. H. J.: Gastroenterology 48:312, 1965.
†Mean ±2 S.E.
‡Ten subjects only.

zymes can be tested by infusing cholecysto-kinin intravenously or by perfusing the duodenum with test solutions. When the duodenum is perfused with a glucose solution, there is washout of trypsin, but sustained enzyme output occurs only with perfusion with a mixture of amino acids or with a micellar solution of fat (Fig 9–4). The output attained during perfusion with amino acids is equal to that following the maximal tolerated intravenous dose of cholecystokinin. Only the essential amino acids are effective, and a mixture of phenylalanine, valine and methionine is more effective than the individual amino acids. Lipase, trypsin and amylase are secreted in parallel by the normal human pancreas, but in patients with acute or chronic pancreatitis, there is greater impairment of secretion of proteolytic than of nonproteolytic enzymes.

Rapid hydrolysis of proteins in the duodenum during digestion of a meal liberates amino acids that stimulate pancreatic secretion of enzymes. When three meals are eaten during the day, the rate of pancreatic secretion equals the maximum attained after infusion of cholecystokinin, and the high rate of secretion is sustained for 12 hours.

Fig 9–4.—Rate of trypsin output in normal human subjects during duodenal perfusion with isotonic NaCl, a glucose solution, a micellar solution of fatty acids or a solution of a mixture of essential amino acids. Pancreatic secretion of bicarbonate-containing juice was maintained by continuous intravenous infusion of secretin. (Adapted from Go, V. L. W., Hofmann, A. F., and Summerskill, W. H. J.: J. Clin. Invest. 49:1558, 1970.)

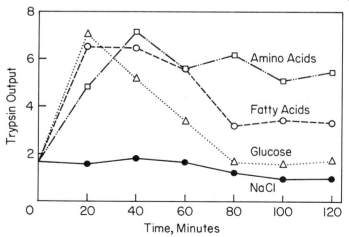

Plasma Amylase and Lipase

Normal human plasma slowly digests starch, but less than one fourth of plasma amylase comes from the pancreas. Elevation of intraductal pressure promotes escape of amylase from the pancreas into the blood; when the pancreas secretes against obstruction, ductules rupture, fluid containing amylase escapes through the capsule, and the amylase is carried by lymph to the plasma. Stimulation of the pancreas by parasympathomimetic drugs, palpation of the gland or eating of a heavy meal accompanied by liberal amounts of alcohol raises the plasma amylase level; and increasingly severe insults raise it much higher. The enzyme also appears in the urine. Renal clearance of pancreatic amylase is rapid, and consequently there may be a large rise in urinary amylase when there is only a small rise in plasma amylase. During severe injury to the pancreas, the gland autolyzes; then plasma amylase rises 15–40 times, and urinary amylase may rise 100 times. The plasma level falls again as recovery ensues or as fibrotic tissue replaces acinar cells.

Lipase is also liberated into the plasma, and its concentration roughly parallels that of amylase. Lipase activated within the gland or released into the peritoneal cavity causes fatty necrosis.

Pancreatic and Biliary Duct Pressures and Pancreatitis

Forceful retrograde injection of bile into the pancreas causes immediate coagulation of lobules; fatty necrosis and dissolution of necrotic cells with hemorrhage follow in 24 hours. Reflux of bile into the pancreas, particularly when their common duct is obstructed, has frequently been blamed for the naturally occurring pancreatitis, which ranges from barely detectable to fatal, complete autolysis. This plausible explanation of the etiology of pancreatitis may not be correct. In order to cause pancreatitis, bile must be injected into the pancreatic duct under a pressure far higher than any naturally occurring. Relative pressures in the biliary and pancreatic ducts have been measured in otherwise undisturbed dogs by means of small plastic cannulas leading from the ducts through the skin. Except occasionally and briefly after feeding, pancreatic intraductal pressure is regularly greater than biliary intraductal pressure. Pancreatic pressure has been found to be higher than biliary pressure in a patient with pancreatitis. In cholecystectomized dogs, the biliary pressure is greater than in intact animals, and sometimes it exceeds the pancreatic pressure. In fasting animals, pancreatic pressure ranges from 3–22 mm Hg. It varies from day to day and from one animal to another, and waves of pressure changes of 2–3 mm Hg without rhythm commonly occur. Pressure is not increased when the pancreas is stimulated by food or secretin. If fresh bile is infused into the pancreatic ducts at a pressure likely to occur naturally—about 3 mm Hg—there is no intrapancreatic activation of proteolytic enzymes, and pancreatitis does not occur. However, infusion of bile mixed with pancreatic juice whose proteolytic enzymes have been activated by enterokinase is capable of producing fatal pancreatitis. It is possible that, under some circumstances, duodenal contents containing activated enzymes may reflux into the pancreas.

Consequences of Loss of Pancreatic Juice

When pancreatic juice flows to the outside through a fistula, dehydration and metabolic acidosis are the consequences. It is difficult to support life by maintaining water and salt balance; even in the more expert hands, dogs with pancreatic fistulas die after about 6 months. The condition of the subject, man or dog, can be improved by the feeding of raw pancreas.

Obstruction of the pancreatic duct, with subsequent fibrotic destruction of acinar cells, also results in deficiency of digestive enzymes unaccompanied by fluid and acid-base difficulties. Surgical removal of the pancreas deprives the subject of insulin and glucagon as well as of pancreatic enzymes. The digestive consequences of deprivation are considered in Chapter 17. In the dog, but not in man, there is progressive decrease in body weight and blood lipid concentration together with development of a fatty liver with impaired function. Instead of the normal 5–7% fat (half phospholipids, one-fourth neutral fats and the rest cholesterol and cholesterol esters), the liver may contain 20–40% fat. The increase is chiefly in glycerides and cholesterol esters. Fatty liver can be prevented by the feeding of adequate amounts of choline or its precursors. The major choline precursor is methionine, from which the methyl groups of choline are derived by transmethylation. Because fatty liver in pancreatectomized animals can also be prevented by the feeding of raw pancreas, fresh pancreatic juice or crystalline trypsin, current opinion is that lack of pancreatic enzymes results in deficient methionine absorption, and hence deficient choline synthesis, for the reason that not enough methionine is liberated from dietary protein. The feeding of raw pancreas corrects this by providing proteolytic enzymes.

REFERENCES

Babkin, B. P.: *Secretory Mechanism of the Digestive Glands* (2d ed.; New York: Paul B. Hoeber, Inc., 1950).

Banks, P. A.: Acute pancreatitis, Gastroenterology 61:382, 1971.

Brunner, H., Northfield, T. C., Hofmann, A. F., Go, V. L. W., and Summerskill, W. H. J.: Gastric emptying and secretion of bile acids, cholesterol, and pancreatic enzymes during digestion, Mayo Clin. Proc. 49:851, 1974.

Case, R. M., Harper, A. A., and Scratcherd, T.: Water and electrolyte secretion by the pancreas, in Botelho, S. Y., Brooks, F. P., and Shelley, W. B. (eds.): *Exocrine Glands* (Philadelphia: University of Pennsylvania Press, 1969), pp. 39–56.

Gregory, R. A.: *Secretory Mechanisms of the Gastrointestinal Tract* (London: Edward Arnold & Co., 1962).

Hallenbeck, G. A.: Biliary and pancreatic intraductal pressures, in Code, C. F. (ed.): *Handbook of Physiology:* Sec. 6. *Alimentary Canal,* Vol. II (Washington, D.C.: American Physiological Society, 1967), pp. 1007–1026.

Harper, A. A.: Hormonal control of pancreatic secretion, in Code, C. F. (ed.): *Handbook of Physiology:* Sec. 6. *Alimentary Canal,* Vol. II (Washington, D.C.: American Physiological Society, 1967), pp. 969–996.

Janowitz, H. D.: Pancreatic secretion of fluid and electrolytes, in Code, C. F. (ed.): *Handbook of Physiology:* Sec. 6. *Alimentary Canal,* Vol. II (Washington, D.C.: American Physiological Society, 1967), pp. 925–934.

Makhlouf, G. M., and Blum, A. L.: An assessment of models for pancreatic secretion, Gastroenterology 59:896, 1970.

Mangos, J. A., McSherry, N. R., Nousia-Arvantakis, S., and Schilling, R. F.: Transductal fluxes of anions in the rat pancreas, Proc. Soc. Exp. Biol. Med. 146:321, 1974.

Palade, G. E.: Intracellular aspects of the process of protein synthesis, Science 189:347, 1975.

de Reuck, A. V. S., and Cameron, M. P. (ed.): *The Exocrine Pancreas* (London: J. & A. Churchill, Ltd., 1962).

Rothman, S. S.: Protein transport by the pancreas, Science 190:747, 1975.

Scratcherd, T., Dreiling, D., Waller, S. L., and Salmon, P. R.: Symposium on diagnosis of pancreatic disease, Gut 16:648, 1975.

Swanson, C. H., and Solomon, A. K.: Micropuncture analysis of the cellular mechanism of electrolyte secretion by the in vitro rabbit pancreas, J. Gen. Physiol. 65:22, 1975.

Webster, P. D., III: Hormonal control of pancreatic and acinar cell metabolism, in Botelho, S. Y., Brooks, F. P., and Shelley, W. B. (eds.): *Exocrine Glands* (Philadelphia: University of Pennsylvania Press, 1969).

Wormsley, K. G., and Goldberg, D. M.: The interrelationships of the pancreatic enzymes, Gut 13:398, 1972.

10

Secretion of the Bile

BILE IS CONTINUOUSLY secreted by the liver into bile capillaries from which it flows into the hepatic ducts. There are two major components of bile. A *bile-acid independent fraction* has a composition similar to that of pancreatic juice, and its rate of secretion is largely controlled by the hormones secretin and cholecystokinin. The *bile-acid dependent fraction* contains newly synthesized bile acids and bile acids returned to the liver in portal blood. These bile acids are secreted in micelles, which also contain sodium, lecithin and cholesterol. The rate of secretion of the bile-acid dependent fraction is governed by the rate of return of bile acids to the liver. In man, the two components together are secreted at the rate of 250–1,100 ml per day. During the interdigestive period in man and other animals having a gallbladder, the resistance to bile flow offered by the sphincter at the duodenal termination of the common bile duct shunts much of the bile into the relaxed gallbladder. Bile acids, bile pigments and other organic constituents of the bile are concentrated as much as 20 times by the gallbladder's rapid reabsorption of water and electrolytes. Within 30 min of a meal, chyme in the duodenum releases the hormone cholecystokinin, which causes the gallbladder to discharge its contents into the duodenum. Bile acids assist in emulsification, hydrolysis

and absorption of fats. Some bile acids are deconjugated and dehydroxylated in the small intestine, and some are absorbed by passive diffusion the length of the small intestine. Conjugated bile salts, and some deconjugated bile acids, are actively reabsorbed in the terminal ileum, and all reabsorbed bile acids are extracted from blood by the liver and are again secreted in bile. Deconjugated bile acids are reconjugated before being secreted.

Pressure of Bile Secretion

The secretion of bile is an active process requiring expenditure of energy by liver cells; it is not simply ultrafiltration. If the bile flows into a vertical tube, its rate of flow is constant until a pressure *above* the pressure of blood in the liver is reached; then flow stops abruptly. In man, this limiting pressure is about 23 mm Hg.

After bile has ceased to flow against a high pressure, a further increase in pressure distends the biliary tree and results in irreversible loss of bile from the tree until the pressure of bile returns to the limiting pressure. The loss occurs because bile leaks out of the bile capillaries; and the rate of leakage is higher, the greater the bile pressure. Therefore the limiting pressure is the one at which continuous secretion of bile into the capillaries equals the rate at which bile leaks from them.

Pressure in the portal vein of a man with a normal liver is about 10 mm Hg; pressure in the unobstructed hepatic vein, measured by means of a catheter threaded into it by way of the antecubital vein and inferior vena cava, is 3 – 4 mm Hg. Pressure in liver sinusoids must lie between these two values. An estimate of sinusoidal pressure is obtained by wedging a catheter with an end-tip into the hepatic vein until it obstructs a small branch. Because the hepatic vein is a valveless end-vein distal to freely communicating hepatic sinusoids, a catheter with an end-hole, wedged into and blocking a division of the hepatic vein, measures pressure in the sinusoids. There is, then, direct connection between the tip of the catheter and the sinusoid bed by way of a static column of blood, but blood can continue to flow through the sinusoids because it can escape into other branches of the hepatic vein. The "wedge pressure" so obtained is 4 – 5 mm Hg. Direct measurement of pressures within the liver vessels of the rat, made by inserting a microneedle attached to a manometer, has shown the mean pressure in the intralobular portal venules to be 4 – 5 mm Hg and in the central veins to be only 1 – 2 mm Hg.

Bile flow depends on liver blood flow chiefly because the oxygen supply to liver cells, and therefore their energy supply, depends on blood flow. If abundant oxygen is assured by raising the quantity of oxygen in blood going to the liver, bile flow becomes nearly independent of blood flow.

Electrolytes of the Bile

The electrolyte composition of human bile secreted at basal rate, except for the presence of bile salts, resembles an ultrafiltrate of plasma (Table 10–1). Stimulation of bile flow by secretin raises the concentration of bicarbonate and lowers that of chloride. In this respect the response of the liver to secretin is similar to that of the pancreas. The carbonic anhydrase inhibitor acetazolamide reduces secretin-stimulated bicarbonate output and volume flow of bile as it reduces those of pancreatic juice. The intrahepatic bile ducts and canaliculi, rather than the parenchymal cells of the liver, are probably the sites at which the major fraction of water and electrolytes of the bile are secreted, and their cells are the ones that respond to secretin and that are inhibited by acetazolamide.

Secretin stimulates bile flow without increasing bile salt secretion, and all stimuli for secretin release elicit bile secretion. Large doses of secretin augment bile secretion to about twice the basal rate; during a meal bile flow increases 3 – 6 times, so secretin is not the only important regulator of bile flow. Bile secreted in response to secretin contains a high concentration of bicarbonate (Fig 10 – 1).

The intravenous injection of pure cholecystokinin or gastrin evokes secretion of bile

TABLE 10–1.—MEAN AND RANGES OF BASAL FLOW AND ELECTROLYTE COMPOSITION OF HUMAN BILE OBTAINED FROM 7 SUBJECTS COMPARED WITH FLOW AND COMPOSITION AT THE PEAK OF RESPONSE TO INTRAVENOUS SECRETIN*

	FLOW (ML/10′)	HCO_3 (MEQ/L)	CL^- (MEQ/L)	NA^+ (MEQ/L)	K^+ (MEQ/L)	BILE SALTS
Basal:						
Mean	3.5	26	102	151	4.6	27
Range	2.7 – 4.1	19 – 29	83 – 112	144 – 159	3.6 – 5.2	19 – 55
Secretin:						
Mean	11.9	46	95	146	4.0	9
Range	5.8 – 28.5	31 – 81	62 – 111	140 – 153	3.7 – 4.7	3 – 12

*From Waitman, A. M., et al.: Gastroenterology 56:286, 1969.

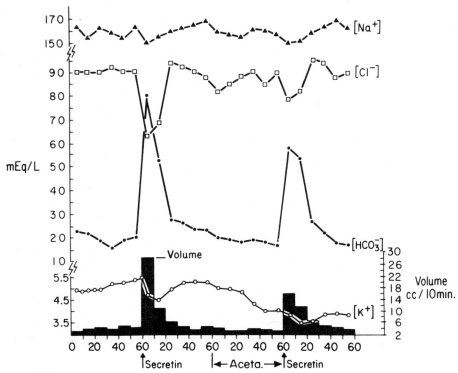

Fig 10–1.—Rate of flow and composition of hepatic bile in a human subject. At the vertical *arrows* secretin (1 unit per kg of body weight) was rapidly injected intravenously. During the third hour the carbonic anhydrase inhibitor acetazolamide was infused in a dose of 50 mg per kg of body weight. (From Waitman, A. M., *et al.*: Gastroenterology 56:286, 1969.)

whose maximal rate is about half that achieved by secretin injection. Gastrin acts by two routes. Its intravenous injection into a gastrectomized dog stimulates bile flow by direct action on the liver. In a normal man or dog, gastrin also acts indirectly by stimulating secretion of acid by the stomach, which in turn releases secretin from the duodenal mucosa. Histamine also stimulates bile flow directly by action on the liver and indirectly by stimulating acid secretion.

Stimulation of the vagus, as by insulin hypoglycemia, increases bile secretion, and the response is blocked by anticholinergic drugs. After bilateral vagotomy, the increase in bile secretion following feeding is lost, demonstrating that vagally mediated reflexes control much of the normal response to a meal. Since vagal stimulation releases gas-

trin, some of the effect of vagal reflexes must be mediated by that hormone.

Afferent nerves from the gallbladder and biliary tree are stimulated by tension in the walls of those vessels; these nerves mediate pain, which is referred to the epigastrium with radiation around the costal margins. Stimulation of the splanchnic nerves to the liver causes vasoconstriction and decreases bile flow. Systemically released epinephrine has two effects: (1) direct vasoconstriction of hepatic vasculature, causing increased impedance; and (2) a redistribution of cardiac output, producing greater blood flow in the splanchnic area with a corresponding increase in pressure gradients through the liver. The latter effect is the more important.

Bile secretion is not stimulated by calomel or by the proprietary preparations of vegetable drugs sold as liver pills. The flow of the bile-acid independent fraction is increased by phenobarbital.

In normal man, the bile-acid independent fraction is at least one third of the total bile secreted.

Other Constituents of the Bile

Bilirubin, the metabolic product of hemoglobin porphyrin degraded in the reticuloendothelial system, is normally present in plasma at about $0.2 - 0.8$ mg per 100 ml. It is conjugated in the liver with glucuronic acid and secreted into bile at up to 1,000 times this concentration. The pigment is further concentrated in the gallbladder. In the intestine, bilirubin is converted by bacteria first to the colorless mesobilirubinogen and to urobilinogen, which, with other breakdown products, colors the stool brown. A small fraction of intestinal urobilinogen is absorbed and secreted into the bile or excreted into the urine at the rate of $0.5 - 2$ mg per day. The capacity of the liver to secrete urobilinogen is low, with the result that in mild liver damage the urinary urobilinogen may rise to 5 mg per day. Liver bile contains $280 - 410$ mg per 100 ml protein, a small fraction being mucoprotein and the rest plasma proteins.

Bile also contains up to 3 gm per liter of neutral fat, and many other substances including products of drug and hormone metabolism and metals such as lead, zinc and copper are secreted in the bile.

Sulfobromophthalein Secretion into Bile

Liver cells secrete many synthetic compounds into the bile. Sulfobromophthalein (Bromsulphalein, BSP) is removed from the blood by liver cells, concentrated within them, to some extent metabolized and secreted into the bile. From two to four derivatives of BSP (the number depending on the species), in addition to unchanged BSP, appear in the bile. The handling of BSP by the liver can be divided into two separate processes: (1) removal of BSP from the plasma

at a rate that depends on its plasma concentration, and (2) secretion of BSP into the bile. The latter process is also a function of plasma concentration; it is a rate-limited process that becomes constant when the plasma concentration of BSP is greater than 3 mg per 100 ml. The maximal rate of BSP secretion into the bile therefore has a T_m (transport maximum) analogous to that of the renal tubules. In 12 men without liver disease, the T_m for BSP secretion was found to be 9.5 mg per min, with a standard deviation of 1.9. In 7 women who were also without liver disease, the value was 7.1 ± 0.8. The T_m is reduced, sometimes nearly to zero, in patients with liver disease.

If BSP is infused into the blood at a rate greater than its T_m, it is accumulated and stored in the liver cells; this function, too, has a maximum that is reduced by liver disorders. The uptake of BSP by extrahepatic tissues is negligible; likewise, absorption of BSP from the intestine is small, being less than 5% of the amount secreted in the bile. Consequently it can be assumed that most of the BSP removed from the blood is taken up and secreted by the liver. This is the basis of a crude test of liver function, in which a dose of 5 mg per kg of body weight is given intravenously and a single plasma sample is taken 45 min later. In normal persons, no more than 5% of the dose is found in the plasma at 45 min.

Other compounds secreted by the liver into the bile include tetraiodophenolphthalein and tetrabromophenolphthalein, which, because of their content of heavy atoms, are relatively impervious to x-rays. These compounds are concentrated in the gallbladder, and the x-ray shadow they cast reveals gallbladder form and function.

Liver Blood Flow

Liver blood flow in man is measured by injecting BSP continuously into a vein at the rate of Q mg per min, and the rate is adjusted so that the arterial plasma concentration of the dye remains constant, as determined by frequent analysis. This rate Q is, of couse, below the T_m of the liver. A catheter is inserted, so that a sample of blood draining the liver through the hepatic veins can be secured. It is assumed that this sample is representative of the total venous drainage of the liver. It is also assumed that the concentration of BSP in all the plasma reaching the liver by way of the hepatic artery and portal vein is represented by the concentration in the systemic artery from which blood is drawn.

If the concentration of BSP in arterial plasma is A mg per liter and if the estimated hepatic plasma flow is $EHPF$ in liters per min, then the rate at which BSP is brought to the liver is $A(EHPF)$ mg per min. The rate at which BSP leaves the liver in the same volume of blood is $V(EHPF)$ mg per min, where V mg per liter is the concentration of BSP in hepatic venous blood. The rate at which BSP is removed by the liver is equal to the difference between that entering and leaving, or $(EHPF)(A-V)$ mg per min. If the only BSP that leaves the plasma is that removed by the liver, the rate at which the liver removes it is equal to the rate Q at which it must be injected to keep the plasma level constant. Then $Q = (EHPF)(A-V)$, or $EHPF = Q/(A-V)$. The estimated hepatic blood flow $(EHBF)$ is determined by dividing this value by the fraction of plasma in arterial blood. With this method, normal values ranging from 1,000–1,800 ml per min per 1.73 m² body surface area are obtained; the mean is about 1,500. This is about 25% of the cardiac output at rest.

In unanesthetized dogs, the hepatic artery supplies about one third of the total hepatic blood flow. There are no reliable estimates of the proportion in man. Blood reaching the liver by the hepatic artery and the portal vein is apparently thoroughly mixed within the liver sinusoids. Of the total oxygen uptake of 40–50 cc per min by a 1,500 gm liver, about half is supplied through the hepatic artery.

Lymph flow from a 1,500 gm liver ranges from 0.4–0.8 ml per min, and its protein content is almost equal to that of plasma.

Bile Acids and Bile Salts*

Bile acids are synthesized in the liver from cholesterol (Fig 10–2). The limiting reaction is hydroxylation catalyzed by 7-α-hydrolase, and this step in synthesis is the one controlled in the negative feedback regulation of bile acid synthesis.

The major bile acids formed in the human liver are the trihydroxy cholic acid (Greek *chole* = bile) and the dihydroxy chenodeoxycholic, or chenic, acid (Greek *chenos* = goose). Because these are the ones synthesized, they are the *primary bile acids*.

Primary bile acids are dehydroxylated by bacteria in the digestive tract. The major products are deoxycholic acid, a dihydroxy acid derived from cholic acid, and lithocholic acid (Greek *lithos* = stone), a monohydroxy acid derived from chenodeoxycholic acid. These are *secondary bile acids*.

Steroid chemists amuse themselves by identifying many minor bile acids such as ursodeoxycholic acid (Latin *ursus* = bear), an isomeric form of chenodeoxycholic acid in which the hydroxyl at the 7-position has the beta configuration.

The *bile acid pool* is the total amount of primary and secondary bile acids present in the body at any one time.

Primary and secondary bile acids are secreted as conjugated bile salts by the liver. Either taurine (Greek *tauros* = bull) or glycine is added in peptide bond to the carbonyl

---

Bile acids is the generic term for the steroid compounds first isolated in protonated form by organic chemists. The papers by O. Rosenheim and H. King (The ring system of sterols and bile acids, J. Soc. Chem. Industry 51:464 and 554, 1932) in which their structure was first correctly described are masterpieces of armchair chemistry. *Bile salts* is a generic term referring to the conjugates of bile acids with glycine or taurine in peptide linkage or bile alcohols linked in ester bond with sulfate. The term bile salts is used here to name the conjugated compounds, whether or not they are ionized. The term bile acids is used as an all-inclusive name for all compounds containing the cholane nucleus, whether or not they are conjugated or ionized. The nomenclature of these compounds is not settled, and the reader can expect to encounter changes. For example, taurine is linked to cholic acid in a peptide bond, and the name for the compound that is likely to be adopted is *cholyltaurine*, replacing *taurocholic acid*.

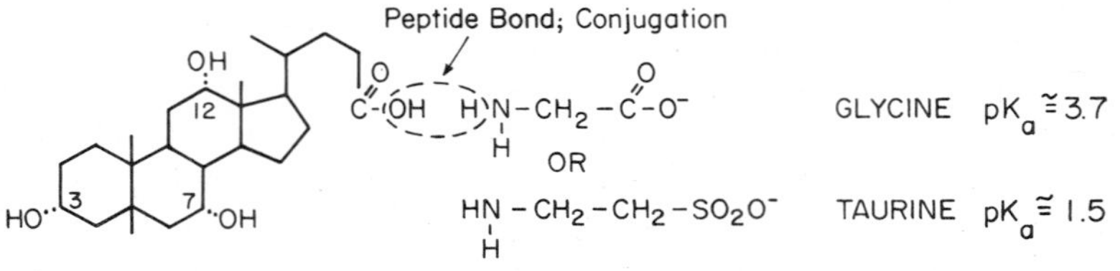

Peptide Bond; Conjugation

GLYCINE $pK_a \approx 3.7$

TAURINE $pK_a \approx 1.5$

CHOLIC ACID (Primary Bile Acid)

NO − OH on 12 = Chenodeoxycholic Acid (primary)

NO − OH on 7 = Deoxycholic Acid (secondary)

NO − OH on 7 and 12 = Lithocholic Acid (secondary)

Fig 10–2.—The structure of the common bile acids. The bile acid is conjugated with either glycine or taurine by elimination of water to form a peptide bond. The approximate ionization constants of glycocholic and taurocholic acids are given on the right. (From Davenport, H. W.: *A Digest of Digestion* [Chicago: Year Book Medical Publishers, Inc., 1975]. Adapted from Hofmann, A. F.: Gastroenterology 48:484 1965.)

group of the side chain of the cholane nucleus, making the corresponding taurocholic or glycocholic bile salts. These occur in human bile in the ratio of about 1:3, because taurine is in short supply. If taurine is fed, the fraction secreted as taurocholic acids rises, and when taurine is lost as the result of imperfect reabsorption, the ratio falls to 1:20.

Bile acids are planar molecules. Hydrophobic groups project on one side of the molecule, and hydrophilic groups project on the other side. The hydrophilic groups are the hydroxyl groups of the cholane nucleus, the peptide bond on the side chain and the carbonyl or sulfate group of glycine or taurine. Bile acids accumulate at oil-water interfaces and decrease interfacial tension. The effectiveness of individual bile acids in reducing interfacial tension depends upon the number and character of their hydrophilic groups. Thus, taurocholic acid emulsifies fat and forms micelles far more readily than does unconjugated lithocholic acid.

Taurine conjugates of bile acids are anions at the pH (6.0–7.7) of bile and intestinal contents, and consequently they always exist as salts. Glycine conjugates of bile acids are less ionized at intestinal pH, and consequently glycocholates in the intestine are a mixture of ionized and un-ionized molecules. The charges of bile anions are balanced by cations, chiefly sodium. The osmotic pressure, measured directly by the freezing point depression method of 67 samples of bile drawn from the common ducts of cholecystectomized dogs, was found to be 299 ± 11 mOsm; plasma osmolality of the same dogs was 303 ± 4 mOsm. However, the sum of the cations and anions of bile is always greater than 299 mOsm. The reason for this discrepancy is that the bile salts form osmotically inactive micelles within which some cations are sequestered. In samples of dog bile whose sodium concentration determined chemically was 245–270 mN, the sodium activity, measured with a sodium-sensitive electrode, was only 148–186 mN. The potassium concentration of the same samples was 8.8–10.2 mN, but the potassium activity was 2.7–3.9 mN. Osmotic coefficients of sodium and potassium in bile are even lower than their activity coefficients.

Ionized bile salts, being large, negatively charged ions at the pH prevailing in the intestinal lumen, are not absorbed by passive diffusion in the duodenum or jejunum. Consequently, their concentration remains high until fat digestion and absorption is complete. Some small fraction of the un-ionized glycocholates are absorbed by diffusion in the duodenum and jejunum; but the remainder, plus the ionized bile salts, are absorbed by active transport in the lower ileum. Upon reaching the liver, all bile salts in the portal

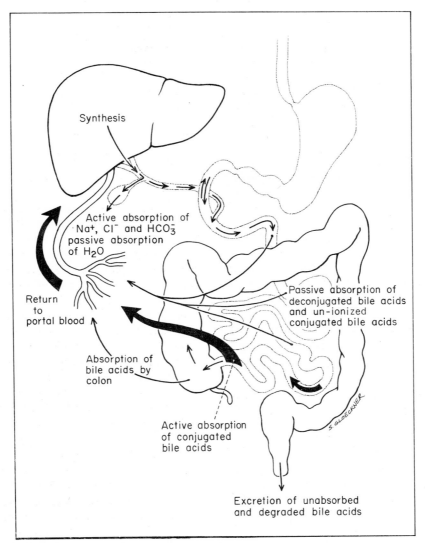

Synthesis

Active absorption of
Na+, Cl− and HCO$_3^-$
passive absorption
of H$_2$O

Return
to
portal blood

Absorption of
bile acids by
colon

Passive absorption of
deconjugated bile acids
and un-ionized
conjugated bile acids

Active absorption
of conjugated
bile acids

Excretion of unabsorbed
and degraded bile acids

Fig 10–3.—The enterohepatic circulation. (From Davenport, H. W.: *A Digest of Digestion* [Chicago: Year Book Medical Publishers, Inc., 1975].)

blood are immediately taken up by liver cells and again secreted into the bile. This cycle is called the *enterohepatic circulation* (Fig 10–3).

Bile-Acid Dependent Bile

A substance increasing bile flow is called a *choleretic.* Bile salts themselves are powerful choleretics, the most effective being dehydrocholate. Bile salts act directly on the parenchymal cells of the liver, and their intravenous injection is promptly followed by their secretion into the bile. The active secretion of bile salts is accompanied by an apparently passive flow of water and electrolytes into the bile, with the result that the volume, as well as bile salt concentration, of the bile increases. When bile salts are continuously infused, the rate of bile salt secretion is equal to, or only slightly greater than, the rate of infusion. When the rate of infusion is in-

TABLE 10–2.—BILE ACID AND
CHOLESTEROL SECRETION IN BILE OF
NORMAL SUBJECTS FED A LOW-CALORIE
AND HIGH-CALORIE DIET*

	20 CAL/KG	40 CAL/KG
Bile acid secretion, gm/day	16.0–37.0	19.0–72.0
Cholesterol secretion, gm/day	1.6– 3.0	2.4– 3.7
Bile acid cycles/day	3.4– 5.7	5.0–13.5

*Adapted from Brunner, H., Hofmann, A. F., and Summer-skill, W. H. J.: Gastroenterology 62:188, 1972.

creased, the rate of flow of bile likewise goes up. The secretion of bile salts is interrupted by their storage in the gallbladder during the interdigestive period.

In normal man, the secretion of the bile-acid dependent fraction is governed by the rate of return of bile acids to the liver, and this in turn is controlled by stimuli for bile secretion and gallbladder contraction. The data in Table 10–2 show that doubling the caloric intake increases bile acid output. Increasing the number of enterohepatic cycles

Fig 10–4.—Schematic representation of a micelle in longitudinal section. Bile acids for a cylindrical shell with their hydrophilic groups outward. The core consists of interdigitating lipid molecules. During digestion these are free fatty acids and 2-monoglycerides, as depicted here. In the bile, the lipids are lecithin and neutral fats. Cholesterol is dissolved in the lipid phase of the micelle. The micelle is surrounded by cations, not shown here, which balance the anionic charges on the bile acids.

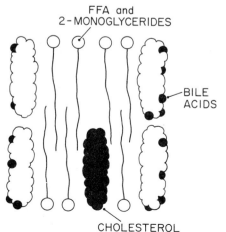

FFA and
2-MONOGLYCERIDES

BILE ACIDS

CHOLESTEROL

makes a small pool of bile acids repeatedly available for digestion.

Bile acids, lecithin and cholesterol are secreted together by the liver. Bile acids and lecithin form micelles in which cholesterol, which is almost totally insoluble in water, is dissolved (Fig 10–4). The concentrations of bile acids and lecithin in bile determine the amount of cholesterol that can be carried in micellar solution. When bile contains more cholesterol than can be carried in micellar solution, the extra cholesterol is in the form of microcrystals, and the bile is supersaturated with respect to cholesterol.

The amount of bile acids and lecithin secreted into the bile, and therefore the bile's ability to dissolve cholesterol, depends upon the enterohepatic circulation. During an overnight fast, little bile acid is returned to the liver from the intestine, and most of the bile acids secreted is sequestered in the gallbladder. The enterohepatic circulation is diminished. In normal subjects, bile tends to be supersaturated with cholesterol at night when bile acid secretion is low. Immediately after a meal, bile acids are discharged from the gallbladder into the intestine, and the enterohepatic circulation of bile acids is accelerated. During the day, and especially just after meals, bile is not usually saturated with cholesterol.

Bile Acid Pool Size

The method of measuring bile acid pool size is illustrated in Figure 10–5. A nasogastric tube is passed with its opening in the subject's duodenum so that freshly excreted

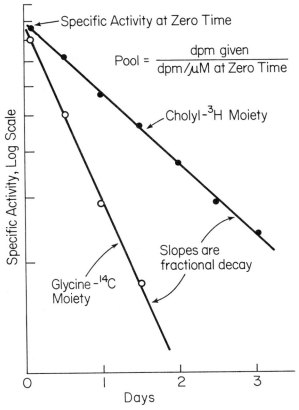

Fig 10–5. — Measurement of bile acid pool size, fractional decay and deconjugation. In this example, cholylglycine (glycocholic acid) labeled with ^{3}H in the cholane nucleus and with ^{14}C in the glycine was given. Dpm = disintegrations per minute.

bile can be sampled. A small amount of isotopically labeled bile is given. In this example, glycocholic acid doubly labeled with ^{3}H on the cholic acid and ^{14}C in the glycine was used. The administered radioactivity in terms of disintegrations per min (dpm) of each isotope is accurately known. Bile samples are collected over 168 hours. The radioactivity and the cholic acid concentrations are determined, and the specific activity of the cholic acid in the sample is calculated. This is dpm per milligram or per millimole of cholic acid. Likewise, the specific activity of the glycine conjugated to cholic acid is calculated. This is dpm per milligram or per millimole of glycine.

The rate of disappearance at any instant of each of the labeled components of glycocholic acid is proportional to the amount present at that instant. This is a first order reaction, and the logarithm of the specific activity of the serial samples plotted against time gives a straight line with negative slope. The line extrapolated to zero time gives the specific activity of the bile acid pool if mixing of the administered bile acid with the rest of the pool were instantaneous. The zero-time specific activity, dpm per milligram or millimole, divided into the number of dpm's originally administered gives the size of the pool of glycocholic acid in milligrams or millimoles.

Other components of the pool can be measured in the same way. Alternatively, if all the components of the pool are measured in an early sample, their individual pool sizes can be calculated once the size of the pool of any component is known.

Representative data on the size and composition of the bile acid pool in normal subjects are given in Table 10–3. The lower limit of the normal pool in adult men is about 1.8 gm. Pool size in patients with cholelithiasis and in cholecystectomized patients is usually below the normal limit.

The slope of the lines plotted in Figure 10–5 is the fractional decay, or rate constant, of the labeled compound. The data given for the cholyl moiety show that half the activity disappeared in 1.56 days. The rate constant is 0.321 days^{-1}. If the pool size remains constant, the amount of cholic acid disappearing each day is replaced by newly synthesized cholic acid. With this rate constant, 0.33 gm a day replaces the cholic acid lost.

During its enterohepatic circulation, some cholic acid is converted to deoxycholic acid by 7-α-dehydroxylation. The newly produced secondary bile acid enters a pool of deoxycholic acid and participates in the enterohepatic circulation.

Both pool size and rate constant of chenodeoxycholic acid are smaller than those of cholic acid. Synthesis of about 0.18 gm a day maintains a pool size of about 0.68 gm. Bacterial dehydroxylation of chenodeoxycholic

TABLE 10–3.—REPRESENTATIVE FIGURES FOR COMPOSITION OF BILE ACID POOL IN NORMAL ADULT HUMAN SUBJECTS

	GM	RANGE, GM
Primary bile acids		
Total cholic conjugates	1.45	0.55 – 1.90
Cholylglycine	1.05	
Cholyltaurine	0.40	
Total chenic acid conjugates	0.68	0.13 – 0.77
Secondary bile acids		
Total deoxycholic conjugates	0.35	0.01 – 0.72
Total lithocholic conjugates	0.05	0.01 – 0.07
Total pool size	2.53	1.86 – 3.24

acid produces lithocholic acid, which is discussed below.

The damaged liver of cirrhotic patients produces cholic acid at 25% of the normal rate as the result of impaired conversion of cholesterol to cholic acid. The rate of production of chenodeoxycholic acid is 70% of normal. As a result, the bile acid pool of patients with cirrhosis is about half that of normal persons, and its fraction of cholic and deoxycholic acids is reduced.

The enterohepatic circulation of cholic, deoxycholic and chenodeoxycholic (or chenic) acids is summarized in Figure 10–6.

Fig 10–6.—Enterohepatic circulation of the major primary and secondary bile acids in man. The circulation of lithocholic acid, which is normally less than 5% of the total bile acid pool, is shown in Figure 10–7. (Adapted from Hoffman, N. E., and Hofmann, A. F.: Gastroenterology 67:887, 1974.)

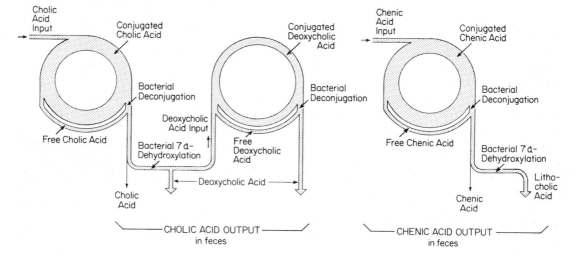

Deconjugation During Enterohepatic Circulation

Bacteria in the small intestine deconjugate bile acids during their enterohepatic circulation, and bacteria in the colon deconjugate bile acids that have escaped reabsorption in the terminal ileum. The concentration of deconjugated bile acids rises from zero in the duodenum to about 1 mM in the lower ileum during digestion of a meal.

The data in Figure 10–5 were obtained when doubly labeled glycocholic acid was given; the ^3H label on the cholic acid and the ^{14}C in the glycine were chemically linked by a peptide bond. Nevertheless, the fractional decay of the glycine label is about 3 times as fast as the cholic acid label. Eighteen percent of glycocholic acid is deconjugated during each enterohepatic cycle. The deconjugated bile acid is reabsorbed to the extent of 95% and, being reconjugated in the liver, is again secreted in the bile. Glycine liberated from the bile acid enters the large glycine pool of the body, and its carbons are almost entirely oxidized to carbon dioxide. Carbon dioxide is eventually breathed off in the lungs, and this is the basis of a simple test of the rate of deconjugation. A bile acid conjugated with ^{14}C-glycine is given, and the rate at which ^{14}C-CO_2 appears in the breath is a measure of the rate of deconjugation. Replacement of glycine oxidized after deconjugation requires 450 mg of glycine a day.

The rate of deconjugation of taurine conjugates is less than half that of glycine conjugates. A small fraction of the taurine removed from bile acids is reabsorbed and enters the taurine pool; little is reincorporated into conjugated bile acids. The taurine not absorbed is degraded by bacterial action, and its sulfur appears in the urine as sulfate.

Lithocholic Acid

Lithocholic acid is derived from chenodeoxycholic acid by 7-α-dehydroxylation in the intestinal lumen. In normal man, 60–150 mg of chenodeoxycholic acid, or one third to one half of the chenodeoxycholic acid synthesized each day, is converted to lithocholic acid. Newly formed lithocholic acid is absorbed into portal blood and rapidly extracted by the liver. In the liver, it is all conjugated, chiefly with glycine.

Sixty percent of the lithocholic acid reaching the liver is sulfated by addition of sulfate to the hydroxyl group in the 3-position. Sulfated lithocholic acid is secreted in the bile. It is poorly absorbed by the intestinal epithelium, and 80% of that secreted by the liver is excreted in the stool (Fig 10–7). Some of the unsulfated fraction of lithocholic acid secreted in the bile is reabsorbed, and 60% of this

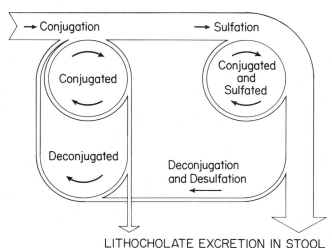

LITHOCHOLATE EXCRETION IN STOOL

Fig 10–7. — The enterohepatic circulation of lithocholic acid and its clearance from the bile acid pool by sulfation. (Adapted from Cowen, A. E., Korman, M. G., Hofmann, A. F., Cass, O. W., and Coffin, S. B.: Gastroenterology 69:67, 1975.)

is sulfated when it returns to the liver. Therefore, lithocholic acid is cleared from the bile acid pool by sulfation and excretion in the stool during each cycle of its enterohepatic circulation. Although a substantial amount of lithocholic acid enters the bile acid pool each day, lithocholic acid constitutes less than 5% of the pool.

Lithocholic acid in high concentration is toxic to the liver. When chenodeoxycholic acid is fed to patients with cholelithiasis, the amount of lithocholic acid formed each day increases. However, rapid excretion of sulfated lithocholic acid keeps the fraction of lithocholic acid in the bile acid pool low, and toxic effects are avoided in man. Primates are less able to sulfate lithocholic acid, and when they are fed chenodeoxycholic acid, their lithocholic acid concentration may rise to toxic levels.

Intestinal Absorption of Bile Acids

Bile acids are absorbed in the terminal ileum by active transport, and they are absorbed the whole length of the small intestine by passive diffusion. More than 95% of the bile acids are absorbed in the small intestine. Some of the remaining bile acids are absorbed in the colon by passive diffusion, and the rest, after bacterial deconjugation and modification, is excreted in the stool. Excretion averages 500 mg a day.

Conjugated and ionized bile acids are rapidly absorbed by active transport by a mechanism confined to the terminal ileum. The transport process exhibits saturation kinetics, and there is competition among bile acids for absorption. A small amount of deconjugated bile acids may also be absorbed by active transport in the terminal ileum.

Because bile acids have molecular weights greater than 500, their passive absorption depends chiefly upon their solubility in the lipid portion of the membrane of intestinal epithelial cells. Therefore, factors decreasing water solubility and increasing fat solubility promote passive absorption. Glycine conjugates, because their pK_a is high, are only slightly ionized in intestinal contents. They are more readily absorbed by passive diffusion than are taurine conjugates whose pK_a is low and which are ionized in intestinal contents. Deconjugation increases passive absorption by 9 times, and dehydroxylation increases passive permeation by more than 4 times.

The fraction absorbed by each process is unknown for man, but abolition of active absorption in the monkey by removal of the last third of the small intestine reduces total bile acid absorption by 45%. Perhaps in man, half of the absorption is active, half passive.

In man, resection or disease of the terminal ileum reduces absorption of bile acids, and diarrhea may ensue for one of two reasons. If bile acid malabsorption is mild, synthesis of bile acids by the liver is able to maintain pool size, and no defect in fat digestion and absorption occurs. However, unabsorbed bile acids escape into the colon where they or their degradation products inhibit sodium and water absorption. This form of diarrhea can be prevented by feeding cholestyramine, a resinous polymer of aminated styrene, which absorbs and thereby sequesters bile acids. If bile acid malabsorption is severe, digestion and absorption of long-chain triglycerides is defective, and unabsorbed fatty acids or their derivatives cause diarrhea. This form can be prevented by substituting medium-chain for long-chain triglycerides.

Bile Acids in Blood

Absorbed bile acids are carried from the intestine to the liver in the plasma of portal blood. Dihydroxy bile acids are bound to plasma proteins to the extent of 96–98%, and trihydroxy acids are bound 83–91%. Bile acids are rapidly removed from portal blood by the liver, and the rate of removal is slightly less for those more tightly bound to plasma proteins. The concentration of bile acids in peripheral plasma is very low, being less than 2 μM. The concentration in the plasma of normal persons reaches a peak about 2 hours after a meal when the acids secreted in response to the meal reach the terminal ileum (Fig 10–8). In patients with bile acid malabsorption resulting from terminal ileal disease, the peaks reached following a meal are much lower. In such patients, the peak occurring after breakfast is higher than subsequent peaks, because it reflects absorption of bile acids synthesized overnight and stored in the gallbladder. The plasma concentration of bile acids in cholecystectomized patients is much more constant. Peripheral plasma concentration of bile acids in a patient with portacaval shunt may be 100 times the normal value.

Removal of bile acids from the blood can be used as a test of liver function. Following intravenous injection, glycocholic acid disappears from the plasma of normal persons very rapidly. The disappearance curve has two components, one with a half-time of 1.7–3 min and the other with a half time of 7–16 min. Patients with liver disease, whose other liver function tests may give normal results, have greatly prolonged bile acid disappearance times.

Control of Bile Acid Synthesis

Under normal circumstances, 500–900 mg of bile acids is lost each day, and pool size is kept constant by equal synthesis of new bile acids. Rate of synthesis is governed by the rate at which bile acids return to the liver, and control is exercised by feedback inhibition of 7-α-hydrolase. If the rate of return is increased by feeding bile acids, pool size rises and rate of synthesis falls. If the rate of return is reduced, rate of synthesis is increased up to the maximum capacity of the liver. In the monkey, the liver can synthesize bile acids up to 10 times the normal rate. The same is probably true for the human liver.

If bile acids are drawn off through a fistula or if reabsorption is reduced, the rate of synthesis rises. In the monkey, the rate of synthesis can keep up with loss if no more than 20% of the bile acids secreted fails to be reabsorbed. However, if 33% of the bile

Fig 10–8. – The concentration of conjugated bile acids in the plasma of normal persons, cholecystectomized subjects and bile-acid malabsorbing subjects over 24 hours. The subjects were given liquid meals at the times indicated by the *arrows*. (Courtesy of A. F. Hofmann.)

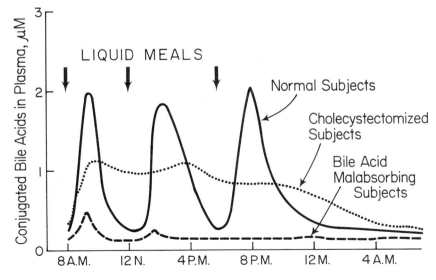

acids secreted is lost, synthesis cannot increase further, and pool size falls. With greater losses, pool size falls further. If the enterohepatic circulation is abolished in man by total failure of reabsorption, the rate of secretion of bile acids falls to 3–5 gm per day; this rate of secretion is equal to the maximal capacity of the liver to synthesize bile acids. Thus, in patients incapable of reabsorbing bile acids, only 3–5 gm of bile acids reaches the duodenum each day, in contrast with the normal 16–72 gm per day. In such patients, most of the bile is delivered in the morning when the gallbladder evacuates bile it has stored overnight.

Function of the Gallbladder

In animals possessing a gallbladder* that concentrates and stores bile, the entry of bile into the duodenum does not parallel its secretion by the liver. In the interdigestive pe-

---

*In animals without a gallbladder (e.g., pigeon, rat, pocket gopher and horse), the sphincter of the hepato-pancreatic ampulla has little or no resistance, and the liver secretes large amounts of bile. Some animals (e.g., guinea pig, rabbit and bush rat) whose gallbladders have a low ability to concentrate bile also have a large continuous secretion. The pig, goat, sheep and cow have gallbladders with low concentrating capacity, but their livers secrete relatively small amounts of bile. The duck, man, mouse, chicken, dog, cat and striped gopher secrete bile at low rates; but their gallbladders have high concentrating capacity, with the result that the volume of bile reaching the intestine is small.

riod, the sphincter of the ampulla offers resistance to bile flow. The gallbladder is relaxed, and liver bile enters it. In man, the bladder's capacity ranges from 14–60 ml, with a mean of 33. Not all bile is diverted to the gallbladder, for the duodenum contains bile in the interdigestive period.

Rapid reabsorption of fluid by the mucosa of the gallbladder prevents a rise in pressure in the biliary tree. Water and electrolytes are reabsorbed, leaving behind bile salts, bile pigments and cholesterol, which become 5–20 times more concentrated in bladder bile than they are in liver bile (Table 10–4). The primary process responsible for concentrating bile is the active transport of sodium by the mucosa from lumen to blood. Chloride and bicarbonate follow sodium, and water follows the ions in isotonic proportions. Of water-soluble substances, only the small ones whose molecular diameter is less than 6–8 Å can penetrate the gallbladder mucosa. The concentration of K^+ tends to rise as well as that of bile salts; but because it can penetrate the mucosa, K^+ diffuses down its concentration gradient from lumen to blood. The mucosal surface of the gallbladder is negative by about 18 mV with respect to the serosal surface, and the final luminal concentration of K^+ is that which is in electrochemical equilibrium with K^+ in interstitial fluid. Probably Ca^{2+} is also at its electrochemical equilibrium in gallbladder bile.

TABLE 10–4.–COMPARISON OF LIVER AND GALLBLADDER BILE*

	UNITS	LIVER BILE	GALLBLADDER BILE
Na^+	mEq/L	140 –159	220 – 340
K^+	mEq/L	4 – 5	6 – 14
Ca^{2+}	mEq/L	2 – 5	5 – 32
Cl^-	mEq/L	62 –112	1 – 10
Bile salts	mEq/L	3 – 55	290 – 340
pH		7.2 – 7.7	5.6 – 7.4
Cholesterol	mg/100 ml	60 –170	350 – 930
Pigment	mg/100 ml	50 –170	200 –1,500

*Compiled in part from data on human bile contained in Waitman, A. M., et al.: Gastroenterology 56:286, 1969, and Dittmer, D. S. (ed.): *Blood and Other Body Fluids* (Washington, D.C.: Federation of American Societies for Experimental Biology, 1961) and from data on canine bile contained in Ravdin, I. S., et al.: Am. J. Physiol. 99:317, 1932, and Wheeler, H. O., and Ramos, O. L.: J. Clin. Invest. 39:161, 1960.

The concentration of Cl⁻ is determined by two opposing processes: active transport out of the gallbladder and diffusion back along its electrochemical gradient. The high total concentration of Na⁺ in gallbladder bile is accounted for by the fact that a large fraction of the ion is held in association with bile salts in osmotically inactive micelles. The gallbladder mucosa secretes mucin, the concentration of which in gallbladder bile is about 1.6%.

The mucosa of the gallbladder consists of a single layer of tall, columnar epithelial cells, each bound to its neighbors at the apical end by tight junctions. When the mucosa is reabsorbing fluid, long lateral channels appear between the cells (Fig 10–9). The human gallbladder, unlike that of rabbit or fish, develops a potential difference of about 8 mV, serosal surface positive, when it is absorbing, and it can maintain a short circuit current of 136 μa. The current is not present if choline or potassium is substituted for sodium, it is abolished by ouabain and it is almost independent of anions. Absorption is apparently effected by an electrogenic sodium pump. Chloride ions follow sodium passively. Transport of ions into the lateral channels raises the local osmotic pressure, and water flows along its osmotic gradient from the lumen through the cells into the lateral channels. Solute moves down the channels by diffusion and by being swept along by the water stream. When fluid transport is in a steady state, there is a standing osmotic gradient within the lateral channels, high at the apical end and nearly or exactly isotonic at the basal end, where the fluid crosses the basement membrane to enter interstitial fluid and to be removed by the capillaries. Secretion of acid by the gallbladder epithelium removes bicarbonate from the contents and slightly acidifies the bile.

The organic constituents of bile are concentrated by being left behind during absorption of water and electrolytes. About 10% of the lecithin in the gallbladder is absorbed, and a small amount of deconjugated bile

Fig 10–9.—Scheme showing mechanism of fluid transport by the gallbladder. (Adapted from Dietschy, J. M.: Gastroenterology 50:692, 1966.)

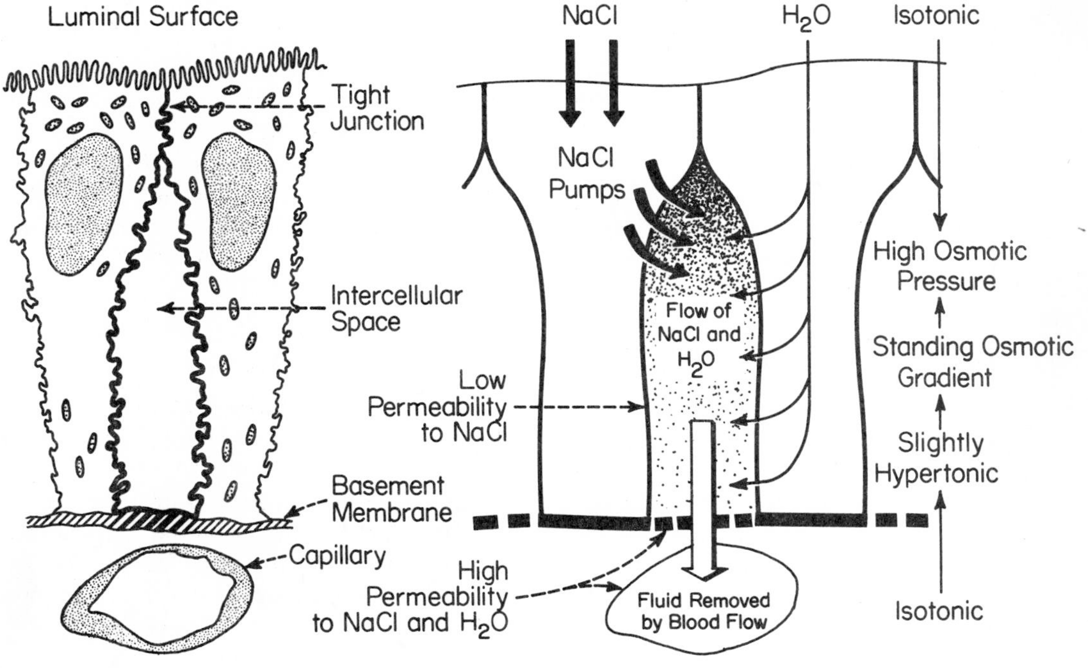

acids, when present in gallbladder bile, is absorbed by passive diffusion. Iodine-containing, synthetic organic compounds are secreted by the liver into bile, and they are not absorbed by the gallbladder epithelium. Consequently, they become concentrated in gallbladder bile. Because they are radiopaque, they serve as contrast media.

Contraction of the Gallbladder

Within the first half hour after a meal, the human gallbladder begins to contract. Emptying occurs irregularly over a period of 20–105 min, and it is seldom complete. In 15 normal human subjects whose gallbladder size was measured by x-ray photographs taken in two directions, the greatest volume expelled was found to be 27 ml and the least 8 ml. In another 15 subjects, the percentage emptying ranged from 51–99 and averaged 84. Gallbladder contraction raises the pressure of bile within the ducts. When bile pressure reaches a certain level, the ampullary sphincter yields, allowing bile to spurt into the duodenum. As bile pressure falls, resistance of the ampullary sphincter rises again, and flow of bile into the duodenum is cut off until intraductal pressure again becomes high enough to make the sphincter yield.

A substance promoting discharge of gallbladder bile is a *cholecystagogue,* and the effect is mediated by two pathways—nervous and humoral. The efferent nerves to both gallbladder and sphincter are in the vagus; their mediator is acetylcholine. Gallbladder contraction is part of the cephalic phase of digestion. Besides being activated by the process of eating, this reflex is subject to influences having emotional concomitants. The drinking of olive oil by an Italian may be promptly followed by gallbladder contraction, whereas the gallbladder of an Irishman, to whom olive oil may be repulsive, may be unaffected by the same stimulus. Afferent nerves from the duodenum and other organs may carry impulses arousing or inhibiting vagally mediated gallbladder and sphincter

movements, but such reflexes have not been fully studied.

The gallbladder is under humoral control, for an extrinsically denervated gallbladder contracts in response to chyme in the duodenum. The most effective stimuli are fat in any form, egg yolk and meat. The hormone released from the duodenal mucosa is cholecystokinin. This hormone has the same C-terminal sequence of amino acids (Trp-Met-Asp-Phe-NH_2) as gastrin and caerulein. When compared with cholecystokinin on a molar basis, the tetrapeptide itself has 1/143d the potency of cholecystokinin in stimulating gallbladder contraction. Gastrin I and II are equal and are 1/22d as effective. Because caerulein is 16 times more effective than cholecystokinin, it has been used clinically as a cholecystagogue.

In patients with gluten enteropathy, the release of cholecystokinin is reduced, and bile remains sequestered in the gallbladder. The rate of synthesis of bile acids is normal, but their half-life is prolonged. As a result, the bile acid pool size increases and may become as large as 14 gm.

The gush of gallbladder bile into the duodenum early in gastric emptying raises the concentration of bile salts in duodenal contents to 10–46 mM. Thereafter, as gastric contents empty into the duodenum, bile salts remain nearly constant at 2–7 mM throughout the period of intestinal digestion and absorption. Therefore, bile salt concentration in duodenal contents is always above the critical micellar concentration.

The Lithogenic Index: Percentage Saturation of the Bile

Bile is a four-component system of water, phospholipids, bile acids and cholesterol. Because the water content of bile as it is secreted by the liver is nearly constant at 90%, the chief variables determining whether bile is lithogenic is the relation among phospholipids, bile acids and cholesterol. These three are considered in terms of their mole fractions in bile. The amount of each component in a sample of bile is determined. The sum of the three components, in moles, is set equal to 1, and the fraction that a particular component contributes to the sum is its mole fraction.

Phospholipids and bile acids form micelles in which cholesterol dissolves (see Fig 10–4). There are three possible relations between the cholesterol concentration in bile and the phospholipid-bile acid micelles.

1. The amount of cholesterol present is not enough to saturate the micelles. Unsaturated bile does not form cholesterol gallstones; it is *nonlithogenic*. Nonlithogenic bile, in fact, dissolves cholesterol gallstones, and to make the liver secrete nonlithogenic bile is the aim of dietary modification of bile composition.

2. Micelles in bile are saturated with cholesterol, but there is no additional cholesterol

in the bile sample. The relation between the mole fraction of cholesterol in saturated bile and the bile's content of phospholipid and bile acids is shown in Figure 10–10 as the experimentally determined curved line. The abscissa is the ratio: (mole fraction of phospholipid)/(sum of mole fractions of phospholipid and bile acids). Points toward the right along the abscissa represent mixtures with higher fractions of phospholipid, and points to the left represent mixtures with higher fractions of bile acids.

3. Supersaturated bile contains micelles saturated with cholesterol and, in addition, microcrystals of cholesterol. Such bile is *lithogenic,* and it is represented by points above the line of saturation.

The *lithogenic index* of a particular sample of bile is the ratio of the mole fraction of cholesterol in the sample divided by the concentration of cholesterol at saturation. Two

Fig 10–10.—Determination of the lithogenic index of bile by means of a graph. The ordinate (y) is the mole fraction of cholesterol in a sample, and the abscissa (x) is the ratio: *PL/(PL + BA)* (mole fraction of phospholipid/sum of mole fractions of phospholipid and bile acids). The *curved line* is the composition of bile

saturated with cholesterol, and its equation is $y = 3.082 - 0.804x + 117.05x^2 - 204.94x^3$. The percentage saturation of the lithogenic bile is 375%, and that of the nonlithogenic bile is 93%. (Adapted from Thomas, P. J., and Hofmann, A. F.: Gastroenterology 65:698, 1973.)

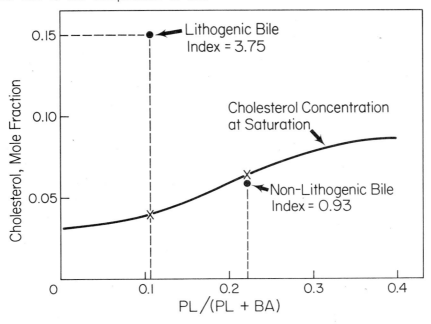

TABLE 10-5.—COMPOSITION OF LITHOGENIC AND NONLITHOGENIC BILE, AND CALCULATION OF LITHOGENIC INDEX AND PERCENT SATURATION*

SAMPLE	CHOLESTEROL[†]	BA[†‡]	PL[†‡]	CHOLESTEROL AT SATURATION[†]	$\frac{PL}{PL + BA}$	LITHOGENIC INDEX	% SATURATION
1	0.154	0.76	0.09	0.041	0.106	3.75	375
2	0.060	0.073	0.21	0.065	0.223	0.93	93

*Adapted from Thomas, P. J., and Hofmann, A. F.: Gastroenterology 65:698, 1973.
[†]In mole fractions.
[‡]BA = bile acid. PL = phospholipid.

examples are given in Figure 10-10 and Table 10-5, one of supersaturated bile and one of unsaturated bile. The ratio of the mole fraction of phospholipid to the sum of the mole fractions of phospholipid plus bile acids is calculated, and this value is marked on the abscissa. A vertical line from that point cuts the line of saturation at the mole fraction of cholesterol with which that bile would be saturated. If the actual mole fraction of cholesterol lies above the line, the bile is supersaturated; if it lies below, the bile is unsaturated.

The *percent saturation* is obtained when the lithogenic index is multiplied by 100.

Gallstones

Many compounds precipitate in the gallbladder to form stones, and although individual stones may be chiefly composed of cholesterol, all stones have several components. In one series of 331 stones from seven countries, cholesterol was 71% of the total crystalline material, and it existed in three crystal forms. Three forms of calcium carbonate comprised 15% of the solids, and calcium palmitate another 6%. Apatite, sodium chloride, palmitic acid and calcium phosphate were present in small amounts. Stones also contain pigments, usually the calcium salt of bilirubin. The composition of stones varies with country of origin, and there are sex differences. Women tend to have a higher proportion of cholesterol in their stones and men more calcium palmitate.

Crystals of calcium salts are important for nucleation and growth of stones, but stones occur because some component of gallblad-

der bile is present in supersaturating concentration. Supersaturation may be the result of physical or chemical changes occurring in the gallbladder, or it may be a characteristic of hepatic bile delivered to the gallbladder.

The secondary bile salt, lithocholic acid, is converted to 6β-OH-lithocholic acid by the liver and secreted into the bile conjugated with taurine or glycine. The taurine conjugate is soluble, but the calcium salt of the glycine conjugate is highly insoluble. Conjugated bilirubin, the form of the pigment secreted by the liver, is soluble in bile. *Escherichia coli* contains strong beta-glucuronidase activity, and in a patient whose gallbladder is infected with the microorganism conjugated bilirubin is hydrolyzed to free bilirubin and glucuronic acid. The calcium salt of free bilirubin precipitates as a pigment stone. Such a precipitate may be the nucleus around which cholesterol crystals aggregate.

In normal persons, the composition of bile varies diurnally. At night when digestion is abated, much of the bile acid pool is sequestered in the gallbladder. The rate of return of bile acids to the liver is low, and bile secreted at night has a relatively low concentration of bile acids. Therefore, bile secreted at night is frequently saturated or supersaturated with cholesterol. During the day, when the enterohepatic circulation of bile acids is brisk, bile secreted by the liver of normal persons tends to be unsaturated with cholesterol.

Persons with gallstones, before as well as after cholecystectomy, have a bile acid pool about half that of normal persons, and they secrete bile saturated with cholesterol night and day. The limiting enzyme for cholesterol synthesis is 3-hydroxy-3-methylglutaryl-

SECRETION OF THE BILE

CoA reductase, and the concentration of this enzyme may be increased in the liver of a person with gallstones. On the other hand, the activity of the limiting enzyme for bile acid synthesis, 7-α-hydrolase, may be subnormal, so that cholesterol synthesis is increased and cholesterol conversion to bile acids decreased.

Because persons without gallstones, as well as persons with them, often secrete bile saturated with cholesterol, the fact that hepatic bile may be saturated with cholesterol is not a sufficient explanation for the occurrence of gallstones. When bile is concentrated in the gallbladder by absorption of electrolytes and water, the proportional relation among lecithin, bile acids and cholesterol is unchanged; when concentrated, unsaturated bile remains unsaturated, and saturated bile continues to be saturated. Why saturated bile in some persons forms large cholesterol crystals that aggregate into radiolucent stones and in other persons does not is unknown.

Obstruction of the bile ducts by stones or tumors results in bile deficiency and its effect on digestion. Absence of bile results in incomplete activation of pancreatic lipase and in reduced absorption of fat. Because the absorption of vitamin K requires bile salts, the prothrombin content of plasma is low and hemostasis becomes inadequate. In the absence of bile pigments, the stool is clay-colored. Although secretion of bile by the liver continues, the high biliary pressure caused by obstruction results in dilatation of the ducts and leakage of the bile back into the circulation. In effect, the liver fails to clear the blood of bile acids and bilirubin, and jaundice results.

Dietary Modification of Bile Composition

If bile composition can be favorably modified by diet, the liver will secrete bile that is less than saturated with cholesterol. Unsaturated bile does not form cholesterol gallstones, and it slowly dissolves cholesterol stones already present in the gallbladder.

Feeding lecithin does not increase the leci-

thin content of bile. The components of lecithin—glycerol, fatty acids, phosphate and organic bases—enter their respective metabolic pools and have other fates. Only a very small fraction of the choline fed in lecithin reappears in biliary lecithin. The only effect upon bile of feeding lecithin is to increase the cycling frequency of bile acids.

When safflower oil or triolein is taken in large amount, biliary lipid secretion and bile acid pool size increase, but the composition of the bile is unchanged.

Feeding chenodeoxycholic acid in the amount of 1–2.25 gm a day increases the bile acid pool size and reduces hepatic synthesis of cholesterol. Therefore, the cholesterol concentration in bile falls below saturation. Consequently, feeding chenodeoxycholic acid is an effective, if tedious and expensive, medical treatment of cholelithiasis. Other bile acids are not effective, because they do not reduce cholesterol synthesis.

When chenodeoxycholic acid is fed to a patient with cholelithiasis, the bile acid pool expands, and the composition of the pool changes. The pool of cholic acid, and consequently the pool of deoxycholic acid, shrinks, and chenodeoxycholic acid becomes 90% of the pool. In some persons, a substantial amount of ursodeoxycholic acid appears in the bile acid pool.

Because chenodeoxycholic acid is converted to lithocholic acid, production of lithocholic acid increases greatly in persons fed chenodeoxycholic acid. However, the pool of lithocholic acid remains small in most persons, because lithocholic acid, after being sulfated in the liver, is excreted in the stool.

Unsaturated bile remains unsaturated when it is concentrated in the gallbladder, and gallstones slowly dissolve when bathed by unsaturated bile. Dissolution of stones is a complex and poorly understood process.

REFERENCES

Bell, G. D.: The present position concerning gallstone dissolution, Gut 15:913, 1974.
Cowen, A. E., Korman, M. G., Hofmann, A. F., and Thomas, P. J.: Metabolism of lithocholate

in healthy man, Gastroenterology 69:77, 1975.

Diamond, J. M.: Transport mechanisms in the gallbladder, in Code, C. F. (ed.): *Handbook of Physiology:* Sec. 6. *Alimentary Canal,* Vol. V (Washington, D.C.: American Physiological Society, 1968), pp. 2451–2482.

Dowling, R. H.: The enterohepatic circulation, Gastroenterology 62:122, 1972.

Hallenbeck, G. A.: Biliary and pancreatic intraductal pressures, in Code, C. F. (ed.): *Handbook of Physiology:* Sec. 6. *Alimentary Canal,* Vol. II (Washington, D.C.: American Physiological Society, 1967), pp. 1007–1026.

Hoffman, N. E., and Hofmann, A. F.: Metabolism of steroid and amino acid moieties of conjugated bile acids in man, Gastroenterology 67:887, 1974.

Hofmann, A. F.: Functions of bile in the alimentary canal, in Code, C. F. (ed.): *Handbook of Physiology:* Sec. 6. *Alimentary Canal,* Vol. V (Washington, D.C.: American Physiological Society, 1968), pp. 2507–2533.

Hofmann, A. F., and Poley, J. R.: Role of bile acid malabsorption in pathogenesis of diarrhea and steatorrhea in patients with ileal resection, Gastroenterology 62:918, 1972.

Northfield, T. C., LaRusso, N. F., Hofmann, A. F., and Thistle, J. L.: Biliary lipid output during three meals and an overnight fast, Gut 16:12, 1975.

Weiner, I. M., and Lack, L.: Bile salt absorption; enterohepatic circulation, in Code, C. F. (ed.): *Handbook of Physiology:* Sec. 6. *Alimentary Canal,* Vol. III (Washington, D.C.: American Physiological Society, 1968), pp. 1439–1456.

Wheeler, H. O.: Water and electrolytes in the bile, in Code, C. F. (ed.): *Handbook of Physiology:* Sec. 6. *Alimentary Canal,* Vol. V (Washington, D.C.: American Physiological Society, 1968), pp. 2409–2432.

11

Intestinal Secretion

THE INDISPENSABLE FUNCTION of the small intestine is absorption of water, salts and foodstuffs. Net amounts of approximately 9 liters of water with accompanying salts plus the digestion products of several hundred grams of food move each day across the intestine from the mucosal side to the blood. In the case of water and salts, this net transfer is the resultant of very large flow in the direction of serosal to mucosal side, offset by a still larger flow in the opposite direction. Transfer in both directions during absorption is described in Chapter 14. In this chapter, only the volume and composition of secretions delivered into the empty bowel will be considered.

Secretory Structure of the Small Intestine

The small intestine is divided functionally into three parts—duodenum, jejunum and ileum—but the microscopic structures of the mucosa of all three parts are similar. The surface, which is thrown into folds, is covered with finger-like or ridge-shaped villi 0.5–1.5 mm long, at a density of 10–40 per mm². The surface of the villi is a layer of columnar epithelial cells whose free border is composed of hundreds of minute processes called microvilli. Mucus-containing goblet cells are scattered among the columnar epithelial cells.

At the base of the villi, and lined with epithelium continuous with them, are the crypts, simple tubes 0.3–0.5 mm deep. There are three crypts to every villus.

Epithelial cells are formed by mitosis in the crypts. New cells migrate in columns up a villus as old cells are extruded from the tip (Fig 11–1), and the entire epithelium is replaced in 3–6 days. Cells are shed at the rate of more than 100 million a minute, and in this process 30–50 gm of endogenous protein, together with other cell constituents, is delivered to the lumen every day. In the steady state, cells are formed in the proliferative zone at the same rate they are shed. Cell production is decreased by x-radiation, methotrexate, folic acid and vitamin B_{12} deficiencies, starvation and old age. Cell production is increased during lactation, by cortisol administration and by bacterial overgrowth. The rate of cell production is also increased in gluten enteropathy.

Cells change as they leave the crypts. They are sensitive to radiation while in the crypts but not after migration to the villi. In the crypts, the cells have low concentrations of phosphatases, esterase and succinic dehydrogenase, but they acquire the enzymes on leaving; there is a sharp line at the base of the villus where precursor cells lacking phosphatases abut on cells rich in those enzymes.

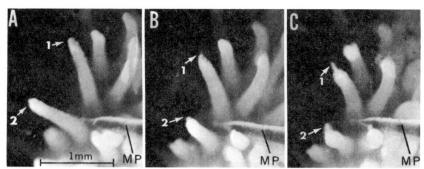

Fig 11–1. — Photomicrographs showing cell extrusion from the tip of a villus of the dog's intestine. **A,** a piece of mucosa has been bathed in physiological salt solution for 10 min. Villi *1* and *2* were separated by a micropipette *(MP)*. No cell extrusion occurred. **B,** after 20 min the cell extrusion in villi *1* and *2* became visible *(arrows)*. **C,** after 40 min there was a large accumulation of extruded cells at the tips of villi *1* and *2*, and extruded cells are visible at the tips of other villi. Villi *1* and *2* also show contraction. (From Lee, J. S.: Am. J. Physiol. 217:1528, 1969.)

Other enzymes develop more gradually, and differentiation continues during migration to the apex. Morphologic as well as chemical differentiation occurs as the cells migrate. Free ribosomes disappear as endoplasmic reticulum is formed. Microvilli increase in number, elongate and become thinner. The apical surface of the cell increases 7–8 times, and this change is accompanied by an increase in the concentration of brush-border enzymes. A terminal web develops, and the fuzzy coat, or glycocalyx, is added to the microvilli.

Apical cells have the highest capacity for absorption.

If the mucosa is destroyed in one spot, epithelialization occurs first by migration of cells from adjacent areas followed by increased rate of cell division in the crypts at the margin of the injured area; later, new crypts form on its floor and villi are regenerated. If the injury penetrates the muscularis mucosae, that layer is not re-formed during repair.

Duodenal Glands

In the duodenum, there is a further anatomical division, for special glands extend from an irregular border at the pyloric ring down the duodenum. In more than 85% of men, the distal border of the area of the glands is marked by islands of glands below the infrapapillary section of the duodenum; and in nearly one third of the men, glandular islands extend into the jejunum. Among other animals, the carnivores have the shortest distribution of these glands, the omnivores a medium one and the herbivores the longest.

The duodenal glands consist of branched and coiled tubules arranged in lobules lying in the submucosa. Their ducts pierce the muscularis mucosae to empty into the bottoms or sides of the crypts. In most species, the glands are composed of one type of cuboidal cell, and they stain deeply for mucus.

Duodenal Secretions

The secretion collected from the upper duodenum is a mixture of that from the duodenal glands and the crypts, but the major components appear to come from the duodenal glands. The juice is highly viscid, being similar to egg white, although its mucin content is only 0.5%. The viscosity of the juice declines rapidly on standing, and there is an increase in nonprotein nitrogen, both effects probably being caused by a proteolytic enzyme contained in the juice. Most samples of juice from the upper duodenum contain amylase, peptidases, enterokinase and other enzymes. Most or all of the enzymes, with the exception of enterokinase, are derived from desquamated cells.

The aqueous portion of the secretion is probably isotonic with the blood. Its mean electrolyte composition is Na^+ 145 (range 136–150), K^+ 6.3 (4.5–8.0), Cl^- 136 (130–140) and HCO_3^- 17 (14–22) mEq per liter. Pepsin is present in low concentration.

Rate of secretion by duodenal glands during fasting is very low, but secretion increases after feeding. The response to food persists after extrinsic denervation. Gastrin, secretin, cholecystokinin and glucagon, all of which are liberated during digestion of a meal, stimulate duodenal secretion when given in large doses, but whether these or other hormones are responsible for the observed increase in secretion following feeding is unknown.

The small volume of duodenal secretion gives negligible neutralization of gastric contents as they pass through the duodenum, but the mucin lubricates the area and is a mechanical barrier holding neutral or alkaline fluid against the mucosal surface. The upper duodenum is more resistant to attack by acid chyme than is the jejunum; if in the dog the duodenum is bypassed surgically so that gastric contents empty directly into the jejunum, ulceration of the jejunal mucosa almost invariably occurs. However, duodenal ulcers in man occur exclusively in the duodenal gland area.

Intestinal and Colonic Secretion

Most digestive enzymes of the small intestine are integral parts of the membrane of the brush border; they are not secreted. When epithelial cells desquamate, intracellular enzymes as well as the enzymes of the brush border are shed into the lumen, and they are solubilized by bile. Alkaline phosphatase and enterokinase are among the few enzymes actually secreted, and their secretion is stimulated by the hormones secretin and cholecystokinin.

During digestion and absorption, there is a large exchange of water and electrolytes across the intestinal mucosa; this is described in Chapter 14. In the interdigestive phase there is little or no secretion by the intestinal mucosa.

Some animals—e.g., the racoon, black bear and skunk—have no ileocecal valve; their intestine is continuous, without interruption, from duodenum to rectum. This emphasizes the fact that the secretory and absorptive behavior of the colonic mucosa is only quantitatively different from that of the rest of the intestine. In the intestine of man, villi do not occur below the ileocecal valve, and the colonic mucosa consists of crypts, with the free surface between crypts covered by columnar epithelial cells. Crypts and epithelium have a high density of mucus-containing goblet cells, and colonic secretion is rich in mucus. The aqueous secretion is scanty. Reabsorption predominates in the colon; when colonic contents remain stagnant, the aqueous phase is absorbed, leaving only an inspissated mass.

The small amount of fluid secreted by the colon is alkaline, for bicarbonate is secreted as chloride is reabsorbed. Bicarbonate neutralizes the products of fecal fermentation, for the surface of a fecal mass is neutral while the pH of its center may be 4.8. The secretion contains no enterokinase, invertase, lipase or trypsin; lysozyme is present.

Slimy, viscid mucus comprises only 0.4% of fresh colonic secretion. Its electrolyte composition is K^+ 146–200, Na^+ 3–10 and HCO_3^- 87–155 mEq per liter. When the goblet cell is stimulated, concentrated mucus leaves through the cell's stoma. As the mucus becomes free, the cell's nucleus becomes round and its goblet part less bellied out. More of the cytoplasm is revealed; and, as mucus is lost, the cell comes to resemble a simple columnar epithelial cell. At this stage, a few granules of mucus may remain in the cytoplasm just above the nucleus. With continued stimulation, the cell loses its cytoplasm and becomes first cuboidal and finally squamous.

The cells do not readily desquamate even under strenuous stimulation by surface irritants. Surface irritants (mustard oil and the like) stimulate secretion, and so does the mechanical irritation caused by the colon's rubbing against itself or over the fecal mass. This secretion occurs independently of extrinsic innervation. Secretion is evoked by stimulation of peripheral ends of the nervi erigentes and by central stimulation of one cut nervus erigens with the other remaining intact. Acetylcholine is the mediator; and parasympathomimetic drugs, as well as pilocarpine, cause secretion, while atropine blocks it. Parasympathetic stimulation increases motility of the colon, but concomitant increase in secretion is not merely secondary to movement. Blood flow through the mucosa increases; and, in general, movement, secretion and blood flow are parallel one with another. Stimulation of the sympathetic supply, which relaxes colonic movement and causes vasoconstriction, also reduces any ongoing secretion. No hormonal control of colonic secretion has been identified.

Intestinal Secretion in Cholera

The toxin produced by *Vibrio cholera* during its growth phase stimulates the jejunum and to a lesser extent the ileum of man and animals to secrete massive amounts of fluid. The toxin has no direct effect upon the colon, but because the colon's reabsorptive capacity is overwhelmed, fulminant diarrhea results. Enterotoxins elaborated by noninvasive strains of *E. coli* also stimulate intestinal secretion, and they cause many of the cases of severe, acute diarrhea suffered by travelers.

The fluid secreted by the intestine is protein free, and its major cations are sodium and potassium. Its concentration of bicarbonate approaches 80 mM, and its concentration of chloride is correspondingly low. Loss of many liters of this fluid results in dehydration and metabolic acidosis. Secretion of aldosterone is stimulated, and reabsorption of sodium and secretion of potassium by the colon are increased. If dehydration and metabolic acidosis can be corrected by adequate intravenous infusions, the disease is painless and self-limiting. Administration by mouth of electrolyte fluids not containing glucose or amino acids is ineffective, for the administered fluid merely adds to that secreted by the intestine and gushes from the anus.

The cholera toxin binds to intestinal epithelial cells and stimulates them to secrete by activating adenyl cyclase and increasing the cells' content of cyclic AMP. Fluid shed by the mucosa is actively secreted, not filtered; there are no changes in capillary or mucosal permeability and no changes in intestinal hemodynamics that can account for the effects of the toxin. Although the toxin, when injected intradermally, locally increases capillary permeability, it does not do so in the intestine for the reason that it is not absorbed. It produces little or no cytotoxic effects, and it does not cause desquamation.

The toxin probably stimulates secretion of bicarbonate and chloride ions. Sodium ions to maintain electric neutrality and water to maintain isotonicity follow passively. Some persons believe that a neutral sodium pump is also stimulated. Bicarbonate secretion is actually the result of secretion of hydroxyl ions, which, upon entering the lumen, combine with ubiquitous carbon dioxide. This is demonstrated by the fact that the partial pressure of carbon dioxide in the luminal fluid falls when bicarbonate secretion is stimulated by the toxin. Potassium is also actively secreted, for its concentration in the fluid is about 3 times that which can be accounted for by its electrochemical gradient. Reabsorption of sodium may be simultaneously reduced. However, reabsorption of sodium can be strongly stimulated by glucose or amino acids present in the luminal fluid. When sodium is reabsorbed, anions and water accompany it, and, consequently, administration by mouth of electrolyte fluids containing glucose or amino acids mitigates the diarrhea. A combination of glucose and glycine is more effective than either alone.

Other Stimuli of Intestinal Secretion

Some non-β-cell tumors of the pancreas contain and secrete hormones that stimulate secretion of bicarbonate-containing fluids by the pancreas, liver and small intestine, and the resulting diarrheal state is called *pancreatic cholera*. The hormone mainly responsible is vasoactive intestinal peptide (VIP, see Chapter 12), and the concentration of VIP in the plasma may be very high. In some instances, secretin or a secretin-like hormone may also be secreted by the tumor, and other hormones including 5-hydroxytryptamine have been found in tumor tissue. The tumors do not secrete gastrin, and secretion of acid is not increased. VIP and secretin stimulate secretion of bicarbonate-containing fluid by the pancreas and liver, and the gallbladder of a patient with pancreatic cholera may be grossly distended with dilute bile. However, the volume of fluid passing the duodenal-jejunal junction is normal, and therefore the diarrhea must result from excessive secretion of fluid by the jejunum and perhaps the ileum. Reabsorption of fluid by the colon is increased as the result of secondary aldosteronism.

The process of absorption reduces the volume of intestinal contents essentially to zero. A major constituent of the diet of herbivorous animals is cellulose, which is digested by microorganisms in the cecum. As it passes through the small intestine, it contributes nothing to the osmotic pressure of luminal contents. To provide adequate volume, the small intestine of herbivorous animals regularly secretes fluid into the lumen. In man, intestinal secretion is stimulated by oleic and ricinoleic acid, and reabsorption is inhibited by bile acids; in this way adequate luminal volume is maintained during digestion and absorption of fat.

The Appendix

Among animals having a cecal appendix, only in man, the rabbit and the chimpanzee does the appendix secrete a significant volume of fluid spontaneously or in response to pilocarpine; the appendix of the dog, red fox, howling gibbon and tiger does not. If the secretion drains freely, secretion can continue indefinitely; but if the lumen is obstructed, continued secretion results in a rise in intraluminal pressure, which in 4 hours may come to equal systolic blood pressure. Focal ischemic necrosis occurs, followed by rupture, with infection in the scattered contents. Lymphoid tissue of the entire small intestine, the colon and perhaps the stomach is capable of forming antibodies, but the lymphoid tissue of the appendix appears to be an especially lively site of antibody production.

REFERENCES

Alpers, D. H., and Kinzie, J. L.: Regulation of small intestinal protein metabolism, Gastroenterology 64:471, 1973.

Banwell, J. G., and Sheer, H.: Effect of bacterial enterotoxins on the gastrointestinal tract, Gastroenterology 65:467, 1973.

Cooke, A. R.: The glands of Brunner, in Code, C. F. (ed.): *Handbook of Physiology:* Sec. 6. *Alimentary Canal,* Vol. II (Washington, D.C.: American Physiological Society, 1967), pp. 1087–1095.

Field, M.: Intestinal secretion, Gastroenterology 66:1063, 1974.

Gregory, R. A.: *Secretory Mechanisms of the Gastrointestinal Tract* (London: Edward Arnold & Co., 1962).

Lipkin, M.: Proliferation and differentiation of gastrointestinal cells, Physiol. Rev. 53:891, 1973.

Loehry, C. A., and Creamer, B.: Three-dimensional structure of the human small intestinal mucosa in health and disease, Gut 10:6, 1969.

12

Control of Secretion

CONTROL OF THE digestive tract is exerted through extrinsic nerves, intrinsic nerves and hormones, and all three systems are complexly integrated. Impulses in afferent nerves in the head, relayed through the vagal nucleus and efferent fibers, excite and sensitize secretory cells in the stomach, pancreas and liver by releasing acetylcholine. At the same time, other vagal fibers releasing other mediators cause receptive relaxation of the stomach. Vagal impulses to the pyloric glandular mucosa liberate the hormone gastrin whose major function is to stimulate secretion of acid by the oxyntic glandular mucosa. This is the *cephalic phase* of secretion.

Distention of the stomach by food excites secretion through local reflex arcs in the intrinsic plexuses and through long reflex arcs in the vagus nerves. Protein digestion products in the stomach directly stimulate the oxyntic glandular mucosa to secrete acid and, acting together with distention, stimulate release of gastrin from cells of the pyloric glandular mucosa. This is the *gastric phase* of secretion.

Chyme entering the duodenum stimulates the release of hormones from the duodenal mucosa. Properties of chyme stimulating release of hormones are its bulk, its osmotic pressure, its acidity and its content of protein-digestion and fat-digestion products. The hormone cholecystokinin stimulates pancreatic secretion of enzymes and contraction of the gallbladder. Secretin, acting in conjunction with cholecystokinin, stimulates secretion of bicarbonate-containing juice by the pancreas and liver. These hormones, together with others not yet completely characterized, participate in the regulation of gastric emptying so that the duodenum receives no more chyme than it can deal with in preparation for intestinal digestion and absorption. Hormones liberated from the duodenal mucosa and impulses carried in the intrinsic plexuses and in the vagus nerves both stimulate and inhibit secretion of gastric juice. This is the *intestinal phase* of secretion. In addition, hormones and nervous impulses from the intestine regulate food intake by acting on the satiety mechanism in the hypothalamus.

The quantities of gastric, pancreatic, hepatic and intestinal secretions are nicely adjusted to the quantity and quality of food through multiple actions of hormones and nerves. Thus, gastric secretion of acid is regulated in part by the buffering power of the food, for after the buffers of gastric contents have been acidified, acid chyme in contact with the pyloric glandular mucosa inhibits release of gastrin. At the same time, excitatory impulses in the vagus nerves die out, and inhibitory impulses and hormones from the duodenum stop acid secretion. The process of control then uses another negative feedback loop, for acid upon entering the duodenum stimulates secretion of bicarbonate-containing fluids that neutralize acid.

Questions to Be Answered

The first step in the elucidation of hormonal control of a physiological process is the demonstration that a specific stimulus acting on a specific receptor organ releases a specific chemical messenger, which, traveling through the blood, causes a specific target organ to give a specific response. For gastroenterology, this was begun in the afternoon of January 16, 1902.* Then Bayliss and Starling found that when acid was placed in an extrinsically denervated loop of upper small intestine, the pancreas responded by secreting. This could not be a nervous reflex, and Starling exclaimed: "Then it must be a chemical reflex!" He then showed that a crude extract of the intestinal mucosa, when given intravenously, also stimulated pancreatic secretion. Acid, the specific stimulus, acting on the upper intestinal mucosa, releases the hormone secretin, which causes the pancreas to secrete bicarbonate-containing juice. This is a negative feedback loop in which the stimulus evokes a response that ends in the elimination of the stimulus. It is a long distance from this concept to our current, but incomplete, understanding of the control of pancreatic secretion, and along the way many questions have been posed.

The next step is the isolation, purification, structural analysis and chemical synthesis of the messenger. Because gastrointestinal hormones are polypeptides, this took 60 years or more; peptide chemistry had to mature. Then the pure hormone is administered to many animals, including man, and a catalog of its actions is compiled. These are the *pharmacological* actions of the hormone. But

one asks: Is this hormone actually the one released by the stimulus? Are there species differences, and are there molecular variants? If there are variants, which is released in what quantity? What is the spectrum of action of each, its potency and its half-life? From what cells is the hormone released, and what determines the competence of the cells to respond to the stimulus? Are there other stimuli, and what conditions, nervous, humoral, nutritional, affect the release? Are there synergistic or antagonistic hormones or conditions? Is the amount of hormone released by the natural stimulus adequate to produce the observed response? Are there subtle and long-term effects of the hormone, perhaps trophic actions? When these questions are answered, we have the *physiological* properties of the endocrine system.

How is the system disordered in disease? Have the cells releasing the hormone disappeared, become uncontrolled or displaced? Is the hormone not released at all, or is it discharged in such an amount that its pharmacological properties become important? Is there a pathologic variant of the normal hormone that has other actions? What about the rate of destruction of the hormone, the response of the target organ and the influence of other hormones? The answers to these questions comprise the *pathology* of gastrointestinal endocrinology.

The endocrinology of the digestive tract is a rapidly developing science. Within the lifetime of this edition, old questions will be answered afresh, old answers will be found to be wrong and new questions will demand new answers. Nothing is still, all flows; and the student should beware.

Direct Vagal Stimulation Through the Cephalic Phase

Gastric secretion of acid and pepsinogen follows stimulation of afferent nerves in the head; hence this part of the response is called the cephalic phase. The final common path is the vagus nerve, for the cephalic phase is completely abolished by vagotomy. Vagal

*This is the birthday of endocrinology. See Martin, C.: Br. Med. J. 1:900, 1927. The date is not to be confused with June 16, 1904, which is Bloomsday.

impulses reach the stomach by way of the anterior and posterior vagal trunks arising from the esophageal vagus plexus formed by right and left vagus nerves on the surface of the esophagus just above the diaphragm. They are distributed to the anterior and posterior aspects of the stomach and end in synapse with postganglionic cells in the myenteric plexus. This plexus, lying between longitudinal and circular smooth-muscle layers, sends connecting branches to the submucous plexus. In man, the submucous plexus consists of a network of nonmedullated fibers in the submucous region. The network contains very few cell bodies. From this network, fibers penetrate the mucosa where they innervate the secretory cells. The mediator is acetylcholine, and it directly stimulates oxyntic and chief cells to secrete. Long-lived parasympathomimetic drugs also stimulate secretion, and the action of these drugs and of acetylcholine is blocked by atropine. Acetylcholine also sensitizes the secretory cells

to gastrin, and the magnitude of the response to acetylcholine and gastrin together is far greater than is the sum of the responses to each alone.

Stimulation of gastric secretion through the cephalic phase is demonstrated by sham feeding. A dog is provided with an esophagostomy so that the food it eats falls back into its feeding dish instead of reaching the stomach. Copious gastric secretion occurs during and after sham feeding, and the response is abolished by vagotomy. Sham feeding can be demonstrated in a patient whose esophagus does not empty into his stomach. If such a patient chews food, his stomach secretes (Fig 12–1). The effective stimuli may originate from taste and smell receptors or by conditioned reflexes from other receptors. In man, tasting, smelling or chewing palatable food, but not indifferent substances such as paraffin wax, is followed by gastric secretion. The results shown in Figure 12–1, obtained during sham feeding, show a

Fig 12–1.—Rate of acid secretion by the stomach of a woman, 24, with complete stenosis of the esophagus. In her usual manner of eating, the food was tasted, chewed, partly swallowed, regurgitated, expectorated and then placed in the stomach through a gastrostomy. In the two instances recorded here, the food was not placed in the stomach after it had been tasted and chewed. *Meal I* (mean of 4 experiments) consisted of 8 oz of cereal gruel,

eaten dutifully but with obvious distaste. *Meal II* (mean of 12 experiments) was composed of the subject's unrestricted choice: fresh vegetables, salad with dressing, 3 slices of white bread with butter, 2 glasses of milk, potatoes, half a fried chicken or 2 broiled lamb chops or fried ham steak or 2 fried eggs, ice cream and cake. (Adapted from Janowitz, H. D., *et al.*: Gastroenterology 16:104, 1950.)

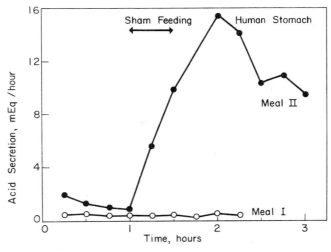

strong response to a meal of choice but none to a hospital meal dutifully chewed but not swallowed.

In addition to its effects in releasing acetylcholine near secretory cells, stimulation of the vagus nerve by sham feeding releases gastrin from the pyloric glandular mucosa and thereby stimulates acid secretion. This effect can be demonstrated only when the pyloric glandular mucosa is protected from contact with acid, for acid inhibits the release of gastrin (Fig 12–2).

The cephalic phase of gastric secretion may not be important in man, because entry of food into the stomach also stimulates secretion. A steak meal, homogenized and fed through a nasogastric tube, evokes the same acid secretion as does a steak meal eaten with relish.

In man, reduction of blood glucose to about 45 mg per 100 ml, or about half the fasting level, by insulin administration strongly

Fig 12–2. – Acid secretion by an extrinsically denervated test pouch of the oxyntic gland area of the dog stomach in response to sham feeding before and after vagal denervation of a pouch of the pyloric gland area. The antrum is separated from the stomach by a mucosal bridge to protect it from acid, but its innervation is at first intact. **Left,** sham feeding stimulates secretion by the test pouch. **Right,** no secretion follows sham feeding after denervation of the antral pouch. At the end of this experiment, the dog was fed 300 gm of meat with the esophagostomy held closed so that the food passed into the stomach. (Adapted from Pe Thein, M., and Schofield, B: J. Physiol. 148:291, 1959.)

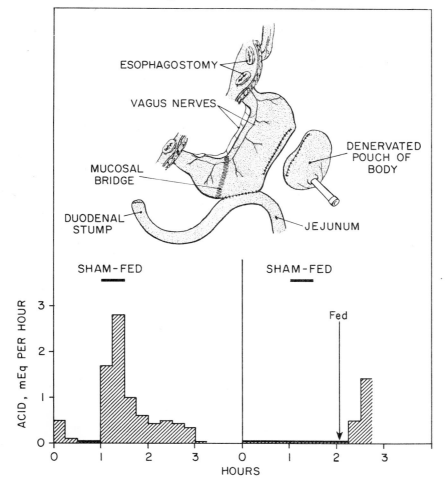

stimulates acid and pepsinogen secretion. Intravenous infusion of insulin at 0.1 units per kg·hr gives a maximum secretory response.

The effect of insulin is prevented by maintaining the blood glucose concentration with glucose infusion. The fall of blood glucose is sensed by cells in the hypothalamus, and excitation is relayed to the stomach along the vagus nerve. The response does not occur in completely vagotomized subjects, and therefore the absence of gastric secretion following insulin-induced hypoglycemia is a test for vagal denervation. Administration of the nonmetabolized sugar 2-deoxy-D-glucose in an intravenous dose of 50–200 mg per kg of body weight is also a powerful stimulant to gastric secretion by way of the vagus nerve. The sugar probably acts by competing with

glucose in the glucose-sensitive cells in the medial forebrain bundle of the hypothalamus.

If the acid secreted by a man in response to insulin hypoglycemia is prevented from reaching the gastric antrum by intragastric neutralization, plasma gastrin rises by about 30% above basal level 45 min after insulin is given. When atropine in a dose of 0.015 mg per kg is given 20 min before insulin is administered, plasma gastrin rises to twice the basal level. There appears to be a cholinergic inhibition of gastrin release, which is blocked by atropine (see also Fig 12–3).

Insulin hypoglycemia is used as a test of the completeness of vagotomy, and positive results are usually attributed to incomplete vagotomy or regrowth of vagal fibers. However, gastrin release following insulin hypoglycemia is not abolished by vagotomy. Gastrin release is stimulated by epinephrine, and the general sympathetic discharge occurring during hypoglycemia may be responsible for gastrin release, and in turn for acid secretion, in some subjects with complete vagotomies.

Gastric Phase of Secretion

At the end of sham feeding, gastric secretion elicited by the cephalic phase dies away; but if food enters the stomach, gastric secretion continues at a high rate for about an hour and then dies out as the contents of the stomach become acid. A total of about 800 ml is secreted in response to a meal of grilled steak. The acid juice contains a high concentration of pepsin. The gastric phase of secretion is stimulated by the volume and chemical composition of gastric contents.

Distention stimulates gastric secretion of acid and pepsin in the dog by five means:

1. Distention of a separated pouch of the antrum releases gastrin from G cells in the pyloric glandular mucosa. Distention of the whole stomach, body as well as antrum, does not release gastrin in dog or man.

2. Distention of the antrum stimulates secretion by the oxyntic glandular mucosa through a cholinergic reflex. Integrity of the

Fig 12–3.—Plasma gastrin concentration in a normal subject and in a subject with truncal vagotomy in a control period and following ingestion of a protein meal. (Adapted in part from Korman, M. G., Hansky, J., and Scott, P. R., Gut 13:39, 1972, and in part from Malagelada, J-R., Longstreth, G. F., Summerskill, W. H. J., and Go, V. L. W., Gastroenterology 70:203, 1976.)

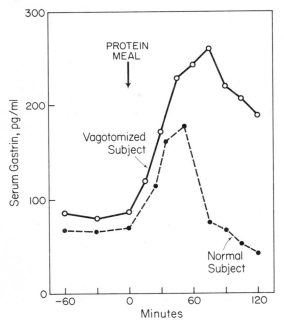

vagus nerve is essential for this reflex. Distention stimulates receptors in the wall of the stomach, which send afferent impulses to the central nervous system, and in return impulses are sent to the stomach. Local plexuses also participate in this reflex. Distention of the whole stomach in man also stimulates secretion through a similar reflex, which is abolished by vagotomy.

3. Distention of the body of the stomach stimulates acid secretion through a reflex whose afferent and efferent pathways are in the vagus nerve. This reflex occurs in man.

4. Distention of the body of the stomach stimulates acid secretion through a cholinergic reflex confined to the intramural plexuses. Experiments necessary to demonstrate this reflex cannot be done in man.

5. Distention of the body of the stomach stimulates release of gastrin from the pyloric glandular mucosa. The vagus nerve is necessary for this oxynto-pyloric reflex. Whether the reflex occurs in man is unknown.

Chemical compounds in gastric contents stimulate acid secretion by acting directly upon the oxyntic cells or by stimulating the release of gastrin.

Protein digestion products in neutral solution (but not intact proteins) stimulate acid secretion when they bathe the oxyntic glandular mucosa. This direct effect upon the oxyntic mucosa is enhanced by concurrent distention of the body of the stomach, and when topical application of protein digestion products is combined with distention, the resulting rate of acid secretion equals that occurring in response to a maximal dose of exogenous gastrin. If the protein digestion products are acidified to pH 2.0, they do not stimulate secretion. The protein digestion products appear to act directly upon the oxyntic cells in a way that is not understood. Their effect is only slightly decreased if the mucosa is treated with a local anesthetic or if the responsiveness of the oxyntic cells to acetylcholine or to histamine is abolished by intravenous injection of atropine or an H_2 antagonist. Both atropine and H_2 antagonists suppress the response to a meal in the in-

tact stomach, and consequently it is unlikely that the direct effect of protein digestion products upon the oxyntic cells has an important role in the normal response to a meal.

Ethanol stimulates isolated pieces of the oxyntic glandular mucosa to secrete acid by becoming a source of energy for the cells. In man, a solution of ethanol is no better stimulant of acid secretion than is the same volume of Adam's ale.

Release of Gastrin

A totally denervated or transplanted pouch of the oxyntic gland area secretes acid and pepsinogen when secretagogues are in contact with the mucosa of the pyloric gland area or when the antrum is simply distended. The same response is obtained if the antrum itself is separated from the stomach and transplanted under the skin to deprive it of extrinsic innervation. Therefore neither afferent nerves from the source of stimulation nor efferent nerves to the target organ are required. Intravenous injection of secretagogues does not stimulate secretion, and so absorption of secretagogues into the blood is not the humoral influence. The conclusion is that a hormone, gastrin, is produced and released by the pyloric glandular mucosa into gastric venous blood. The hormone, after release from the mucosa, passes through the liver and thence into the general circulation, where, on reaching the oxyntic gland area of the stomach, it stimulates secretion. An insignificant amount of gastrin travels in lymph draining the antrum.

Gastrin is synthesized and stored in G cells in the pyloric glandular and duodenal mucosa. Gastrin is released from basal storage granules of the G cells into interstitial fluid and thence into portal blood. G cells are flask-shaped; their necks project to the lumen of the glands, and the exposed surface is covered with short microvilli, which, one supposes, receive stimuli for release of gastrin.

The major stimulus for gastrin release is a

neutral solution of L-amino acids, or their polypeptides, bathing the pyloric glandular mucosa. Native proteins are ineffective. Solutions of calcium salts, including milk, also release gastrin. Decalcified milk has only a small effect. Hypercalcemia resulting from intravenous infusion of calcium salts releases gastrin, but a patient with hypercalcemia caused by hyperparathyroidism does not have hypergastrinemia unless he has a gastrinoma, a tumor secreting gastrin.

In the dog, gastrin is also released from a surgically prepared pouch of the antrum by distention, but it is not released by distention of the whole, intact stomach. Distention of the stomach is not an important releaser of gastrin in man.

Epinephrine releases gastrin. In man, the effect is antagonized by a beta-adrenergic blocking agent.

Aliphatic alcohols, of which ethanol is the most potent, release gastrin in the dog, a confirmed teetotaler, but in man, ethanol, if it releases gastrin at all, has only a trivial effect. Caffeine does not release gastrin, but decaffeinated coffee does so and stimulates acid secretion, an effect attributable to peptides in the brew.

When injected intravenously in man, the peptide bombesin, extracted from frog skin, releases gastrin. This effect is not abolished by acidification of the antrum. Infusion of liver extract into the small intestine of a dog causes a rise in plasma gastrin concentration, and acid secretion is stimulated. This effect is abolished by surgical removal of the antrum, which is therefore the source of the incremental gastrin, but it is not abolished by acidification of the antrum. Perhaps digestion products in the duodenum release a bombesin-like hormone, which in turn releases gastrin.

Gastrin release is inhibited by acid in contact with the pyloric glandular mucosa. Release is completely suppressed when the pH of antral contents is 1.0. When a solution of amino acids is placed in the stomach of a normal man, the increment in plasma gastrin is the same when the pH of the solution is 3.0, 4.0 or 5.5, but when the pH of the solution is 2.5, the rise in plasma gastrin is very small. A patient with pernicious anemia has achlorhydria, and his fasting plasma gastrin concentration is high. If such a patient drinks acid, preferably through a straw so that he does not dissolve his teeth, his plasma gastrin concentration falls steeply within 15 min. When he eats a meal, his plasma gastrin rises to several times the fasting level, because gastrin release is not inhibited by acid.

The effect of acid upon the pyloric gland area profoundly influences the secretory response to food. During the interdigestive period, the pH of the antrum is low and gastrin release is inhibited. However, the volume of gastric contents is small, and as soon as food enters the stomach, the acid is neutralized and diluted. The pH of the material in contact with the pyloric gland area rises, and gastrin is secreted in response to vagal impulses of the cephalic phase and to distention and secretagogues of the gastric phase. Acid is promptly secreted, and secretion continues at a high rate until the buffering power of the food is exhausted. Then acidification of the antrum terminates the gastric phase; the pH of the contents drops to 1.5, gastrin release is inhibited and secretion declines.

Protein is the chief buffer of food, and Figure 12−4 shows that acid secretion is directly related to the protein content of a meal. Six dogs with vagally innervated pouches of the oxyntic gland area were fed on many occasions 100-calorie meals of broiled ground round of beef so that the response to the meal was well known. The same dogs were then fed 100-calorie meals of 29 common foods, ranging from haddock fillets to canned peaches, and the secretory responses were carefully measured. The results were expressed as the gastric secretory equivalent, that is, the total acid secreted in response to 100 calories of food divided by the acid secreted by the same dog in response to 100 calories of beef, the quotient being multiplied by 100. The correlation coefficient between secretory equivalent and protein content was 0.97, and there was also a good cor-

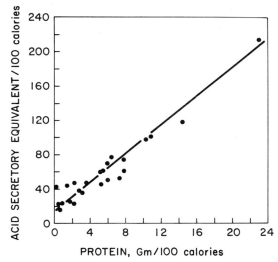

Fig 12–4.—Acid secretory equivalent of 29 common foods in relation to their protein content. The total acid secreted by a dog's vagally innervated pouch in response to a 100-calorie meal of any of the 29 foods divided by the total acid secreted by the same dog's pouch in response to a 100-calorie meal of broiled ground beef, the quotient multiplied by 100, is the acid secretory equivalent per 100 calories. The foods used ranged from canned peaches, containing the least protein per 100 calories and giving the smallest acid secretory equivalent, to haddock, with the most protein per 100 calories and the strongest effect on secretion. (Adapted from Saint-Hilaire, S., *et al.*: Gastroenterology 39:1, 1960.)

relation (r=0.82) between secretory equivalent and the amount of acid required to reduce 100-calorie portions of the food to pH 2.0.

Addition of sodium bicarbonate or other alkalinizing agents to food increases the secretory response because it neutralizes acid in the stomach but does not in itself stimulate secretion.

The hormones secretin, gastric inhibitory polypeptide (GIP), vasoactive intestinal peptide (VIP), glucagon and calcitonin, when injected in pharmacological doses, inhibit gastrin release, but they probably do not have a physiological role in regulating the plasma concentration of gastrin.

The concentration of gastrin in human plasma following a protein meal is shown in Figure 12–3. In the fasting state, the gastrin concentration in the plasma of a patient with duodenal ulcer is usually not significantly higher than that in the plasma of a normal person, but following the same protein meal it rises higher and remains elevated longer.

The plasma gastrin concentration in a normal person who has been given atropine usually rises higher following a protein meal than it does when he is in a control state. The response of a vagotomized patient to a meal is also exaggerated (see Fig 12–3), and a large response occurs in a patient with highly selective vagotomy as well as in a patient with truncal vagotomy. The supranormal release of gastrin in these circumstances is probably the result of hypochlorhydria caused by atropinization or vagotomy, for with reduced secretion of acid, inhibition of gastrin release by acidification of gastric contents is delayed.

The Gastrins

Gastrins are a family of straight-chain peptides. The structures of three are given in Table 12–1. The hormone synthesized in the G cells contains 34 amino acids, and it is called *big gastrin*. The amino acid at its N-terminus is pyroglutamic acid (Glp), and the phenylalanine at its C-terminus is amidated. These terminal substitutions prevent destruction of the hormone by aminopeptidases and carboxypeptidases. Big gastrin is only one of the forms of gastrin in plasma. Another form is a heptadecapeptide, *little gastrin*, which is derived from big gastrin by cleavage after two basic amino acids. Little gastrin is the most abundant form of gastrin stored in the G cells, and it too is released into plasma. A smaller molecule containing 14 amino acids called *mini-gastrin* is found in some samples of plasma.* A very small amount of a much larger and still uncharacterized mole-

---

*On account of the difficulty in measuring tryptophan in peptides, mini-gastrin was first thought to contain only one tryptophan residue and to be a peptide of 13 amino acids.

TABLE 12–1.—STRUCTURE OF SOME GASTRINS, CHOLECYSTOKININ AND CAERULEIN

Human big gastrin, HG–34–II*

```
 1    2    3    4    5    6    7    8    9   10   11   12   13   14   15   16   17
Glp—Leu—Gly—Pro—Gln—Gly—His—Pro—Ser—Leu—Val—Ala—Asp—Pro—Ser—Lys—Lys—
18   19   20   21   22   23   24   25   26   27   28   29   30   31   32   33   34
Gln—Gly—Pro—Trp—Leu—Glu—Glu—Glu—Glu—Glu—Ala—Tyr—Gly——Trp—Met—Asp—Phe—NH2
                                                    |
                                                   HSO3
```

Human little gastrin, HG–17–II

```
 1    2    3    4    5    6    7    8    9   10   11   12   13   14   15   16   17
Glp—Gly—Pro—Trp—Leu—Glu—Glu—Glu—Glu—Glu—Ala—Tyr—Gly——Trp—Met—Asp—Phe—NH2
                                              |
                                             HSO3
```

Human mini-gastrin, HG–14–II

```
 1    2    3    4    5    6    7    8    9   10   11   12   13   14
Trp—Leu—Glu—Glu—Glu—Glu—Glu—Ala—Tyr—Gly——Trp—Met—Asp—Phe—NH2
                                   |
                                  HSO3
```

Porcine little gastrin, PG–17–II

```
 5
—Met—
```

Feline little gastrin, FG–17–II

```
 5                  10
—Met—              —Ala—
```

Canine little gastrin, CG–17–II

```
 5        8
—Met—    —Ala—
```

Ovine little gastrin, OG–17–II

```
 5
—Val—
```

Pentagastrin

```
C(CH3)3—OCO—NH—CH2—CH2—CO——Trp—Met—Asp—Phe—NH2
```

Cholecystokinin

```
                      26   27   28   29   30   31   32   33
(25 more)—Asp—Tyr—Met—Gly——Trp—Met—Asp—Phe—NH2
               |
              HSO3
```

Caerulein

```
Glp—Glu—Asp—Tyr—Thr—Gly——Trp—Met—Asp—Phe—NH2
             |
            HSO3
```

*Gastrin Is are not sulfated.

174

cule called *big big gastrin* occurs in the void volume of plasma samples run through a Sephadex column. In all forms of gastrin, the sixth amino acid from the C-terminus is tyrosine, and this tyrosine may or may not be sulfated. Sulfated and nonsulfated gastrins occur in approximately equal quantities, and they are equally potent as hormones.

Gastrins are conventionally named according to their species source, their amino acid content and their sulfation. Thus: HG – 17 – II is *H*uman heptadecapeptide *G*astrin, sulfated. The unsulfated form is numbered I. A more immediately intelligible nomenclature, not generally acceptable, is to omit designation of the unsulfated form and to label the sulfated one =S. Because sulfation has no effect upon the potency of any gastrin, designation of sulfation is unimportant and will be ignored here.

The entire spectrum of physiological activity is exhibited by the last four amino acids at the C-terminus of the gastrins. Unfortunately, biochemical convention is to number peptides from the N-terminus, and therefore the active fragment of big gastrin resides in amino acids 31 – 34, of little gastrin in amino acids 14 – 17 and of mini-gastrin in amino acids 11 – 14, all the same amino acids in the same sequence.

Little gastrin obtained from the tissues of pig, cat, dog and sheep differs from human little gastrin in having one or two amino acid substitutions in the middle of the peptide chain (see Table 12 – 1). Each substitution is the result of a single base change in the codon triplet, and because the substitutions occur in the nonspecific part of the molecule, they have no evolutionary significance.

A synthetic product called *pentagastrin* consists of the C-terminal tetrapeptide to which a substituted beta-alanine has been added. This compound has the advantages of being stable, water-soluble and patentable. It is the commercially available product having all the physiological properties of the natural gastrins.

Only 10% of the gastrin contained in G cells of the antrum is big gastrin; the rest is little gastrin. Big gastrin is the larger fraction of gastrin extractable from the proximal duodenum. There is only a minute amount of any gastrin in the oxyntic glandular mucosa, and the amount extractable from the normal pancreas is so small as to be debatable. Big gastrin is twice as abundant as little gastrin in most plasma samples; after feeding, it rises more rapidly and remains higher than does little gastrin.

Big gastrin has a half-life in man of 42 min, and little gastrin has a half-life of 7 min. The difference in half-lives accounts for the fact that the plasma concentration of big gastrin is higher than is that of little gastrin.

Injection of equimolar amounts of big gastrin and little gastrin causes equal secretions of acid. However, big gastrin, having a much longer half-life, rises to a higher plasma concentration under this circumstance, and it is therefore much less potent, at the molecular level, than is little gastrin in stimulating acid secretion.

The kidney is the chief site of inactivation of gastrins. Gastrin molecules or analogs containing 10 or fewer amino acids are inactivated in the liver by deamidation; gastrins containing 11 or more amino acids pass through the liver intact.

Gastrin concentrations are measured by radioimmunoassay, and the normal fasting plasma concentration is 50 – 100 pg per ml (a pg or picogram is 10^{-12} gm). The upper limit of normal is about 200 pg per ml. Except in the most expert hands, the assay does not differentiate among the various forms of the hormone or between active and inactive forms. Consequently, most reported assay results tell only whether the concentration is high, medium or low, and reported values are most useful when they show changes occurring as the result of some procedure.

Actions of Gastrin

The major short-term action of gastrin is to stimulate secretion of acid. (The major long-term action is a trophic one discussed below.) In man, when the plasma concentration of gastrin rises in response to a meal, acid secretion rises by 1.8% of the stomach's capacity to secrete for every picogram per milliliter rise in plasma gastrin concentration.

Gastric blood flow increases when gastrin stimulates acid secretion. Pepsinogen secretion also accompanies acid secretion, but secretion of pepsinogen is not entirely a direct consequence of the action of gastrin upon chief cells. When acid, whose secretion has been stimulated by gastrin, flows over the surface of the mucosa, it excites receptors, which, through a cholinergic reflex, stimulate pepsinogen secretion.

Many other responses can be detected when gastrin is injected intravenously (Table 12–2). It is difficult to decide which response is physiological and which pharmacological. If a response to gastrin administration occurs during the normal process of digestion and if the same response is elicited by a plasma concentration of gastrin similar to that occurring during digestion, the response may be a physiological one. If the response is produced only by a plasma concentration of gastrin far greater than that which stimulates acid secretion at half-

TABLE 12–2.—ACTIONS OF GASTRINS

Physiological actions:
Gastrins stimulate
Acid secretion
Pepsinogen secretion
Gastric blood flow
Contraction of the circular muscle of the stomach
Growth of gastric and small intestinal mucosa
and the pancreas
Pharmacological (and possibly physiological) actions:
Gastrins stimulate
Water, bicarbonate and electrolyte secretion by
pancreas, liver and small intestinal mucosa
Enzyme secretion by pancreas and small intestinal
mucosa
Contraction of the lower esophageal sphincter,
gallbladder, small intestine and colon
Gastrins inhibit
Contraction of the pyloric sphincter, ileocecal
sphincter and sphincter of the hepatopancreatic
ampulla
Gastric emptying
Absorption of glucose and electrolytes in the small
intestine
Gastrins release
Insulin, glucagon and calcitonin

maximal rate, the response may be a pharmacological one. In addition, other factors influencing the response, nervous and humoral, must be identified and evaluated before a firm distinction between physiological and pharmacological actions can be made.

Intestinal Control of Gastric Secretion

In the dog, the cephalic and gastric phases of secretion can be bypassed by placing chyme or other solutions directly into the duodenum through a fistula. When this is done, the oxyntic glandular mucosa of the animal secretes acid. The response is mediated by hormones and not by a nervous reflex, for an extrinsically denervated pouch of the body of the stomach secretes acid when liver paste is infused into the duodenum.

In the dog, intestinal stimulation of acid secretion may occur through three mechanisms:

1. A hormone is released from the intestinal mucosa that directly stimulates acid secretion. The hormone is not gastrin, for there is no rise in plasma gastrin concentration. The rate of acid secretion is strongly augmented by endogenous or exogenous gastrin, and therefore the hormone must be synergistic with gastrin. This rules out cholecystokinin, for that hormone inhibits gastrin-stimulated acid secretion. These properties place the hormone among the candidate hormones of the gut, and it has been tentatively baptized *entero-oxyntin* to show its origin and its target.

2. The hormone released by chyme in the duodenum augments the response to concurrently released gastrin.

3. Chyme in the intestine may release a bombesin-like hormone, which in turn releases gastrin from the pyloric glandular mucosa.

The intestinal phase of acid secretion can be studied in man by observing the response to feeding in a patient whose antrum has been removed and who therefore has no gastric source of gastrin. Feeding such a patient causes a rise in plasma gastrin, most of it big gastrin. Apparently, chyme in the small intestine of man does release gastrin.

Acid secretion declines during the second half of the digestion of a meal, in part because cephalic and gastric stimuli die out and in part because chyme in the duodenum inhibits acid secretion. Acid and fat- and protein-digestion products in chyme are the stimuli for inhibition, and they act through hormones and nerves.

Early in the digestion of a meal, the pH of chyme delivered to the duodenum is 3.0 or above. Later, when acid has titrated gastric contents, 30 mEq of acid may be delivered to the duodenum in an hour at a pH between 1.0 and 3.0. However, duodenal neutralization of acid is so rapid and effective that only a short segment of the duodenum is acidified for a brief time (see Fig 13–6). Acid in the duodenum does not liberate enough secretin to account for inhibition of acid secretion. Gastric inhibitory polypeptide (GIP) is liberated by fat digestion products, but whether the amount released is adequate to account for inhibition of acid secretion has not yet been determined. Cholecystokinin is also liberated by fat- and protein-digestion products; although the amount liberated does account for inhibition of gastric emptying at the end of digestion, cholecystokinin is probably not the physiological inhibitor of acid secretion. Synergism among several hormones may contribute to inhibition, and there may be an as yet uncharacterized candidate hormone, *bulbogastrone*, which mediates acid inhibition by chyme in the duodenum.

Years ago, extracts of the duodenal mucosa were found to inhibit acid secretion, and the putative agent in the extracts was called *enterogastrone*. Such extracts contain secretin, GIP, VIP and cholecystokinin, all of which, when given in pharmacological amounts, as enterogastrone was given, inhibit acid secretion. There is no one hormone that can be called enterogastrone, and the name should be dropped.

Two peptides isolated from normal male human urine, when injected in small doses, inhibit histamine- and pentagastrin-stimulated acid secretion. These are called *urogastrone*. One of these peptides contains 53 amino acids, and the other contains the same ones except for arginine at the C-terminus. Urogastrone appears to be identical with a peptide isolated from the submaxillary glands of male mice that promotes epidermal growth. The biologic functions of urogastrone and epidermal growth factor are unknown.

Perfusion of the duodenal bulb of a dog with acid at the rate of 8 mEq an hour, a rate well within the normal range, inhibits pentagastrin-stimulated acid secretion by the dog's stomach. Inhibition cannot be attributed to secretin, for pancreatic secretion is only slightly stimulated. Injection of pure secretin in an amount necessary to produce the same inhibition causes far greater pancreatic secretion. The effect is not mediated by another hormone, for pentagastrin-stimulated secretion by an extrinsically denervated pouch is unaffected. Separation of the duodenal bulb from the stomach abolishes the inhibition that follows perfusion of the bulb with acid. These results demonstrate that physiological quantities of acid in the duodenum can inhibit acid secretion through a local reflex.

Inhibition of Acid Secretion by Exogenous Compounds

Carbonic anhydrase is part of the oxyntic cell's metabolic machinery, and very high concentrations of carbonic anhydrase-inhibiting drugs inhibit acid secretion. Thiocyanate and related compounds having an unshared nitrogen atom inhibit acid secretion, as do many compounds that block energy production in the oxyntic cells. Such inhibitors have no practical importance.

Acetylcholine stimulates acid secretion, and it acts synergistically with gastrin. Therefore, atropine and atropine-like drugs inhibit acid secretion stimulated by gastrin and histamine as well as that stimulated by cholinergic reflexes or drugs.

Most actions of histamine are inhibited by a class of antihistaminic drugs that are, in general, substituted ethylamines. Histamine combines with receptors on effector cells, and those receptors that are blocked by this class of antihistaminic drugs are called H_1 receptors. These drugs do not block the stimulation of acid secretion by histamine; oxyntic cells do not have H_1 receptors. Another class of drugs, substituted imidazoles, does block the effect of histamine upon oxyntic cells. These are called H_2 blocking agents, and oxyntic cells are said to have H_2 receptors. One drug of this class, metiamide, not only inhibits histamine-stimulated acid secretion, but it also inhibits gastrin- and acetylcholine-stimulated acid secretion. The large and continuous acid secretion of a patient with a gastrinoma is 90% inhibited by metiamide. Metiamide has an unfortunate side-effect, and it is not available for general clinical use. Other and apparently safer drugs, of which cimetidine is one, are also H_2 blockers.

Prostaglandins of the E class inhibit acid secretion stimulated by histamine or pentagastrin. The compounds are effective in very small doses when given orally or parenterally. A substituted prostaglandin, 16,16-dimethyl prostaglandin E_2, in an oral dose of only 15 μg per kg or less reduces basal acid secretion in man by 99% and inhibits histamine-stimulated acid secretion by 79%. Acid secretion stimulated by a meal also is inhibited. Inhibition lasts several hours. The drug has no effect upon pepsin secretion.

Secretin and glucagon in pharmacologic amounts inhibit acid secretion. A clinical trial of secretin therapy in duodenal ulcer disease gave unpromising results.

Stimulation of Pancreatic and Biliary Secretion

Chyme entering the duodenum contains protein digestion products and acid more or less buffered by gastric contents. Digestion of fat begins in the duodenum, and soon after chyme enters, fat digestion products accu-mulate. Flow of pancreatic juice containing bicarbonate and digestive enzymes begins, and within a few minutes enzyme activity in duodenal contents is high. Bile flow is stimulated and the gallbladder contracts. The concentration of bile acids in the duodenum rises above the critical micellar concentration.

Some of the flow of pancreatic juice and bile is stimulated by the cephalic phase through the action of vagal impulses, and some is stimulated through liberation of gastrin during the gastric phase of digestion. The major part of pancreatic enzyme secretion and contraction of the gallbladder is stimulated by cholecystokinin released into the blood from the duodenal mucosa by protein- and fat-digestion products, and most of the flow of bicarbonate-containing fluid from pancreas and liver is stimulated by the joint action of cholecystokinin and secretin, the latter released into the blood from the duodenal mucosa by acid.

Cholecystokinin

The fact that fat- and protein-digestion products in the duodenum cause contraction of the gallbladder by means of a hormone was demonstrated by the classic means of endocrinology in 1928, and the hormone responsible was named cholecystokinin. Years later, the fact that fat- and protein-digestion products in the duodenum stimulate enzyme secretion by the pancreas through mediation by a hormone was established by similar means, and the hormone responsible was named pancreozymin. Still later, a polypeptide isolated from the duodenal mucosa proved to have the properties of both hormones; cholecystokinin and pancreozymin are the same molecule. For convenience, the molecule is given its earlier name, but its property of stimulating the pancreas is still called pancreozymin.

Cholecystokinin is a polypeptide containing 33 amino acids. Its partial structure is given in Table 12–1. The last five amino acids in its C-terminal sequence are identical with those in gastrin. Since the last four of these constitute the active fragment of gastrin, both hormones have the same physiological actions. They differ quantitatively, not qualitatively. In gastrin, the sulfated tyrosyl residue is separated from the active C-terminal sequence by a single amino acid, and whether or not the tyrosyl residue is sulfated makes no difference to the hormone's physiological activity. In cholecystokinin, the sulfated tyrosyl residue is separated from the C-terminal sequence by two amino acids, and this tyrosyl residue must be sulfated for the molecule to have cholecystokinin activity.

The C-terminal octapeptide of cholecystokinin has all the hormone's actions, and it is used as a substitute for cholecystokinin in experimental work.

Cholecystokinin is present in cells in the mucosa of the duodenum and jejunum but not the ileum. It is released by the L-forms of essential amino acids. The D-forms and the nonessential amino acids are ineffective. The most potent stimulus is L-tryptophan, and the optimal concentration of tryptophan is 50 mM. In decreasing order of potency are phenylalanine, lysine, methionine, arginine, valine, isoleucine, histidine, leucine and threonine. The maximal response to tryptophan is 80% of that to a maximal dose of cholecystokinin, and the rate of secretion of pancreatic enzymes when the duodenum is irrigated with a mixture of essential amino acids is nearly the same as that following intravenous administration of a maximally tolerated dose of cholecystokinin.

Perfusion of the duodenum with acid at a physiological rate releases cholecystokinin as well as secretin, but the amount of cholecystokinin released is only one fifth that of secretin.

The actions of cholecystokinin are listed in Table 12–3.

In the dog, gastric emptying is slowed when cholecystokinin or its octapeptide is injected or when cholecystokinin is released by perfusion of the duodenum with tryptophan. The amount of cholecystokinin required to inhibit gastric emptying is the same

TABLE 12–3.—ACTIONS OF CHOLECYSTOKININ

Physiological actions
  Cholecystokinin
    Stimulates contraction of the gallbladder and relaxation of the sphincter of the hepato-pancreatic ampulla
    Stimulates secretion of enzymes by the pancreas
    Weakly stimulates secretion of bicarbonate-containing juice by the pancreas and liver but is synergistic with secretin
    Slows gastric emptying
    Is a trophic hormone for the pancreas
Pharmacological (and possibly physiological) actions
  Cholecystokinin
    Weakly stimulates acid secretion in absence of gastrin
    Competitively inhibits gastrin-stimulated acid secretion
    Stimulates pepsinogen secretion
    Inhibits the lower esophageal sphincter
    Stimulates motility of the sigmoid colon
    Stimulates secretion by the duodenal glands
    Releases insulin and glucagon
    Terminates or stimulates feeding behavior

as that required for contraction of the gallbladder and secretion of pancreatic enzymes. The effect of cholecystokinin in slowing gastric emptying is classed as a physiological one. In contrast, the dose of gastrin required to inhibit gastric emptying is far higher than that required to stimulate acid secretion, and that effect of gastrin is classed as a pharmacological one.

Because cholecystokinin has the same C-terminal sequence of amino acids as do the gastrins, cholecystokinin stimulates acid secretion. It is only a weak stimulant, and it is called a *partial agonist*. It competitively inhibits gastrin-stimulated acid secretion by occupying gastrin receptors and denying them to the more potent hormone. It is not known whether the inhibition of acid secretion by cholecystokinin helps to terminate acid secretion at the end of the gastric phase of digestion.

A rat, like other animals, behaves as though it were satiated after eating a certain amount of food. It stops eating, grooms itself and sleeps. If a rat is provided with a gastric fistula that remains closed, and if the rat is fed a liquid diet, the rat performs cycles of eating, grooming and sleeping. If the fistula is opened so that the liquid food drains from the stomach, the rat does not stop eating, but it does stop eating when a small volume of food is directly infused into the duodenum. Termination of feeding behavior appears to be regulated by action messages sent from the duodenum to the hypothalamus. The exact nature of these messages is at present unknown.

In the rat, cat and man, injection of cholecystokinin in pharmacological doses inhibits feeding, but in the cat and man, injection of the hormone in physiological doses enhances food intake. Secretin has no effect.

Caerulein

The skin of the Australian frog *Hyla caerulea* contains a high concentration of a decapeptide, caerulein, whose structure is given in Table 12–1. It has the same C-terminal tetrapeptide as gastrin and cholecystokinin and has a sulfated tyrosyl residue in a position corresponding to cholecystokinin's. Consequently, it shares the properties and potency of the other two polypeptides. In fact, its ability to cause contraction of the gallbladder is so great that it has been used in nanogram per kilogram doses during cholecystography.

Secretin

The structure of the polypeptide secretin is given in Table 12–4. Of the 27 amino acids in secretin, 14 occupy the same position, counting from the N-terminus, as do those in glucagon. Consequently, secretin in pharmacological doses mimics glucagon, and glucagon in pharmacological doses mimics

secretin. There is no minimal active fragment of secretin. The molecule has a helical structure between positions 5 and 13, which may explain why the whole chain of 27 amino acids is required for activity.

The chief action of secretin is to stimulate secretion of bicarbonate-containing juice by the pancreas and liver.

Secretin is present in S cells throughout the duodenum, but it is most concentrated in the duodenal bulb. Over the range of pH from 0 to 3.0, release of secretin is independent of the pH of fluid entering the duodenum; the amount of secretin released is proportional to the amount of acid entering the duodenum, not to the concentration of acid. Above pH 3.0 there is a sharp decline in the amount of secretin released as the pH rises until an absolute threshold is reached at pH 4.5–5.0. When the contents of the stomach are acid, the duodenal bulb is acidified only briefly when a spurt of chyme enters it (see Fig 13–6), and acid is completely neutralized in the first portion of the duodenum. Therefore, only a small quantity of secretin is released into the blood during digestion of a meal, and this amount of secretin *in itself* is not enough to stimulate the actually occurring pancreatic and biliary secretion of bicarbonate.

Synergism of secretin and cholecystokinin accounts for the quantity of bicarbonate and enzymes secreted. The two hormones have the following relations:

1. Cholecystokinin is released by acid in the duodenum at a rate one-fifth that at which secretin is released by the same amount of acid.

2. Secretin is released by acid but not by any other component of chyme.

3. Cholecystokinin alone weakly stimu-

TABLE 12–4.—STRUCTURE OF SECRETIN*

1	2	3	4	5	6	7	8	9	10	11	12	13	14
His	*Ser*	Asp	*Gly*	*Thr*	*Phe*	*Thr*	*Ser*	Glu	Leu	*Ser*	Arg	Leu	Arg—

15	16	17	18	19	20	21	22	23	24	25	26	27
Asp	*Ser*	Ala	*Arg*	Leu	*Gln*	Arg	Leu	Leu	*Gln*	Gly	*Leu*	Val—NH$_2$

*The amino acids in italics occur in corresponding positions in glucagon.

lates secretion of bicarbonate-containing juice by pancreas and liver, but it amplifies the action of secretin.

4. Secretin alone weakly stimulates secretion of enzymes by the pancreas, but it amplifies the action of cholecystokinin.

5. Secretin and cholecystokinin stimulate pepsinogen secretion and inhibit acid secretion.

Because the structure of secretin has nothing in common with the structure of gastrin, secretin might be expected to be a noncompetitive inhibitor of gastrin-stimulated acid secretion. It is in the dog, but it is a competitive inhibitor in man.

Secretin's inhibition of acid secretion is probably only a pharmacological property; so are its effects in stimulating contraction of the gallbladder, inhibiting contraction of the lower esophageal sphincter and gastric and intestinal motility, releasing insulin and causing lipolysis in adipose tissue.

Candidate Hormones

Polypeptides isolated from the gastrointestinal mucosa have gastrointestinal actions when injected intravenously. They cannot be called authentic hormones until it is demonstrated that they are released under physiological conditions and participate in the control of a target organ. On the other hand, some physiological function appears to be controlled by an as yet unidentified hormone. Until the evidence is complete, the polypeptides and the putative hormones remain candidates for hormonal status. Like characters in a Pirandello play, their identities remain uncertain.

A partial list of candidate hormones is given in Table 12-5.

Trophic Actions

Trophic actions of gastrin upon the oxyntic glandular mucosa, the mucosa of the small intestine and the pancreas have been demonstrated in experimental animals. Synthesis, storage and release of gastrin depend upon frequent use of the digestive tract. Gastrin concentrations in plasma and antrum fall by more than two-thirds when a rat is fasted, and they return to normal on refeeding. Normal levels are maintained on a liquid but

TABLE 12-5.—SOME CANDIDATE HORMONES OF THE
DIGESTIVE TRACT

NAME	FUNCTION
Chymodenin	In duodenal mucosa; stimulates secretion of chymotrypsinogen by the pancreas
Gastric inhibitory polypeptide (GIP)	In duodenal mucosa; inhibits gastric motility, releases insulin, stimulates intestinal secretion
Motilin	In duodenal mucosa; stimulates gastric motility
Vasoactive intestinal peptide (VIP)	In duodenal mucosa; vasodilator, inhibits acid and pepsinogen secretion, stimulates intestinal secretion, mimics glucagon
Urogastrone	In human male urine; inhibits gastric secretion, probably identical with epidermal growth factor
Gastrone	In mucous fraction of gastric juice; inhibits gastric secretion
Bulbogastrone	In mucosa of duodenal bulb; released by acid, and inhibits gastric secretion
Duocrinin and enterocrinin	In duodenal mucosa; stimulate intestinal secretion
Enteroglucagon	In duodenal mucosa; mimics glucagon
Incretin	In duodenal mucosa; releases insulin
Villikinin	In duodenal mucosa; stimulates contraction of intestinal villi
Bombesin	In frog skin and in antral and duodenal mucosa of frogs, perhaps in mammalian duodenal mucosa; releases gastrin

not on a bulky, non-nutritive diet. When a rat or dog is completely supported by parenteral nutrition, the weights of the oxyntic mucosa, the small intestinal mucosa and the pancreas fall, but the fall can be prevented by administration of pentagastrin. On the other hand, administration of gastrin to the rat or dog stimulates protein synthesis and uptake of tritiated thymidine in the three tissues named but not in liver or muscle.

In man, there is an association between average plasma gastrin concentrations and acid secretory capacity of the stomach. There is a decrease in secretory capacity after antrectomy, which removes a major source of gastrin. On the other hand, a patient with duodenal ulcer often has a secretory capacity greater than normal and a supranormal gastrin release during digestion. A patient with chronic hypergastrinemia resulting from a gastrinoma usually has an acid secretory capacity far above normal. It is a reasonable but as yet unproved hypothesis that the trophic action of gastrin accounts for these facts.

Cholecystokinin is a trophic hormone for the pancreas. Hormones of the anterior pituitary also control gastrointestinal mucosal growth. The mucosa atrophies in hypophysectomized animals, and administration of hormones of the anterior pituitary partially restores the mucosa. It is likely that hormones of the anterior pituitary and releasing or release-inhibiting hormones of the hypothalamus influence gastrointestinal function in an as yet undetermined fashion.

Endocrine Adenomatosis

Early in embryologic development, cells break from the neuroectoderm and, migrating to the gut and its appendages, become parents of hormone-secreting cells. The hormones are all polypeptides, and many of them, gastrin and cholecystokinin, secretin and glucagon, are closely related in structure and function. Because the daughter cells have similar histochemical properties, they are called members of the APUD series, a name that could well be discarded with no harm done.

A tumor, such as an insulinoma, arising from the daughter cells may secrete only one hormone, or it may, like a gastrinoma, secrete a family of similar hormones. Other tumors, such as those responsible for pancreatic cholera, secrete several different hormones. In the case of multiple endocrine adenomatosis, a number of different hormone-secreting tumors may develop simultaneously. Thus, gastrinomas and tumors or hyperplasia of the parathyroid glands are often associated.

There are two classes of hypergastrinemia: one associated with hypochlorhydria in which feedback inhibition of gastrin release is diminished or absent, and the other associated with hyperchlorhydria in which increased acid secretion is driven by increased plasma gastrin. Hypergastrinemia in the latter instance is usually caused by a gastrinoma. The association between a tumor and hypersecretion of acid was first recognized by Zollinger and Ellison,* and one form of the syndrome bears their name. Gastrinomas occur as discrete or diffuse tumors, often in the pancreas but occasionally in other digestive organs, and they usually have many slowly growing metastases. Gastrinomas contain more little gastrin than big gastrin, but in the plasma big gastrin is far more abundant than little gastrin. The difference is explained by the longer half-life of big gastrin. Gastrinomas also secrete a form of little gastrin consisting of the N-terminal 13 amino acids. Because this fragment lacks the four amino acids at the C-terminus, it has no biologic activity. However, it cross-reacts with antibodies raised to the N-terminus of little gastrin, and it gives falsely high values for the plasma concentration of gastrin in radioimmunoassay.

Gastrinomas are not under the negative feedback control exercised by acid, and they liberate gastrin continuously at a high rate. Patients with gastrinomas have a very high acid secretory capacity, perhaps on account

---

*Ann. Surg. 142:709, 1955.

of the trophic action of gastrin. Because this capacity is continuously used, the basal rate of acid secretion approaches maximal secretory capacity. The consequences of uncontrolled and continuous acid secretion are multiple duodenal and jejunal ulceration and inactivation of digestive enzymes secreted by the pancreas. Steatorrhea, azotorrhea and diarrhea are frequent results.

Gastrin release by G cells of the antral mucosa is depressed when secretin is injected in pharmacological doses, but secretion of gastrin by gastrinomas is enhanced by secretin. Consequently, a rise in plasma gastrin concentration, rather than a fall, following secretin injection, is evidence that much of the plasma gastrin comes from a gastrinoma. Gastrinomas, as well as G cells in the antral mucosa, are stimulated to release gastrin by a rise in plasma calcium concentration. Gastrinomas are often accompanied by tumors, or perhaps only by hyperplasia, of the parathyroid glands, and the hypercalcemia caused by hyperparathyroidism drives gastrinomas to secrete gastrin.

In very rare instances, hyperplasia of G cells in the gastric antrum may also be responsible for high plasma concentrations of gastrin. Abnormally large G cells appear not to be inhibited by acid bathing the mucosa, and plasma gastrin concentration is high during fasting and feeding.

In some patients, the small intestinal mucosa continuously secretes abnormally large amounts of bicarbonate-containing fluid. This condition may or may not be accompanied by pancreatic and hepatic secretion of alkaline juice and by gastric hypersecretion of acid. It results in diarrhea, metabolic acidosis, shrinkage of extracellular fluid volume and a fall in total exchangeable potassium. It is often called *pancreatic cholera,* although the intestinal contribution of fluid is more important than the pancreatic one. In several instances, a tumor associated with pancreatic cholera has been found to contain large amounts of VIP, and there was a high concentration of that compound in plasma. VIP stimulates bicarbonate secretion, and it may be the agent responsible for the hypersecretion of pancreatic cholera. Gastrin, cholecystokinin, a secretin-like hormone and 5-HT also occur in pancreatic cholera-associated tumors.

Nervous Influences

Nervous influences on secretion, mediated by both branches of the autonomic nervous system, are closely correlated with other nervous activity expressed as behavior. Feeding, which is the behavior most intimately connected with secretion, is controlled at the hypothalamic level. During satiety, activity in the ventromedial region of the hypothalamus inhibits the lateral region; but during feeding, the lateral region dominates. The medial and lateral regions of the hypothalamus also control secretion. Stimulation of the medial region in rats inhibits both food intake and gastric secretion, whereas stimulation of the lateral region increases food intake and gastric secretion. Rats made hyperphagic by destruction of the medial satiety center secrete acid and pepsinogen in the basal state at rates more than double those occurring in control periods. The efferent pathway passes from the hypothalamus through the mesencephalon to the medulla and vagal nucleus.

The hypothalamus is also activated by what is crudely designated "stress," and stressful situations may lead to gastrointestinal lesions. If mice are restrained by being wrapped in wire gauze, 92% of them develop ulcers of the gastric mucosa in 24 hours, but only 4% of restrained hamsters do so. Pairs of monkeys have been placed in experimental chairlike restraining apparatus and subjected to electric shocks to the feet delivered automatically every 20 sec for 6 hours. After a 6-hour rest, the shocks were again given for 6 hours, and the cycle was repeated indefinitely. Both monkeys were given hand levers which they could press; the lever to the first or "executive" monkey, when pressed at the correct time, prevented the shocks to both monkeys, but the lever for the second or control monkey had no effect. The executive monkey quickly learned to press the lever appropriately (both monkeys actually received very few shocks), but his partner, after a few trials, ignored his lever. In four experiments, the executive monkeys died after 9, 23, 25 or 48 days of lever-pressing, and in each instance the cause of death was extensive gastrointestinal erosions and hemorrhage. No control monkey died or had gastric ulceration. In similar experiments with rats, the executive rat developed hemorrhagic lesions of the gastric mucosa, but his yoked partner did not.

Activity in higher centers converging on hypothalamic and medullary centers affects secretion. In general, strong physical activity or those aspects of behavior interpreted as expressing rage or the experience of pain inhibit gastric secretion and motility through sympathetic discharge and inhibition of parasympathetic activity. The effect on the stomach may outlast the overt response. In the dog, an innervated pouch of the oxyntic gland area usually secretes when the animal eats. If however, the dog is provoked into a rage, subsequent feeding, even after the dog has calmed, may evoke little or no gastric secretion.

Similar suppression of secretion in rage has been found in man, and literature from the most ancient times contains anecdotes describing cessation of digestion during episodes of sympathetic discharge. In one closely studied human subject whose stomach could be inspected through a fistula, gastric secretion was reduced during depression, and it was subnormal for several months when the dominant psychic state was self-reproach. Increased secretion has been found to accompany aggressive actions. The same fistulous subject, when unjustly reproached, experienced strong feelings of hostility; at the same time, both acid secretion and mucosal blood flow increased about 25% and remained elevated for 2 weeks. The generalization derived from these studies is that, when the affective state is one of fear, sadness or withdrawal, gastric secretion is re-duced, but when the dominant element is aggressiveness or the will to fight back, gastric secretion increases. Interpretation of such studies is deeply penetrated by subjective elements, and no one knows how widely these conclusions apply. Attempts to extend the study of the relation between emotions, behavior and secretion, using techniques of psychoanalysis, hypnosis or operant conditioning, have not yet produced any generally acceptable conclusions.

REFERENCES

Andersson, S.: Gastric and duodenal mechanisms inhibiting gastric secretion of acid, in Code, C. F. (ed.): *Handbook of Physiology:* Sec. 6. *Alimentary Canal,* Vol. II (Washington, D.C.: American Physiological Society, 1967), pp. 865–878.

Babkin, B. P.: *Secretory Mechanism of the Digestive Glands* (2d ed.; New York: Paul B. Hoeber, Inc., 1950).

Code, C. F.: Histamine and gastric secretion: A later look, 1955–1965, Fed. Proc. 24:1311–1321, 1965.

Erspamer, V.: Caerulein, Gut 11:79, 1970.

Gregory, R. A.: *Secretory Mechanisms of the Gastrointestinal Tract* (London: Edward Arnold & Co., 1962).

Grossman, M. I., and Konturek, S. J.: Gastric acid does drive pancreatic bicarbonate secretion, Scand. J. Gastroenterol. 9:299, 1974.

Grossman, M. I., *et al.*: Candidate hormones of the gut, Gastroenterology 67:730, 1974.

Johnson, L. R.: The trophic action of gastrointestinal hormones, Gastroenterology 70:278, 1976.

Jorpes, J. E., and Mutt, V. (eds.): *Secretin, Cholecystokinin, Pancreozymin, and Gastrin* (Berlin: Springer-Verlag, 1973).

Preshaw, R. W.: Duodenal acidification and the release of secretin, in Botelho, S. Y., Brooks, F. P., and Shelley, W. B. (eds.): *Exocrine Glands* (Philadelphia: University of Pennsylvania Press, 1969), pp. 247–252.

Smith, G. P., and Brooks, F. P.: Hypothalamic control of gastric secretion, Gastroenterology 52:727, 1967.

Smith, G. P., Gibbs, J., and Young, R. C.: Cholecystokinin and intestinal satiety in the rat, Fed. Proc. 33:1146, 1974.

Thompson, J. C. (ed.): *Gastrointestinal Hormones, A Symposium* (Austin: University of Texas Press, 1975).

Walsh, J. H., and Grossman, M. I.: Gastrin, N. Engl. J. Med. 292:1324, 1377, 1975.

PART III
DIGESTION AND ABSORPTION

13

Gastric Digestion and Emptying; Absorption

THE BODY OF THE stomach accepts large amounts of food at intervals determined by habit and opportunity. The contents are distributed in layers as they are swallowed. Movements of the body are feeble, and food lodges undisturbed while the central mass undergoes salivary digestion. Some of the food soon empties from the antrum into the duodenum, and the duodenum, sensing the composition of its contents, adjusts gastric emptying so that no more is emptied than the duodenum can handle. Liquids are emptied at a rate proportional to the volume of gastric contents, but emptying of solids is delayed until they can be reduced in size.

At each meal the acidity of the contents of the body falls, remains low for 1–2 hours and then rises steadily as the stomach empties (Fig 13–1). Consequently, much of the meal is emptied into the duodenum before it is acidified. With substantial meals, or with meals that empty slowly, the low acidity of the body's contents may be undisturbed for several hours. On the other hand, the contents of the antrum are thoroughly mixed with acid and pepsin as peristaltic waves pass over the antrum, and antral contents become more acid earlier than do contents of the body. Emptying is seldom complete

between meals during the day, but early in the morning, after an overnight fast, the stomach contains only about 20 ml of gastric juice.

Overnight there is a longer interval between meals, and the stomach is usually empty between 1:00 and 3:00 A.M. At this time, as the data in Figure 13–2 show, the pH of the small volume of contents obtained from 19 subjects sampled by an indwelling tube was between 1.0 and 2.0, with acidity averaging 66 mN but not exceeding 100 mN. In about half the subjects, the pH remained low until breakfast; but in the other half, it approached neutrality, probably because acid secretion stopped altogether.

Gastric Carbohydrate and Protein Digestion

If the starch of the meal has been mixed with salivary α-amylase by chewing, its digestion in the body can continue as long as the pH remains high. In one experiment, two young men who were capable of voluntary regurgitation were fed meals consisting of 150 gm of beefsteak, 150 gm of potato, 50 gm of green peas, 10 gm of butter and 200 ml of water. The men chewed the food according to their normal habits, and at the end of

187

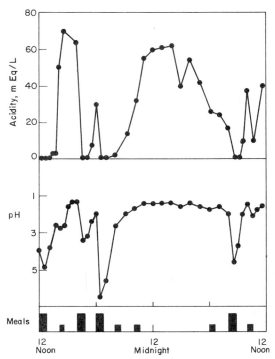

Fig 13-1.—Acidity of material aspirated from stomach of a normal man, age 36, over a period of 24 hours. Results are presented in terms of titratable acidity and pH. Large rectangles represent meals: *lunch*—fish, potato, carrot puree, fruit puree, custard and tea; *tea*—bread and butter, tea, golden syrup; *supper*—fish, potato, custard, bread and butter; *breakfast*—porridge, boiled egg, bread and butter, tea. Small rectangles represent drinks of milk. (Adapted from James, A. H., and Pickering, G. W.: Clin. Sci. 8:181, 1949.)

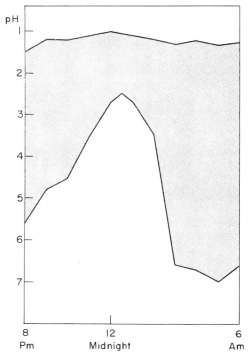

Fig 13-2.—Range of acidity of overnight samples drawn at hourly intervals from stomachs of 19 normal adult subjects who ate their last meal at 6:00 P.M. Data from one additional subject whose pH values ranged from 5.0–7.0 are omitted. (Adapted from James, A. H., and Pickering, G. W.: Clin. Sci. 8:181, 1949.)

30–60 min they vomited at command. The pH of the vomitus was above 3.4. In it, 35–48% of the starch had been hydrolyzed to oligosaccharides, but only a trace of the protein had been digested.

Pepsins, acting on the periphery of the mass in the body and in the antrum, reduce the size of lumps of meat and assist in dispersion of fat and carbohydrate by breaking the walls of animal cells. Although exhaustive digestion of representative proteins by pepsins in vitro breaks them into fragments of all sizes, it liberates no more than 15% of their available amino-nitrogen. Consequent-ly, peptic digestion in the stomach cannot be expected to be complete.

When radioiodinated human serum albumin (RISA) is added to a liquid meal taken by normal men, samples recovered by tube from the region of the pylorus contain 85–90% of the administered ^{131}I in a form precipitable by 5% phosphotungstic acid. This means that the largest part of protein digestion occurs beyond the stomach.

Gastric Digestion and Absorption of Fat

Acid and pepsin tend to break natural or artificial emulsions in the stomach, with the result that fat forms large drops unsuitable for gastric digestion. Liquid or semisolid fats float in gastric contents, and in a man in the upright position the fat in a broken emulsion empties after the rest of the meal has emptied. Acid hydrolysis of ester bonds is negligible in the stomach. Gastric juice contains a lipase distinct from pancreatic lipase; the gastric enzyme partially hydrolyzes triglycerides at pH 6.0–8.0. Lipolytic activity is maximal with tributyrin as the substrate, and it falls off as the chain length of the fatty acids in the triglycerides increases, so that gastric hydrolysis of triglycerides of chain length greater than 10 is negligible in the adult and small in the child. Depot and liver fat of animals and oil of plants contain little or no triglycerides of fatty acids shorter than 10 carbon atoms; and cow's, goat's and human milk contains only 4–9% tributyrin and less than 10% triglycerides with fatty acids shorter than 14 carbon atoms. Only minute amounts of long-chain fatty acids are absorbed through the gastric mucosa. If the water-soluble short-chain fatty acids — acetic, propionic or butyric — are present in gastric contents at pH low enough (pH 4.0 or lower) for an appreciable fraction to be in the un-ionized, fat-soluble form, they rapidly diffuse through the lipoprotein membranes of the mucosal cells and are absorbed.

Gastric Emptying of Liquids

The three major determinants of the rate of emptying of the stomach are the volume of the meal, its osmotic pressure and its chemical composition. The most nearly standardized data describing the time course of emptying of a liquid meal have been obtained by use of the *serial test meal*. In this test, a subject, after his stomach is washed out, is given a known volume of the meal, either by mouth or by tube. At some time, say 15 min later, the stomach is emptied as thoroughly as possible, and the volume and composition of the material recovered are measured.

If a substance not absorbed by the stomach (e.g., phenol red) is present in the meal at a known concentration, the volume of secretions added to the meal can be calculated from its dilution. Then, at some later time, at which the subject's stomach is assumed to be in the same functional state, a similar meal is given. This time, a different interval, say 30 min, is allowed to elapse before the gastric contents are recovered. A large number of replications of the procedure gives a description of the course of emptying. The relation of the volume of gastric contents to emptying has been measured this way, using a meal composed of 20 gm of citrus pectin, 35 gm of sucrose, 60–70 mg of phenol red, NaOH to adjust pH to 6.5 and distilled water to make 1 liter. The osmotic pressure was adjusted by adding sucrose.

The shape of the full stomach is roughly that of a cylinder on top of a cone; and for a cylinder or cone, the transmural pressure P is equal to the tension in the wall T divided by the radius R. Rearranging, we have $T = P \cdot R$. The radius is proportional to the square root of the volume contained: $T \propto P\sqrt{V}$. Since tension in the wall of the stomach is the ultimate driving force for emptying, we expect the rate of emptying to be a function of the square root of the volume within the stomach. The data in Figure 13–3 show that the square root of the volume of gastric contents decreases in a linear fashion with increasing time. The slope of the line, or the rate of emptying, depends upon the composition of the meal. These data show that the concept of "emptying time" or the time taken for the last traces of a meal to leave the stomach is meaningless, for the bulk leaves early.

Although the square root of volume gives the best fit of the data, the data are almost as well fitted if the logarithm of the volume remaining in the stomach is plotted against time. In either case, the data show that with liquid meals, the rate of emptying is greatest when the volume is greatest, and that means that emptying is fastest at the beginning of digestion of a meal. Consequently, the greatest bulk of gastric contents is delivered to the duodenum before much gastric digestion or acidification has occurred.

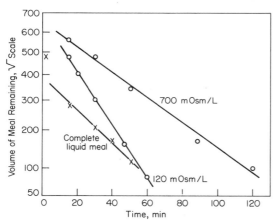

Fig 13–3. — The square root of the volume of gastric contents of a normal man plotted against time, showing that rate of emptying is a linear function of the square root of the volume. Data for two liquid meals are given. Each meal contained 20 gm of citrus pectin per liter and either 35 or 200 gm of sucrose per liter The osmotic pressures are those of the meals determined by freezing-point lowering as the meals were ingested. The complete liquid meal was a mixture of milk product, cream and sugar containing 1.28 cal per ml: fat 40%, protein 15% and carbohydrate 45%. (Adapted from Hunt, J. N., and Spurrell, W. R.: J. Physiol. 113:157, 1951; and Hunt, J. N.: Gastroenterology 45:149, 1963. The square root plot was proposed by Hopkins, A.: J. Physiol. 182:144, 1966.)

The rates of emptying of the three major foodstuffs in liquid form are regulated so that equal numbers of calories are delivered to the duodenum in the same time (Fig 13–4). There are roughly 9 kilocalories in a gram of fat and 4 kilocalories in a gram of carbohydrate or protein. Consequently, an emulsion of 4 gm of fat (or 36 kilocalories) is emptied in the same time as a solution of 9 gm (or 36 kilocalories) of carbohydrate or protein. The digestion products of fat, carbohydrate and protein are the compounds that affect the receptors in the duodenum, which in turn control the rate of emptying. Nevertheless, an emulsion of triglyceride is emptied at the same rate as an isocaloric solution of oleic acid for the reason that the triglyceride is rapidly hydrolyzed in the duodenum by pan-creatic lipase. A gram of starch is emptied at the same rate as 1 gm of glucose, and 1 gm of casein is emptied at the same rate as 1 gm of its constituent amino acids, because hydrolysis of both kinds of macromolecules in the duodenum is extremely fast. Because fat contains the most calories per gram, meals composed largely of fat of any kind — corn oil, cottonseed oil, olive oil, butter, lard or fatty acids — may remain in the stomach 20 hours. A breakfast full of fat, 140 ml of thick cream, two eggs scrambled in butter, two slices of toast and 250 ml of coffee with milk and sugar has been found to remain in the stomach 6 hours; but a leaner meal of two helpings of cold roast pheasant with chutney and mashed potatoes, washed down with beer and coffee, is gone in 4 hours.

A person who chooses an energy-dense diet has an empty stomach sooner than one who takes his food in more dilute form; this may tempt him to eat more. There is some evidence that the choice of an energy-rich diet is associated with obesity.

Heavy exercise up to 71% of maximal oxygen uptake has little or no effect upon the emptying of liquids and none upon intestinal absorption of glucose, sodium, potassium, chloride, bicarbonate or water. There has been no systematic study of the effect of exercise upon the digestion of regular meals. Emperor Frederick II (1215–50), *stupor mundi*, is said to have compared the full stomach of one of his servants, killed after a hard, postprandial ride, with the empty one of another, similarly fed servant, killed after a delightful conversation with the emperor himself, but the experiment is inadequately documented. Severe pain, fear or other stimuli of massive sympathetic discharge delay gastric emptying. Patients with duodenal ulcers empty liquid meals more rapidly than do normal persons.

Water drunk with meals, aside from its effect on volume and osmotic pressure, has no special influence on emptying. The Olympic record for gastric emptying is held jointly by subjects of whole intestinal perfusion; each has had physiological salt solution

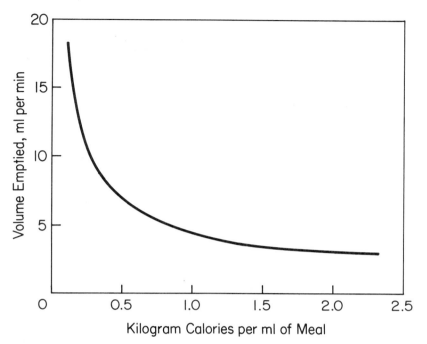

Fig 13–4.—The rate of emptying of a liquid meal in man as a function of its nutritional density. (Adapted from Hunt, J. N., and Stubbs, D. F.: J. Physiol. 245:209, 1975.)

pumped into his stomach through a tube at the rate of 75 ml a minute for well over an hour.

Osmotic Pressure and Gastric Emptying

Gastric contents may have any osmotic pressure, and the osmotic pressure of gastric contents has little or no effect upon gastric function. On the other hand, a major task of the duodenum is to adjust the osmotic pressure of its contents to isotonicity, and to assist in accomplishing this, gastric emptying is regulated by the osmotic pressure of duodenal contents.

A comparison of the rates of emptying of two liquid meals of differing osmotic pressures is given in Figure 13–3. Each meal contained 15 gm of citrus pectin in 750 ml of water. One contained 35 gm of sucrose per liter, giving it a freezing-point depression equal to that of a solution of 120 mOsm. The other contained 200 gm of sucrose per liter; it had a freezing point of − 1.3 C, equivalent

to an osmotic pressure of 700 mOsm. The fluid of higher osmotic pressure emptied more slowly, requiring about twice as long for an equal volume to leave the stomach. Nevertheless, at the beginning, about 350 ml left the stomach within 40 min; emptying was slowed, but not stopped, by high osmotic pressure. Variations in the osmotic pressure of gastric contents caused by sodium chloride also affect emptying. Emptying is fastest at 200 mOsm NaCl, and solutions of lower concentration, including pure water, empty more slowly. Above 200 mOsm NaCl, the rate of emptying is inversely proportional to the salt content.

The signal regulating emptying is probably given by receptors in the duodenal mucosa sensitive to an effective osmotic pressure difference across their walls. Such a difference, with the higher pressure outside, would make them shrink, thereby initiating nerve impulses to be carried along the afferent limb of the enterogastric reflex path. Nonpenetrating solutes — glucose, sorbitol and sodium sulfate — are, on an osmolar basis, approximately equally effective. The rapidly penetrating solute ammonium chloride has no influence upon the rate of gastric emptying. Very high concentrations of the penetrating solutes urea and glycerol slow gastric emptying, but their effect is much less per osmol than that of glucose. Ethyl alcohol solutions up to 1,720 mOsm (about 8%, or between strong beer and light wine) have almost no effect on duodenal osmoreceptors, and hence wine may be expected to flow through the pylorus like water. Strong drinks may delay gastric emptying. In one study of eight men, drinking 120 ml of 100 proof Bourbon whiskey (50% ethanol v/v) just before eating almost doubled the time required for half of a liquid meal to leave the stomach. The fatty canapés often eaten during social drinking slow the emptying of cocktails into the duodenum, but because ethanol is rapidly absorbed through the gastric mucosa, they do little to diminish the effects of the drinks.

The osmotic pressure of gastric contents may be high or low, and adjustment of osmotic pressure is not a necessary prelude to emptying. The osmotic pressure of two glazed doughnuts and a pint of milk is 630 mOsm, whereas that of a meal of steak, bread and butter, tossed salad and tea is 232 mOsm. Gastric contents are diluted in the stomach by the addition of 500 – 1,000 ml of essentially isotonic gastric secretions. Gastric contents that empty early are only slightly diluted, and they have nearly their original osmotic pressure. Thirty minutes after being eaten, a steak meal has been diluted 2 – 6 times, and its remnants after 90 min have been diluted 5 – 14 times. The final osmotic pressure attained by the mixture of food and secretions lies between 160 and 250 mOsm; the reason the mixture is hypotonic is that buffering of each hydrogen ion removes an osmotically active particle from solution.

Gastric Emptying of a Meal

The volume in the stomach and the rate of emptying in the course of digestion of an or-

dinary meal are shown in Figure 13 – 5. After the subject had swallowed tubes that permitted sampling necessary for the determination of volume and rate of emptying, he ate a meal of 90 gm of tenderloin steak, ground cooked and seasoned with 0.1 gm of NaCl and a pinch of polyethylene glycol marked with ^{14}C. The steak was accompanied by 25 gm of white bread covered with 8 gm of butter, and it was followed by 60 gm of vanilla ice cream topped with 35 gm of chocolate syrup. The meal was washed down with 240 ml of water. The volume of an identical meal after homogenization was 400 ml; its pH was 6.0 and its osmotic pressure was 540 mOsm.

The subject responded to the meal by secreting 800 ml of gastric juice, and consequently the volume in the stomach remained high for almost 2 hours despite early rapid emptying.

Duodenal Acidity and Gastric Emptying

The acidity of gastric and intestinal contents is measured by passing a tube to which one or more miniature glass electrodes are attached, and the position of the electrodes is determined fluoroscopically. Immediately after eating, when antral contents are almost neutral, duodenal contents are likewise neutral, and fluctuations in duodenal pH are small and irregular. The pH of gastric contents falls progressively, so that 30 min after a meat meal the pH of chyme within the antrum lies between 2.5 and 3.4. An hour later the pH of antral contents is slightly below 2.0, and it is steady. At this time the pH recorded immediately beyond the pyloric sphincter fluctuates widely (Fig 13 – 6) as chyme is delivered to the duodenal bulb in spurts. At the base of the duodenal bulb the prevailing acid condition is interrupted by neutrality, whereas farther along the prevailing neutral condition is interrupted by acidity. In normal subjects, the average pH of contents of the middle of the second part of the duodenum lies between 5.4 and 7.8, and the pH is below 3.0 less than 1% of this time.

The acid load delivered to the duodenum

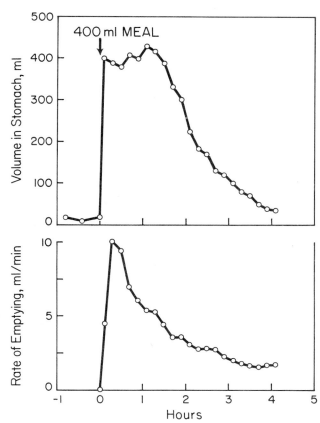

Fig 13–5.—Volume of gastric contents and rate of gastric emptying in a subject eating a 400 ml meal of steak, bread and butter and vanilla ice cream. (Adapted from Malagelada, J-R., Longstreth, G. F., Summerskill, W. H. J., and Go, V. L. W.: Gastroenterology 70:203, 1976.)

Fig 13–6.—The pH in the gastric antrum and at various sites in the duodenum recorded by means of a glass electrode. The records were obtained relatively late in the digestion and emptying of a meal when the contents of the antrum were acid. Fluctuations of pH at the pylorus and beyond occur at multiples of 3 per min, which is the frequency of terminal antral contraction. (From Rhodes, J., and Prestich, C. J.: Gut 7:509, 1966.)

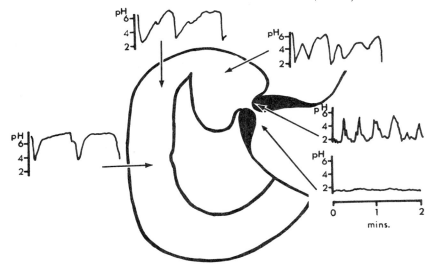

is the product of the acid secreted and the amount emptied. Patients with duodenal ulcer secrete more acid and empty their stomachs more rapidly than do normal persons, and the acid load in the duodenum is greater. However, in duodenal ulcer patients and normal subjects, most of the chyme is delivered to the duodenum at pH 3.0 or greater.

It is not the pH of duodenal contents that determines the rate of gastric emptying, but the neutralizing ability of the duodenum. Acid entering the duodenum both slows gastric emptying and evokes duodenal, pancreatic and biliary secretion to neutralize itself. A steady state is reached in which the rate of emptying of acid from the stomach is matched by the rate at which acid is neutralized in the duodenum.

Gastric Emptying of Solids

Large lumps of meat remain in the stomach as long as 9 hours, until reduced in volume, whereas an equal amount of ground beef leaves quickly.

The size of the lumps prevents them from passing through the narrow pyloric canal. Lumps are pushed toward the pylorus by each peristaltic wave in the antrum, but, denied entrance, they are thrown backward by each retropulsive movement. Eventually, digestion and triturition make them small enough to pass into the duodenum. Whereas pyloroplasty or distal antrectomy has little effect upon the emptying of liquids, those operations remove the barrier to lumps, which then leave the stomach quickly. Presence of solids in the stomach has no effect upon the concurrent emptying of liquids.

Temperature of the Stomach

The empty stomach is at the core temperature of the body. The temperature of swallowed food covers a range of about 80 C. Most persons like their coffee between 50 and 70 C, temperatures above the maximal tolerable temperature of hot water applied to the skin. At the other extreme is a mint julep whose temperature is −7.1 C.

Temperature changes occurring in the stomach and the rate of return to core temperature depend upon the heat load, positive or negative, imposed. A 250-ml liquid meal taken at 57 C raises the temperature of the stomach to 44 C within 2 min, and the gastric temperature returns to core temperature in 18 min. Eating a dish of Philadelphia peach ice cream reduces gastric temperature to 22 C and often inverts the T wave of the ECG by reducing the rate of repolarization of the epicardial surface.

Gastric Absorption of Water-Soluble Substances

In respect to water-soluble substances, the gastric mucosa behaves like a sieve with very small pores. Water molecules can move across the mucosa in both directions, but net movement, aside from secretion, is negligible. Absorption, the difference between flux from lumen to blood and flux from blood to lumen, has been measured by placing a known volume of water, 500–750 ml, in the stomachs of normal men. The volume of secretions added in the experimental period was estimated from the chloride and acid content of the fluid recovered by tube at the end of an hour. A small amount of phenol red, which is not absorbed, was placed in the water. The volume of water lost by passage through the pylorus was calculated as the difference between the amount of phenol red placed in the stomach and the amount recovered. Gastric emptying was delayed by placing olive oil in the duodenum. In eight normal young men, the mean rate of water absorption was only 1.5% of the gastric contents per hour. In this experiment, there was an osmotic gradient of about 300 mOsm in the direction tending to move water from lumen to blood. When a similar gradient in the opposite direction is established by placing hypertonic solutions on the mucosa, only a small net flow of water is found. Water does not move readily through the gastric mucosa along osmotic pressure gradients, and this accounts for the failure of hypo- or hyperosmotic gastric contents to come to isotonicity before a major part of gastric emptying occurs.

Only minute quantities of water-soluble substances—sodium, potassium, glucose, amino acids—are absorbed through the normal gastric mucosa. A trace of radioactive iron was found to be absorbed from a total gastric pouch of a dog. The absorption of small amounts of other substances, such as urea and gases, has been demonstrated; but, in general, gastric absorption of most water-soluble substances, except ethanol, is unimportant.

Fat Solubility and Gastric Absorption

Two well-established principles of general physiology are that membranes of cells behave as though composed of both protein and fat and that fat-soluble substances diffuse rapidly through a membrane that water-soluble substances penetrate slowly, if at all.

These principles apply to the stomach.

A substance of low molecular weight such as ethanol, which is both water and fat soluble, is rapidly absorbed from gastric contents. Its rate of absorption is directly proportional to its concentration. If 350 ml of a 5.6% (w/v) solution of ethanol is placed in the stomach of a normal human subject, 5.4 gm of ethanol is absorbed through the gastric mucosa in 30 min. The rate of absorption of ethanol is entirely independent of the concentration of acid or of drugs (such as aspirin) simultaneously present in gastric contents.

Organic acids and bases may be water soluble at one pH and fat soluble at another. In general, organic acids are ionized and water soluble in a solution whose pH is above the acid dissociation constant (pK_a) of the compound and un-ionized and fat soluble in a solution whose pH is below the pK_a. An example is acetylsalicylic acid (aspirin), whose structure is shown in Figure 13–7. The pK_a of the acid group is 3.5. At neutral pH, all of the compound's molecules are in the ionized, water-soluble form; at pH below 2.0, more than 95% are in the un-ionized fat-soluble

Fig 13–7.—Absorption of aspirin. Aspirin is acetylsalicyclic acid and is fat soluble. If it is ingested as acetylsalicylate, the ionized water-

soluble form, and if it is inadequately buffered, acid secreted by the stomach converts it to acetylsalicylic acid.

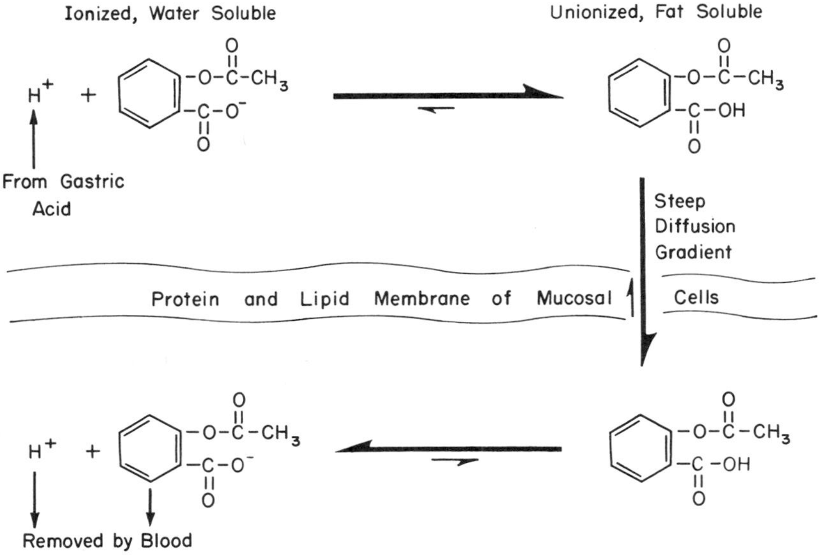

Ionized, Water Soluble   Unionized, Fat Soluble

From Gastric Acid

Steep Diffusion Gradient

Protein and Lipid Membrane of Mucosal Cells

Removed by Blood

form. When acetylsalicylic acid is present in acid gastric contents, its fat-soluble form diffuses across the mucosa into the cells. Because the cell interior is nearly neutral, the compound ionizes as soon as it crosses the membrane. The disappearance of the fat-soluble molecules from the intracellular side of the membrane prevents back-diffusion and maintains a steep gradient from lumen to cell. Eventually the compound escapes from the mucosal cells and is carried away by the blood. In man, acetylsalicylic acid is absorbed from the stomach at the rate of 50% of the amount in gastric contents in 30 min when the pH is 1.0. This is the same rate at which deuterium oxide (D_2O) is absorbed, and faster than that of ethyl alcohol. It is absorbed at about one-tenth this rate from neutral solutions.

Acetylsalicylic acid, once it has entered the surface epithelial cells, damages them. The cells are rapidly desquamated, and the gastric mucosa becomes much more permeable. For example, when the mucosa of the oxyntic gland area is bathed with 20 mM acetylsalicylic acid in 100 mN HCl, a concentration equal to two 5-grain aspirin tablets in 180 ml of gastric juice, the mucosa becomes 5–10 times more permeable to hydrogen and sodium ions. When acid diffuses rapidly into the salicylate-damaged mucosa, frank bleeding often occurs; and hemorrhages, from minor to massive, may follow ingestion of aspirin. The tendency of the salicylate-damaged gastric mucosa to bleed is greatly exacerbated by the simultaneous presence of 4–9% (w/v) ethanol in gastric contents. In contrast, acetylsalicylic acid neutralized and firmly buffered at pH 6.5 is innocuous. Organic bases are uncharged and fat soluble in a solution whose pH is on the alkaline side of their pK_a, but charged and water soluble in a solution whose pH is on the acid side of their pK_a. Therefore, they are absorbed in the stomach more readily from neutral than from acid solutions. Given the same pK_a, absorbability is a function of the solubility of the compound in oil. Barbital, whose pK_a is 7.8, is absorbed very slowly from an acid solution in the stomach, but thiopental, whose pK_a is 7.6 and which is 100 times more soluble than barbital in chloroform, is absorbed 10 times more rapidly. Caffeine is absorbed in the stomach from neutral but not acid solution, and the amount absorbed is directly proportional to the time caffeine solution remains in the stomach. Presence of caffeine has no effect upon the rate of emptying of liquids.

Effects of Gastrectomy

Following partial gastrectomy, gastric digestion is never complete, and gastric absorption is minor; the major parts of digestion and absorption are left to the small intestine. Nevertheless, normal gastric function is necessary for normal nutrition.

In persons with partial gastrectomy, the usual consequence is weight loss, resulting not only from decreased food intake, although appetite may remain good, but from defective intestinal digestion and absorption. The absorption of fat measured in balance studies may be as low as 30% of the amount ingested, whereas normal absorption is over 90%. Both pancreatic secretion, measured by the concentration of pancreatic enzymes in intestinal contents, and biliary secretion, measured by bile concentrations, are reduced in partially gastrectomized subjects. The reason for this probably is that diminished gastric digestion and acid secretion provide subnormal stimulation of pancreatic secretion by way of pancreozymin and secretin. Reduced pancreatic secretion, poor mixing of intestinal contents with enzymes and increased bacterial decomposition of protein in the lower small intestine may account for poor assimilation of protein; digestion of carbohydrate is only slightly affected. Microcytic hypochromic anemia occurring after partial gastrectomy is attributed to faulty iron absorption, a consequence of reduced acid secretion, for acid promotes iron absorption.

The stomach is essential for life, not because of its digestive function but because it secretes the gastric intrinsic factor required for vitamin B_{12} absorption. Complete gastrectomy removes the source of this factor. The normal human liver contains 75,000–225,000 μg of vitamin B_{12}, and at the rate of utilization of 1 μg per day, this is enough to last 2–6 years. This estimate is reasonably accurate, because 4 patients not receiving vitamin B_{12} parenterally were found to develop megaloblastic anemia 4–7 years after total gastrectomy.

REFERENCES

Cooke, A. R.: Control of gastric emptying and motility, Gastroenterology 68:804, 1975.

Dozois, R. G., Kelly, K. A., and Code, C. F.: Effect of distal antrectomy on gastric emptying of liquids and solids, Gastroenterology 61:675, 1971.

Hunt, J. N., and Knox, M. T.: Regulation of gastric emptying, in Code, C. F. (ed.): *Handbook of Physiology:* Sec. 6. *Alimentary Canal,* Vol. IV (Washington, D.C.: American Physiological Society, 1968), pp. 1917–1936.

Hunt, J. N., and Stubbs, D. F.: The volume and energy content of meals as determinants of gastric emptying, J. Physiol. 245:209, 1975.

Rhodes, J., and Prestwich, C. J.: Acidity at different sites in the proximal duodenum of normal subjects and patients with duodenal ulcer, Gut 7:509, 1966.

Schanker, L. S.: On the mechanism of absorption of drugs from the gastrointestinal tract, J. M. Pharm. Chem. 2:343, 1960.

14

Intestinal Absorption of Water and Electrolytes

THE SMALL INTESTINE is the indispensable site of absorption, and its surface area is very large. The macroscopic mucosal area of six human small intestines obtained at autopsy and handled so as to minimize postmortem distortion was found to be $1.9-2.7$ m². Mucosal folds form transverse ridges projecting into the lumen; and because these are less prominent in the distal part of the intestine, the mucosal area per unit length falls off sharply from proximal to distal intestine. Almost half the total mucosal surface occurs in the first quarter of the small intestine. Presence of villi multiply the total absorptive area about 8 times, and microvilli on the absorptive cells increase the area a further $14-24$ times, so that the total absorptive area of the small intestine is from $200-500$ m².

In the small intestine, the columnar epithelial cells cover the villi. One or more central arterioles carry arterial blood to the tip of a villus, and near the tip the arterioles break into a rich capillary network that descends the villus just beneath the bases of the epithelial cells. Thus, blood shoots up the villus as in a fountain and then falls back around the central stream. This arrangement permits countercurrent exchange of absorbed sub-

stances between the descending network and the ascending vessels. The collecting veins drain into the portal vein; all blood and substances absorbed into the portal vein normally pass through the liver before reaching the systemic circulation. Blood flow through the capillaries of the small intestine is about 1 liter per min in the normal man at rest. Flow decreases following sympathetic stimulation, as in exercise; and it increases in the hyperemia that accompanies digestion. Intestinal blood flow is sufficiently fast so that it does not limit the rate of absorption. Every villus also contains a central lacteal emptying into the lymphatic channels, which drain eventually into the left thoracic duct. Total left thoracic duct flow in man is between 100 and 200 ml per hour. The volume of lymph coming from the intestine is not known, but it cannot be more than a few milliliters per minute. Lymph flow increases during absorption, as the result of the pumping action of intestinal contraction and the increase in volume of intestinal interstitial fluid.

There is a gradient of function along the small intestine. Permeability is high in the duodenum and diminishes toward the ileum. Consequently, osmotic equilibrium is quickly established between the contents of the duodenum and the blood. Active absorption of sodium and secretion of bicarbonate increase toward the ileum.

Only a few of the absorptive processes are regulated. The intestine indiscriminately absorbs water, the major electrolytes and the products of digestion of foodstuffs, but absorption of calcium and iron is adjusted to the body's needs. The absorbing capacity of the small intestine is normally far in excess of need, for 50% or more of the intestine can be removed without deleterious effect. Yet in conditions preventing adequate digestion and in some diseases of intestinal mucosa or muscle, the absorption of water, fat or protein may be so deficient as to endanger life.

Water Absorption

Between 5 and 10 liters of water, derived from food and drink and from salivary, gastric, pancreatic, biliary and intestinal secretions, enters the small intestine of the normal adult man in a day. Only about a liter leaves the intestine for the colon; therefore the small intestine absorbs water at an average rate of 200 – 400 ml per hour. This is the minimum rate. The maximum rate, if any upper limit exists, is imposed not by any property of the intestinal mucosa but by the body's ability to bear a water load. The steady state of maximal excretion of water in the urine when water is drunk at quarter-hour intervals is about 1% of the body weight per hour. For a 70-kg man, this equals absorption at the rate of 700 ml per hour. Absorption of a single drink of 1 liter of water occurs within an hour; and because the maximal dilution of plasma during absorption occurs 30 – 60 min after the drink is taken, the peak rate of absorption must be greater than 1 liter per hour. In the cat, rat and rabbit, water can be absorbed so fast that intravascular hemolysis occurs; as much as 3 gm per 100 ml of hemoglobin has been found in portal venous plasma during water absorption in these animals. A dehydrated camel can drink water equal to 30 – 33% of its body weight in 10 min.

In the experimental process of whole intestinal perfusion, a physiological salt solution is pumped through a tube into the stomach of a human subject at the rate of 75 ml a minute. After an hour a steady state is reached in which 200 – 300 ml is passed through the anus at 3- to 4-min intervals. All food particles, fecal masses and 99% of colonic bacteria are flushed out. Water is absorbed from the perfusate at the rate of 600 ml per hour.

Water is absorbed throughout the small intestine, but the chief locus of absorption following a meal is the upper part of the tract. When the unabsorbable substance polyethylene glycol is added to a meal, its concentration at various parts of the digestive tract shows how much the meal has been diluted or concentrated in its passage through the gut. After a meal of steak, bread and butter, salad and tea whose initial volume is 645 ml and which contains a known amount of polyethylene glycol has been eaten, the concentration of polyethylene glycol in samples recovered from the upper duodenum is 20 – 40% of its initial value. The meal, as it enters the duodenum, has been diluted by several times its volume of gastric secretions. The volume passing the mid-duodenum is 1,500 ml; that in the proximal jejunum is 750 ml; and that in the lower small bowel is 250 ml. Since biliary and pancreatic secretions of about 1,000 ml have been added in the duodenum, their volume together with water contributed by the stomach has been absorbed. The net movement of water in the duodenum is in the direction that tends to keep its contents isotonic, and this large absorption of water is the result of absorption of osmotically active components of the meal, chiefly glucose and amino acids, and the neutralization of acid, with consequent reduction of osmotic pressure. Absorption of 55 gm of glucose, or 306 mOsm, would allow 1,000 ml of water to be absorbed. Addition of 1 part of isotonic pancreatic juice to 1 part of isotonic HCl gives, roughly, 1 part of isotonic NaCl and 1 part of free water, which must be absorbed if the duodenal contents are to remain isotonic. By the time the intestinal contents have reached the ileum, the concentration of polyethylene glycol is greater than that in the original meal, showing that all the water added by secretion, as well as some ingested, has been absorbed.

Water Fluxes Across the Intestine

The net flow of water across the intestine is equal to the difference between the unidirectional flux from intestinal lumen to interstitial fluid (and eventually into blood and lymph) and the opposite flux from interstitial fluid to intestinal lumen. When 40 ml of isotonic NaCl solution containing a trace of D_2O was placed in a 16–18 cm length of dog duodenum, the isotope was found to move from lumen to blood at a rate directly proportional to its concentration in the lumen contents. From the measured rate of D_2O movement, the flux of water from lumen to blood was calculated to be 2 ml per min; and because there was no net volume change, the flux from blood to lumen must also have been 2 ml per min. Similar measurements made using 21- to 25-cm lengths of ileum filled with 40 ml of isotonic NaCl showed that the flux from lumen to blood was 2.2 ml per min, and the flux from blood to lumen

was 1.8 ml per min. The net rate of absorption, 0.4 ml per min, was one-fifth the rates of the one-way movements of water.

Fluxes across the human intestine are correspondingly rapid. The results obtained when 50 gm of D_2O was placed in the upper intestine of normal adult men are shown in Figure 14–1, where they can be compared with similar results showing water flux in the stomach. The D_2O was given alone, or it was made isotonic at 154 mN NaCl or hypertonic at 250–1,000 mN NaCl. Fluxes from hypotonic or isotonic solutions were the same; those from hypertonic solutions were slightly slower. Fifty per cent of the D_2O reached the blood in only 2–3 min, and 95% in 3–11 min. These fluxes are much faster than from the stomach, one reason being the larger mucosal area to which water in the intestine is exposed. When barium sulfate was mixed with the water, the x-ray shadow of barium was distributed throughout most of the small intestine within 2–3 min after the fluid was given. An intravenous dose of the parasympatholytic drug methantheline bromide (0.3 mg per kg of body weight) caused the water to remain in a shorter segment of the bowel, and flux from lumen to blood was reduced about half.

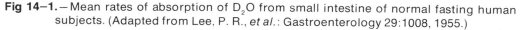

Fig 14–1.—Mean rates of absorption of D_2O from small intestine of normal fasting human subjects. (Adapted from Lee, P. R., *et al.*: Gastroenterology 29:1008, 1955.)

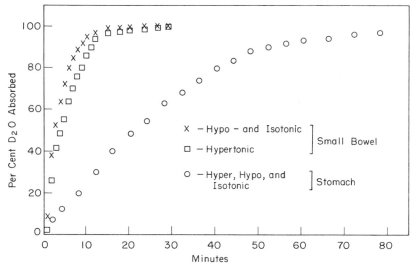

When 1,000 ml of water is drunk, about 50% is absorbed in 30 min. Since 50% of isotopic water enters the blood in 2–3 min, flux from lumen to blood is about 10 times as fast as net absorption. This means that during absorption there must be a flux of water from blood to lumen only slightly slower than that in the opposite direction. Net absorption is therefore the difference between two very large, oppositely oriented fluxes; a small change in either individual flux can cause a large change in net flow.

Intestinal epithelial cells are bound together by tight junctions near their apical borders; and to be absorbed, water must pass through the cells' apical membrane and through the tight junctions. The membrane behaves as though it were a lipoprotein layer traversed by small water-filled pores through which water moves along an osmotic gradient. If the pores are more than figments of mathematical deduction, they occupy less than 0.1% of the mucosal surface, and they will not admit the passage of water-soluble molecules whose molecular weight is greater than 200. Absorbed water mixes with that contained in the cells. It probably leaves by way of the lateral borders of the cells, flowing down the osmotic gradient established by active pumping of small molecules – sodium, glucose and the like. This process is similar to that by which the gallbladder's mucosa absorbs water.

When water containing solutes flows through aqueous channels in a membrane, frictional forces act on the solute; and the solute, especially if it is uncharged, tends to move with the water. This phenomenon is called solvent drag, and it accounts for a fraction of the absorption of urea and other small nonelectrolytes.

Most or all water, and solutes as well, are absorbed by the cells at the tips of the villi.

Intestinal Lymph Flow

Lymph flow from the intestine increases as water is absorbed. The greatest flow occurs during absorption of isotonic NaCl. The percentage of absorbed water transported through lymph channels increases with increasing venous pressure; when venous pressure is below 5 mm Hg, lymph accounts for 10–25% of water transport; but when venous pressure is greater than 20 mm Hg, the rate of lymph flow equals the rate of water absorption. Absorption through lymph channels is assisted by rhythmic contraction of mesenteric lymph ducts and by contraction of the villi.

The lymph flowing in the large ducts is not the actual water and electrolytes absorbed from the intestinal lumen. The reason for this is that water and solute molecules rapidly exchange among capillary blood, interstitial fluid and lymph. If water being absorbed is labeled with tritium, 97% of the tritium is recovered in portal blood and only 3% in lymph. Only fat droplets, once in lymph capillaries, cannot escape.

Osmotic Pressure and Water Movement

Absorption of water is affected by the osmotic pressure and electrolyte composition of the fluid in the lumen. The net flow of water across a segment of dog ileum, together with simultaneously measured unidirectional fluxes, is shown in Figure 14–2. These rates are plotted against the mean concentration of NaCl present in the lumen. When NaCl was 66 mN, less than half isotonic, flux from blood to lumen was high, at 10 ml per 10 min; but flux from lumen to blood was higher, at more than 20 ml per 10 min, so that net absorption was fast. With 150 mN NaCl, the fluid was nearly isotonic; flux from blood to lumen remained the same, but flux from lumen to blood was reduced to a value only slightly higher than the opposing flux. Net absorption was slow. Hypertonic fluid, about 2.5 times isotonic, caused only a small rise in flux from blood to lumen; but it reduced the flux from lumen to blood, with the result that net flow reversed and the volume within the lumen enlarged.

Although water may be absorbed against an osmotic pressure gradient, there is an

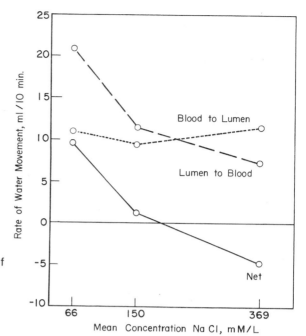

Fig 14–2.—Unidirectional fluxes and net movement of water across mucosa of dog ileum as a function of the concentration of NaCl in the fluid contained in the ileum. (Adapted from Visscher, M. B., et al.: Am. J. Physiol. 142:550, 1944.)

upper limit at which net absorption stops. The line connecting points representing net flow in Figure 14–2 crosses the zero line at about 220 mN NaCl. This concentration is about 130 mOsm greater than isotonicity; at this concentration, net flow would not occur until absorption of NaCl had reduced the osmotic pressure of the lumen's contents. In another series of observations on the water movement across the jejunoileal region of the dog intestine, the osmotic pressure at which the fluxes became equal was found to be 400 mOsm whereas, that of the dog's plasma was 295 mOsm. A similar relation between water movements across the intestine and the total osmotic pressure is obtained if mannitol is used instead of NaCl; therefore, water flow depends on the osmotic pressure and not on the nature of the osmotically active substance.

Solutions in the duodenum and upper jejunum rapidly come into osmotic equilibrium with blood, for the mucosa of the upper small intestine is highly permeable to water and solutes. Chyme, which may have any osmo-lality when it enters the duodenum, is brought quickly to isotonicity and remains isotonic as it moves down the rest of the small intestine. Osmotic equilibrium is attained by absorption of solutes and by net movement of water from blood to lumen. Gastric contents emptied into the duodenum are often hypertonic, and the hydrolytic action of pancreatic enzymes upon protein and starch rapidly produces a large number of small molecules from a few large ones. If the rate of absorption is slower than the rate at which osmotic equilibrium is reached, the volume of fluid in the intestinal lumen increases.

If the solute responsible for the osmotic pressure of intestinal contents cannot be absorbed, sufficient water to dilute it to isotonicity must remain in the lumen. The ions Mg^{2+} and $SO_4^=$ are slowly absorbed, and their presence in the lumen prevents water absorption. The usual dose of 15 gm of Epsom salts ($MgSO_4 \cdot 7H_2O$) retains between 300 and 400 ml of water; this is the basis of the action of this saline cathartic.

The Unstirred Layer

Bulk contents of the lumen of the intestine are mixed by segmentation and peristalsis, and water and solutes are brought to the surface of the mucosa by convection. However, the fluid trapped between the microvilli and within the glycocalyx is stationary, and a series of thin layers, each progressively more stirred, extends from the surface of the epithelial cells to the bulk phase in the lumen. Together, these are the *unstirred layer,* whose effective thickness ranges from 0.01 – 1 mm.

Because there is no convection in the unstirred layer, molecules move within it only by diffusion. The rate at which a substance diffuses is a function of the concentration gradient down which it diffuses and of its diffusion coefficient. The coefficient is, among other things, roughly inversely proportional to the square root of the molecular weight below 450 and to the cube root above that weight. Therefore, during absorption two processes are in series: diffusion from the bulk phase through the unstirred layer to the cell membrane and translation through the cell membrane to the cell's interior. The latter process may be passive or active. If the substance being absorbed is fat soluble or if its diffusion through the membrane is facilitated, its translation into the cell may be so rapid that the substance's concentration at the bottom of the unstirred layer is nearly zero. Then diffusion through the unstirred layer is the rate-limiting step in absorption.

The glycocalyx on the microvilli is a sulfated mucoprotein, and it is negatively charged. The counter ions are cations dissolved in the unstirred layer. If a substantial portion of these cations are hydrogen ions, the acidity of the microclimate within the brush border is greater than the acidity of the bulk phase in the lumen. This may influence the absorption of amphoteric compounds whose fat solubility is increased or decreased in acid as compared with neutral solution.

Absorption of Sodium

Net absorption of sodium results from the difference between two opposing unidirectional fluxes. Rates of movement of sodium from the lumen to blood across segments of the intestine of dogs have been measured by placing isotonic NaCl containing a trace of radioactive sodium (^{22}Na or ^{24}Na) in loops 15 – 20 cm long. The data for jejunum, ileum and colon are shown in Figure 14 – 3 (left). Sodium marked with an asterisk (Na*) is the labeled sodium concentration, which at zero time is the total sodium concentration. In 30 min, there was no change in the total NaCl concentration in the lumen, but the labeled sodium disappeared rapidly; at the end of the period, only 10% of the sodium originally present in the loop was still there. Since the total concentration of sodium was constant, a quantity of sodium equal to that leaving the lumen, minus the net amount of sodium absorbed, must have entered it from the blood. The exchange of sodium was slower in the ileum and still slower in the colon. Movement in the opposite direction was measured by placing unlabeled sodium in the lumen, injecting a trace of ^{24}Na$^+$ into the blood and following the appearance of the isotope in the contents of the lumen. The results given in Figure 14 – 3 (right) show that, although there was a slight fall in total sodium concentration, which indicates net absorption of sodium, there was rapid movement of labeled sodium from blood to lumen.

The relation between the fluxes of sodium across the jejunoileal region of the dog intestine and the concentration of sodium in the lumen is shown in Figure 14 – 4. Flux from blood to lumen is substantially independent of lumen concentration, but flux from lumen to blood increases along with lumen concentration. Net transfer is from lumen to blood at all concentrations down to 40 mN. Consequently a sodium gradient can be established during absorption.

When the jejunum of human subjects is perfused with an isotonic solution of NaCl

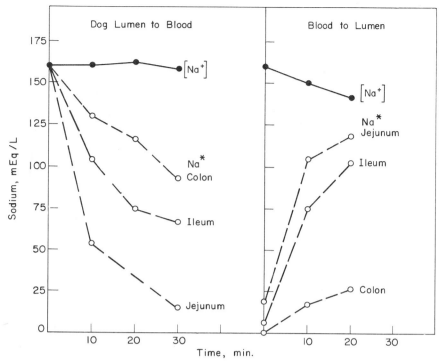

Fig 14–3.—Movement of sodium across intestinal mucosa of unanesthetized dogs with surgically prepared, chronic loops of jejunum, ileum and colon. **Left,** intestinal loops were filled with isotonic NaCl containing a trace amount of ^{24}Na. The total sodium concentration [Na$^+$] remained constant, but the labeled sodium Na* (see text) originally present in the loop rapidly disappeared. **Right,** intestinal loops were filled with approximately isotonic Na$_2$SO$_4$ solutions. A trace amount of ^{24}Na was injected intravenously and its appearance in the contents of the lumen observed. Total sodium concentration fell, but the sodium in the lumen of the loop was rapidly replaced by labeled sodium from the blood. Movement of labeled sodium in both directions is in the following order: jejunum > ileum > colon. (Adapted from Visscher, M. B., *et al.*: Am. J. Physiol. 141:488, 1944.)

(containing no bicarbonate) to which a trace of radioactive sodium (^{22}Na$^+$) has been added, net absorption is found to be 0.29 ± 0.24 mEq/cm per hour. The flux from lumen to blood is 1.04, and that from blood to lumen is 0.75 in the same units. Net absorption in the ileum is 0.16 ± 0.18 mEq/cm per hour with flux from lumen to blood being 0.74 and from blood to lumen 0.58. Rates of absorption are equal in the two parts if the data are calculated on the basis of the estimated surface area per centimeter length of gut.

When 11 normal human subjects were subjected to whole intestinal perfusion with physiological salt solution containing 140 mN sodium, flux from lumen to plasma was found to average 13 mEq per min. Flux from plasma to lumen averaged 11.3 mEq per min, and net absorption was 1.7 mEq per min. In this experiment the volume of fluid in the intestine was 1.8 liters, and mean transit time was 25 min.

Absorption of sodium may occur by passive diffusion when the electrochemical gradient is in the direction from lumen to blood. Absorption also occurs against an electrochemical gradient. In the human jejunum perfused with an isotonic solution of NaCl containing no bicarbonate, sodium is absorbed against a small concentration gradient of 13 mEq per liter and against zero electric gradient. Sodium absorption in the jejunum

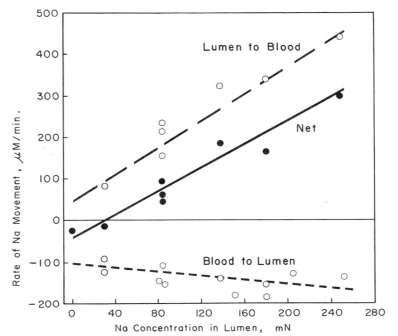

Fig 14–4.—Fluxes of sodium across the intestinal mucosa of dogs provided with chronic loops of the jejunoileal region. Solutions containing NaCl labeled with radioactive sodium were perfused with a single transit through the loops, and fluxes were calculated from the differences in volume and concentration between fluid entering and leaving the loops. The slope of the blood-to-lumen line is not statistically different from zero. (Adapted from Vaughan, B. E.: Am. J. Physiol. 198:1235, 1960.)

is stimulated by the addition of glucose or galactose, both of which are actively absorbed, and by fructose, which is passively absorbed by facilitated diffusion. Absorption of these sugars causes bulk flow of water from lumen to blood, and solvent drag accounts for much of the sodium absorption.

Sodium also participates in active absorption of glucose and galactose, for sodium and the sugars cross the apical membrane of the epithelial cells bound to a common carrier (see Chapter 15). Transport of sodium with a sugar from lumen to blood results in a charge separation; and a potential difference, lumen negative with respect to blood, is established across the mucosa. This potential difference itself opposes active absorption of sodium, and it causes passive movement of potassium down its electrochemical gradient from blood to lumen.

In contrast, active absorption of sodium in the human ileum can occur against a concentration gradient of 110 mEq per liter and an electric gradient of 5–15 mV. The relation between absorption of sodium in the human ileum and the luminal concentration is almost identical with that in the dog's illustrated in Figure 14–4. Sodium absorption in the human ileum is not stimulated by the addition of glucose or galactose to the luminal fluid, nor is it increased by bulk absorption of water from hypotonic luminal contents. Solvent drag probably plays little part in ileal absorption of sodium, and most sodium absorption is accomplished by an active sodium pump. In the ileum of men and dogs, solute absorption can be faster than water absorption, for when the ileum is perfused at a constant rate with isotonic NaCl, the fluid recovered from the lumen has a lower osmolality than that instilled. Under normal circumstances, ileal contents remain in the lumen long enough for osmotic equilibration to occur.

The minimal concentration of sodium reached in the ileal contents of severely salt-depleted human subjects is between 75 and 95 mEq per liter, and the quantity of sodium delivered per day by the ileum to the colon far exceeds sodium intake. During salt depletion, mineralocorticoid secretion by the adrenal cortex increases, and urinary excretion of sodium falls. However, the small intestine exhibits little or no adaptation to sodium load. The variations in sodium absorption brought about by changes in mineralocorticoid levels, if they exist, are very small. In contrast, sodium absorption by the colon definitely increases when aldosterone rises as the result of salt depletion or administration of the hormone.

Countercurrent Flow and Sodium Absorption

Arterial blood is carried up a villus in a central vessel, and venous blood descends in a capillary network surrounding the central vessel (see Fig 4–10). The mean transit time during resting conditions is 4–6 sec. This arrangement permits countercurrent multiplication of diffusible substances absorbed by the cells at the tip of the villus, and the rate of blood flow through the countercurrent system controls the magnitude of multiplication. Blood flow through the intestine of an anesthetized cat averages 29 ml per min · 100 gm of tissue. When the intestine is perfused through the lumen with isotonic NaCl, the concentration of sodium at the tip of the villus is 3–4 times the concentration at the base (Fig 14–5). Addition of glucose, which stimulates sodium absorption, steepens the sodium gradient. When blood flow is increased to 205 ml per min · gm of tissue by infusion of a vasodilator drug, the gradient is washed out. The gradient also disappears during ischemia.

There is no certain method of calculating the partition of sodium at the tip of the villus between extracellular and intracellular water, but the extracellular concentration dur-

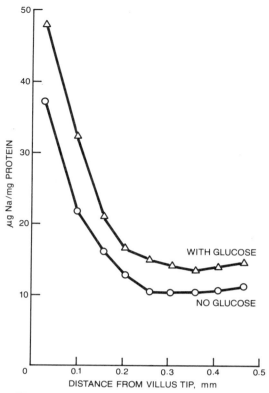

Fig 14–5.—The concentration of sodium along the intestinal villus of an anesthetized cat during perfusion of the lumen with isotonic NaCl solution with and without glucose. (Adapted from Haljamae, H., Jodal, M., and Lundgren, O.: Acta Physiol. Scand. 89:580, 1973.)

ing maximal countercurrent multiplication has been estimated to lie between 300 and 900 mN. Such a concentration is grossly hypertonic, and the osmotic pressure gradient established by countercurrent multiplication facilitates absorption of water.

Absorption of Potassium

As with sodium, the net movement of potassium across the intestinal mucosa is the difference between two opposed unidirectional fluxes. Potassium fluxes are very much smaller than sodium fluxes. Like the flux of sodium, that of potassium from blood to lumen is essentially independent of luminal concentration, and that from lumen to blood increases linearly with increasing luminal concentration. Net movement in jejunum and ileum occurs only down the ion's electrochemical gradient. With a plasma concentration of 4.8 mEq per liter and a potential difference of 5 mV, mucosa negative, the equilibrium concentration is 6 mEq per liter, and absorption occurs when luminal concentration is 1–2 mEq per liter higher than plasma concentration. Potassium absorption is also effected by solvent drag during active absorption of sodium and sugars. There is little difference in behavior of potassium in the jejunum and ileum, and no difference has been identified in man. In the colon, however, potassium is secreted, and the luminal concentration of potassium must be above 25 mEq per liter for net absorption to occur. This accounts for the tendency for potassium deficiency to develop in diarrhea.

Absorption of Chloride and Bicarbonate

The law of electric neutrality of solutions demands that each solution contain an equal number of cations and anions; thus, absorption of sodium cannot occur without an equivalent transport of charge. When anion transport falls behind sodium transport, a small potential difference results from charge separation, and cation flow in the opposite direction along the potential gradient satisfies the requirement for equivalent transfer of charge. A major part of chloride absorption may be merely the consequence of sodium absorption, for chloride absorption closely parallels sodium absorption. In its passive absorption of anions, the intestine does not distinguish among chloride, bromide, iodide, thiocyanate and nitrate; all are absorbed at approximately equal rates.

Nevertheless, the intestinal mucosa of some species, notably the dog and man, can absorb chloride against an electrochemical gradient. The ileum, and to a lesser degree the jejunum, can absorb chloride more rapidly than sodium and can establish a large concentration gradient. If an isotonic mixture of NaCl and Na_2SO_4 is placed in the dog's ileum, the presence of the slowly absorbed sulfate ion keeps the sodium concentration approximately constant, but in 90 min the chloride concentration falls to less than 1 mEq per liter. This concentration is far lower than can be accounted for by any potential difference across the mucosa. Active absorption of chloride in the human and canine ileum is probably accomplished by exchange for bicarbonate secreted into the lumen by the mucosal cells, and it is stimulated by cyclic AMP.

Bicarbonate is rapidly absorbed in the human jejunum if its concentration is above 6 mN. Jejunal contents recovered by tube usually have a bicarbonate concentration of 6–20 mM (mean 8 mM) and a pH of 6.2–6.8 (mean 6.5). Absorption of bicarbonate is apparently effected by secretion of acid by the jejunal mucosa, for during bicarbonate absorption the partial pressure of carbon dioxide in the lumen rises. On the other hand, bicarbonate is secreted into the lumen of the ileum (and of the colon as well) if its concentration in the contents is below 40–45 mM; it is absorbed by the ileal mucosa if its concentration is higher. As with other ions, net absorption or secretion of bicarbonate is the resultant of two opposed unidirectional fluxes; in the ileum the flux of bicarbonate from blood to lumen is independent of luminal concentration, and net movement is governed by changes in flux from lumen to blood. Secretion of bicarbonate is to some extent coupled with chloride absorption in a process of ion exchange, but each is secreted or absorbed by other mechanisms as well. Absorption of chloride and absorption and secretion of bicarbonate are partially blocked by the carbonic anhydrase inhibitor acetazolamide.

Water and electrolytes not absorbed are passed on from the terminal ileum to the colon. Data on water, sodium, potassium and chloride obtained from studies of normal human subjects are given in Chapter 18.

Effects of Obstruction

Obstruction of the small bowel is rapidly followed by distention above the point of obstruction. Distention reduces blood flow through the segment, and this may result in necrosis and rupture. Only a small amount of the distention is caused by the accumulation of fluid delivered from above, for distention inhibits gastric and intestinal motility. Most of the fluid comes from the distended part itself.

If the fluxes across the mucosa are measured by placing isotopically labeled NaCl solutions in the lumen of an obstructed ileal segment of a dog, normal fluxes are found in the direction of blood to lumen, but fluxes of both sodium and water in the direction of lumen to blood are depressed or abolished. This occurs even if the solution in the obstructed segment is hypotonic. As a result, water, sodium and chloride move into the segment but not out of it, distending it still further with fluid having approximately the electrolyte composition of plasma. Sequestration of this fluid in the lumen reduces plasma and interstitial fluid volume, sometimes as much as 35%, with consequent hemoconcentration, renal insufficiency, vascular collapse and death.

Other sequelae are anorexia and vomiting, which exacerbate the fluid loss. These are mediated by afferent nerves, and they occur in otherwise normal dogs following distention of blind pouches made from the jejunum but not in dogs whose jejunal pouches have been denervated. Bilateral splanchnicotomy and excision of the lumbar chain prevents the response of vomiting; but, in addition, bilateral vagotomy is required to abolish anorexia.

Other Factors Influencing Fluxes Across the Intestine

Pentagastrin intravenously infused into dogs in a dose submaximal for acid secretion causes a 15–20% reduction in the rate of absorption of water by both jejunum and ileum. A similar reduction follows release of endogenous gastrin. Intravenous infusion of secretin or cholecystokinin reduces the absorption of sodium, potassium and chloride in the most proximal part of the human jejunum. Free or conjugated dihydroxy bile acids (deoxycholate, chenodeoxycholate and their conjugates) and a free trihydroxy bile acid (cholic) at concentrations of 3–5 mM inhibit net water and electrolyte absorption in the human jejunum, and at higher concentrations, up to 10 mM, stimulate net secretion. Oleic, dodecanoic and decanoic, but not shorter fatty acids, at the concentration of 5 mM in micellar solution inhibit water and electrolyte absorption in the dog's ileum. If these effects of hormones, bile acids and fatty acids have any role in the normal digestive process, it may be to maintain fluidity in the lower small intestine where, otherwise, absorption would dry the lumen.

Dumping Syndrome

It is important that hypertonic solutions become isotonic before they reach the jejunum, for the presence of hypertonic solutions in the jejunum causes nausea, a sense of epigastric fullness with pain, pallor, sweating, vertigo and eventually fainting. This concatenation of signs and symptoms following meals in persons who have been subjected to partial gastrectomy or gastroenterostomy is known as the dumping syndrome, and it results from the rapid arrival of hypertonic fluid in the jejunum. Its physiological expression is very similar to the vasovagal syndrome that may follow venisection or sudden emotional shock: great vasodilatation in the muscles and moderate vasodilatation elsewhere, leading to a profound fall in arterial blood pressure and fainting. Great variability in the dumping syndrome among patients or in one patient from time to time is probably related as much to differences in cardiovascular responses as it is to variations in the cause.

The dumping syndrome may be produced in normal persons by intrajejunal instillation of 250 ml of hypertonic (about 2,000 mOsm) glucose, fructose or amino acids. All these solutions, when placed in the jejunum, increase rapidly in volume; and the fluid entering the intestine is, of course, drawn from extracellular fluid and plasma. If a patient subject to dumping is fed 130–200 ml of a hypertonic meal, his plasma volume begins to fall within 10 min and has fallen 20–30% in 30–40 min. Recovery occurs in 100–120 min. An initially hypotonic diet containing starch has the same effect, for intestinal starch hydrolysis is so rapid that osmotically active oligosaccharides are released within a few minutes. Although distention of the jejunum does not in itself cause a fall in plasma volume, it induces the sense of fullness, nausea and hypermotility leading to diarrhea.

Absorption of Calcium, Magnesium and Phosphate

Calcium is ingested in food and drink of an ordinary diet in the amount of 1,000 mg a day. Calcium also enters the gut as calcium secreted by intestinal cells, as calcium contained in digestive secretions and as calcium contained in desquamated cells. This amounts to approximately 600 mg a day. Consequently, 1,600 mg is available for absorption, and about 700 mg is absorbed, leaving 900 mg to be excreted in the stool. In this example, net absorption is only 100 mg, although 700 mg has actually been absorbed. If the amount ingested is decreased, or if the amount excreted in the stool is increased, negative calcium balance can occur.

In the human small intestine, calcium is actively absorbed when its concentration in the lumen is 1–5 mM. At the highest concentration, the absorbing mechanism is saturated, and at concentrations above 5 mM additional absorption is by passive diffusion.

Calcium is absorbed against a concentration gradient by a two-step active transport mechanism, which uses high energy phosphate. Calcium is first bound to a specific calcium-binding protein in the brush border, and it is transported into the cells. It is dis-

charged from the cells along the lateral and basal membranes.

The calcium-binding protein is under control of 1,25-dihydroxy-vitamin D_3. It is absent from rachitic animals, but it appears within 90 min after a rachitic animal has been given the active form of vitamin D. Parathyroid hormone has a small effect upon calcium absorption by increasing the rate of release of calcium across the basal and lateral borders of the epithelial cells.

The net absorption of calcium is a function of dietary calcium, and it is facultatively regulated to meet the body's needs. Calcium absorption in man is zero or negative when the daily calcium intake is 0.1 mM per kg (or less) of body weight, and it rises linearly to a maximum of about 3 mM per kg a day when dietary intake is high. Rate of absorption is regulated in part by plasma calcium concentration, and intravenous infusion of calcium immediately suppresses absorption from the duodenum and jejunum of a normal subject. Parathyroid hormone secretion, but not that of thyrocalcitonin, may be one of the factors regulating calcium absorption, for malabsorption of calcium occurs in hypoparathyroidism. Age and reproductive status also affect calcium absorption. Young persons absorb calcium more rapidly than do persons over 60.

Bile acids enhance calcium absorption in addition to their effect on the absorption of vitamin D. In bile acid deficiency, unabsorbed free fatty acids are excreted as insoluble calcium salts; this results in negative calcium balance and osteomalacia. A patient whose terminal ileum has been removed often has hyperoxaluria and renal calcium oxalate stones. Such a patient absorbs dietary oxalate 5 times as readily as does a normal subject. The reason is that removal of the terminal ileum reduces bile acid reabsorption, decreases the size of the bile acid pool and increases excretion of calcium as calcium soaps. Consequently, there is less free calcium in the intestinal lumen to precipitate oxalic acid as an unabsorbable calcium salt.

The average daily diet contains about 10

mM of magnesium; of this, 4–5 mM is absorbed. When the magnesium contained in a breakfast was tagged with radioactive magnesium (^{28}Mg), absorption was found to begin after an hour, to continue for 8–12 hours and then to stop, a pattern consistent with absorption through the whole of the small intestine. Magnesium is absorbed by passive diffusion and solvent drag; there is no evidence of facultative control of magnesium absorption. Nor is there evidence that absorption is increased in patients with magnesium deficiency. About 1 mM, or 0.8% of the body's content, of magnesium, appears in the intestinal lumen each day, as the result of either secretion or desquamation.

Phosphate is absorbed by all segments of the small intestine, and absorption occurs both by passive diffusion and active transport. Concurrent absorption of glucose has no effect on that of phosphate, and phosphorylation of sugar is not a necessary step in the absorption of phosphate.

Absorption of Iron and Copper

A carnivorous man eats 15–20 mg of iron in his daily diet; almost all the iron is contained in hemoglobin and myoglobin. Of this he absorbs 0.5–1 mg. A woman subject to loss of iron by menstrual bleeding absorbs 1–1.5 mg per day. In absence of hemorrhage, loss of iron in deciduous cells, chiefly those of the intestinal mucosa, equals the rate of absorption, and the body's load of iron remains constant at approximately 4 gm. The absolute amount of iron absorbed increases with the dose (Fig 14–6) over the short run, but the fraction of the dose absorbed decreases. At the dose of 100 mg (the largest shown in Fig 14-6), 12.6 mg was absorbed. In a normal person the amount absorbed is, over the long run, related to the amount required, for prolonged overloading depresses absorption. Patients with iron deficiency absorb 2–5 times more of a given dose than do normal subjects. After acute bleeding, the iron requirement temporarily increases, but increased absorption does not begin until 3–4 days after hemorrhage.

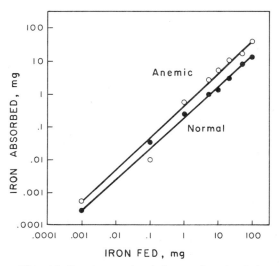

Fig 14–6.—Amount of iron absorbed by fasted normal and anemic human subjects following a single oral dose of iron given with ascorbic acid. The normal subjects had 12.3–16 gm of hemoglobin per 100 ml of blood and 51–190 μg iron per 100 ml of plasma. The anemic subjects had 5–12 gm hemoglobin per 100 ml of blood and 10–25 μg iron per 100 ml of plasma. (Plotted from data of Smith, M. D., and Pannacciulli, I. M.: Br. J. Haematol. 4:428, 1958.)

In man, most iron is absorbed in the heme of hemoglobin and myoglobin, which have been split from their proteins in the lumen. Iron is released within the epithelial cells by heme oxidase, which catalyzes the conversion of heme to bilirubin, carbon monoxide and free iron. Of inorganic iron, ferrous is more rapidly absorbed than ferric. Inorganic iron is absorbed by a saturable active process. Cobalt and manganese are absorbed by the same process, and any one of the metals competitively inhibits absorption of the other two. The mixture of foodstuffs in the intestinal lumen has more influence on absorption of inorganic iron than does the form in which iron is fed. Phosphoproteins and insoluble calcium phosphate bind iron, and ferric iron forms an insoluble, unabsorbable ferric phosphate. By reducing iron, ascorbic acid promotes absorption, and at low pH it forms a chelate with ferrous iron, which remains soluble and available for absorption at the pH prevailing in the small intestine. Neutralized gastric juice mixed with iron salts or with heme promotes absorption; the factor responsible is absent from the gastric juice of patients with pernicious anemia. Salicylates (e.g., aspirin) chelate iron, but they do not reduce absorption. Iron chelated with ethylenediaminetetraacetate is rapidly absorbed, but it is lost equally rapidly by renal excretion. Iron in wine is absorbed no more readily than iron in water, alas! The high iron content of the beer manufactured by the Bantus is said to account for the siderosis prevalent in that race.

The duodenum and jejunum are the sites of iron absorption; only trivial amounts are absorbed in the stomach, lower small intestine and colon. When iron is placed in the duodenum, the initial rate of uptake by mucosal cells is rapid, but the rate falls off to zero in 30–60 min. The ability of mucosal cells to absorb iron is determined by their iron content; and cells that have recently absorbed iron but have not yet delivered it to the plasma temporarily lose their ability to absorb more.

Absorption of heme and inorganic iron into epithelial cells is relatively rapid, but exit into the blood is slow. Within the cells, iron split from heme and absorbed inorganic iron mix with the intracellular pool of iron. Consequently, iron accumulates within the cells, and accumulation is a factor inhibiting further absorption. Iron exists within the cells in at least two forms, one diffusely distributed throughout the cytoplasm. This iron is probably chelated with amino acids. The second is as iron hydroxide micelles in ferritin molecules. Ninety percent of freshly absorbed iron enters ferritin. Some persons think that iron is only temporarily stored in ferritin; others believe it is sequestered within ferritin for the life of the cell. Much of the absorbed iron is actively transported from cell to plasma in 2–4 hours. Transfer of iron from cell to plasma is regulated by the body's need for iron. If the need is high, as during pregnancy or recovery from hemorrhage, a large proportion of the iron contained in the mucosal cells moves to plasma, where it is carried by the globulin transferrin; when the need is low, a small portion leaves the cells. Because of the short life span of the mucosal cells, iron remaining in them is soon lost in the feces.

The iron content of mucosal cells is also determined by the plasma level of iron prevailing at the time the cells are formed within the crypts. If plasma iron is high, the content of the newly formed cells is likewise high; and when these cells have migrated up the villi, their ability to absorb iron is relatively low. If, as after hemorrhage, plasma iron is diverted to the bone marrow to participate in erythropoiesis, the iron content of new mucosal cells is low. When they have migrated up the villi, they absorb iron avidly. Dependence of iron absorption upon the iron content of cells at the time of their formation accounts for the 3–4 day interval between hemorrhage and increased rate of absorption. Iron entering cells from the plasma is lost when they desquamate. Therefore, the intestinal mucosa is a pathway for loss of body iron; and because the cells' iron content reflects the plasma iron concentration, intestinal iron excretion is regulated, to some extent, by the body's iron requirements.

Iron is actively secreted into the lumen by the cells of the rat's jejunum. Whether this occurs in man and whether this is a process regulating the body's store of iron is unknown.

Approximately 40% of the daily intake of copper is absorbed, chiefly in the upper small intestine, and the absolute amount absorbed increases with increasing dose. In normal man, the body load of copper is constant at 60–100 mg; therefore, excretion must equal absorption. Copper is secreted in bile bound to taurochenodeoxycholate, and some of this copper, together with unabsorbed copper, is excreted in the feces. Regulation of absorption and excretion is not understood. In persons with a genetically determined failure to excrete copper at a rate equal to absorption, copper accumulates in many tissues, particularly the liver, central nervous system and the lens. Penicillamine (β,β-dimethylcysteine), by chelating copper, reduces its absorption and promotes its excretion, and penicillamine is used therapeutically and prophylactically in patients who accumulate copper.

Absorption of Vitamin B$_{12}$ and Other Water-Soluble Vitamins

Vitamin B$_{12}$, a highly charged cobalt-containing compound whose molecular weight is 1,357, is probably the largest water-soluble essential nutrient absorbed intact through the intestinal mucosa. Other B vitamins (nicotinic acid and nicotinamide, pantothenic acid, biotin and biocytin and flavine mononucleotide) are small enough to be absorbed in adequate amounts by passive diffusion, but vitamin B$_{12}$ requires a special and still poorly understood system to insure absorption of the 1 μg per day required by an adult man.

In a patient with pernicious anemia who lacks this system, 1–10 mg, or 1,000–10,000 times the daily requirement, must be given by mouth to permit absorption of an adequate amount. Vitamin B$_{12}$ is always present in the diet as one or more of its coenzyme forms bound to protein. All forms are absorbed in the same way. The absorptive process begins with the transfer of the vitamin from dietary protein to a heat-labile mucoprotein, the *intrinsic factor*, secreted by the oxyntic cells of the stomach. Acid facilitates transfer, and in some persons, antibodies to intrinsic factor prevent binding. Two molecules of intrinsic factor appear to be linked by one or two molecules of vitamin B$_{12}$. There are many other compounds in gastrointestinal contents that bind the vitamin but do not promote absorption. In man, intrinsic factor is secreted by the oxyntic glandular mucosa, probably by the parietal cells. When vitamin B$_{12}$ labeled with ^{57}Co is incubated with slices from this area, radioautography shows the isotope to be attached to the parietal cells. Upon infusion of histamine, there is an immediate high peak of intrinsic factor output followed by steady secretion at a low level. In the rat, intrinsic factor is secreted by the chief cells; in the hog, by the pyloric glandular area. Gastric atrophy, which in man results in failure to secrete first acid, then pepsinogen and finally intrinsic factor, leads to pernicious anemia, for the patient fails to absorb enough vitamin B$_{12}$ to allow normal maturation of erythrocytes. The serum of 90% of such patients has within its gamma globulin fraction antibodies against parietal cells. Parenteral administration of vitamin B$_{12}$ bypasses the absorptive defect and prevents macrocytic anemia.

The next step is the uptake of the complex by mucosal cells in the terminal ileum. The combination of intrinsic factor and vitamin B$_{12}$ is exposed to many risks between its sites of formation and absorption. Bacteria residing in diverticula or above intestinal strictures may successfully compete with intrinsic factor for the vitamin and deny it to their host. A tapeworm secretes a compound, which by removing vitamin B$_{12}$ from intrinsic factor makes the vitamin available to the worm but not to the person infested with it. The complex of vitamin and intrinsic factor, upon reaching the terminal ileum, attaches to receptor sites on the microvilli of epithelial cells and is absorbed into the cells. Several hours pass before the vitamin leaves the cells for the plasma, in which it is carried by transcobalamin, a specific transport protein. After a dose of 1 μg of vitamin B$_{12}$ labeled with ^{60}Co is taken by mouth, no isotope enters the blood for 3 hours. Then the radioactive vitamin appears, reaching its peak concentration at 8 hours, and disappears after 12 hours. The vitamin is secreted in the bile and partially reabsorbed in the terminal ileum.

Dietary folate is a mixture of conjugates of reduced folic acid or methyl folic acid. Before absorption, these are hydrolyzed to monoglutamic folate, and this compound appears to be absorbed by a specific, active process in the proximal small intestine. Monoglutamic folate is converted within intestinal cells to reduced monoglutamic methyl folate, and this is the only form of absorbed folate appearing in the blood.

L-Ascorbic acid is absorbed by a sodium-dependent, active process in the human ileum. Thiamine is absorbed by an active process when its concentration in the lumen is 1.5 μM or less, and above that concentration additional thiamine is absorbed by diffusion. Although thiamine is phosphorylated within the mucosal cells, free thiamine is the only form released into the blood.

REFERENCES

Banwell, J. G., et al.: Intestinal fluid and electrolyte transport in human cholera, J. Clin. Invest. 49:183, 1970.

Castle, W. B.: Gastric intrinsic factor and vitamin B$_{12}$ absorption, in Code, C. F. (ed.): Handbook of Physiology: Sec. 6. Alimentary Canal, Vol. III (Washington, D.C.: American Physiological Society, 1968), pp. 1529–1552.

Crosby, W. H.: Iron absorption, in Code, C. F. (ed.): Handbook of Physiology: Sec. 6. Alimentary Canal, Vol. III (Washington, D.C.: American Physiological Society, 1968), pp. 1553–1570.

Curran, P. F., and Schultz, S. G.: Transport across membranes: General principles, in Code, C. F. (ed.): Handbook of Physiology: Sec. 6. Alimentary Canal, Vol. III (Washington, D.C.: American Physiological Society, 1968), pp. 1217–1244.

Dietschy, J. M., Sallee, V. L., and Wilson, F. A.: Unstirred layers and absorption across the intestinal mucosa, Gastroenterology 61:932, 1971.

Fordtran, J. S.: Stimulation of active and passive sodium absorption by sugars in the human jejunum, J. Clin. Invest. 55:728, 1975.

Fordtran, J. S., and Ingelfinger, F. J.: Absorption of water, electrolytes, and sugars from the human gut, in Code, C. F. (ed.): Handbook of Physiology: Sec. 6. Alimentary Canal, Vol. III (Washington, D.C.: American Physiological Society, 1968), pp. 1457–1490.

Glass, G. B. J.: Gastric intrinsic factor and its function in the metabolism of vitamin B$_{12}$, Physiol. Rev. 43:529, 1963.

Haljamae, H., Jodal, M., and Lundgren, O.: Countercurrent multiplication of sodium in intestinal villi during absorption of sodium and chloride, Acta Physiol. Scand. 89:580, 1973.

Ireland, P., and Fordtran, J. S.: Effect of dietary calcium and age on jejunal absorption in humans studied by intestinal perfusion, J. Clin. Invest. 52:2672, 1973.

Love, A. H. G., Rohde, J. E., Abrams, M. E., and Veal, N.: The measurement of bidirectional fluxes across the intestinal wall in man using whole gut perfusion, Clin. Sci. 44:267, 1973.

Rindi, G., and Ventura, U.: Thiamine intestinal transport, Physiol. Rev. 52:821, 1972.

Rosenberg, I. H., and Godwin, H. A.: The digestion and absorption of dietary folate, Gastroenterology 60:445, 1971.

Schultz, S. G., and Curran, P. F.: Intestinal absorption of sodium chloride and water, in Code, C. F. (ed.): Handbook of Physiology: Sec. 6. Alimentary Canal, Vol. III (Washington, D.C.: American Physiological Society, 1968), pp. 1245–1276.

Svanvik, J.: Mucosal blood circulation and its influence on passive absorption in the small intestine, Acta Physiol. Scand. [Suppl.] 385:1, 1973.

Toskes, P. P., and Deren, J. J.: Vitamin B$_{12}$ absorption and malabsorption, Gastroenterology 65:662, 1973.

Turnberg, L. A., et al.: Interrelationships of chloride, bicarbonate, sodium and hydrogen transport in the human ileum, J. Clin. Invest. 49:557, 1970.

Turnberg, L. A., et al.: Mechanism of bicarbonate absorption and its relationship to sodium transport in the human jejunum, J. Clin. Invest. 49:548, 1970.

Van Campen, D.: Regulation of iron absorption, Fed. Proc. 33:100, 1974.

Wilson, T. H.: Intestinal Absorption (Philadelphia: W. B. Saunders Co., 1962).

Wiseman, G.: Absorption from the Intestine (New York: Academic Press, 1964).

15

Intestinal Digestion and Absorption of Carbohydrate

ALTHOUGH MOST of the carbohydrates eaten are oligosaccharides or polysaccharides, monosaccharides are delivered by the intestinal mucosa to the portal blood. Dietary carbohydrate is entirely dispensable; nevertheless, it is 50–60% of the American mixed diet and in many countries is a larger percentage. Carbohydrate intake ranges from 250 to 800 gm per day, and it supplies 1,000–2,500 calories.

The major dietary form of carbohydrate is plant starch composed of straight and branched chains of glucose. The straight chains are held together by 1,4′-α-glycosidic linkages, and at the branch points the linkages are 1,6′-α-glycosidic. The polysaccharide cellulose linked in the 1,4′-β-configuration is not attacked by the enzymes of animals. Although it is digested by microbial flora in the stomach, rumen or cecum of herbivorous animals and in the colon of man, its digestion and absorption in the human small intestine is negligible. Small amounts of other glucose homopolysaccharides, including animal glycogen; some homopolysaccharides of galactose, mannose, arabinose and xylose; and some heteropolysaccharides are eaten. Of the oligosaccharides, the quantitatively most important are su-

crose (glucose-fructose), maltose (glucose-glucose) and lactose of milk (glucose-galactose). Only small amounts of monosaccharides occur in the normal diet. Some hexose alcohols, pentoses and amine derivatives of carbohydrates are ingested as parts of complex molecules.

Hydrolysis of starch is catalyzed by salivary and pancreatic amylases; disaccharides are split at the brush border of mucosal cells as they are being absorbed. Glucose and galactose are the major sugars actively absorbed; fructose is absorbed by diffusion.

Digestion of Starch

Salivary α-amylase attacks dietary starch, and the extent to which salivary digestion proceeds depends upon the failure of the stomach to mix its contents with acid and on the delay of gastric emptying. About 50% of the starch may be broken down in the stomach before it enters the duodenum. There starch mixes with pancreatic amylase in a pH environment favorable for action of the enzyme. Most dietary starch is amylopectin, consisting of long straight chains of glucose molecules linked at α-1,4- points and branches attached to the chains by linkage at α-1,6-

points. Salivary and pancreatic amylase hydrolyze only α-1,4- linkages within the chain; they do not attack terminal α-1,4- linkages, nor do they break the α-1,6- branching links. Consequently, the products of amylytic digestion of amylopectin within the lumen are maltose (glucose-glucose; α-1,4- linkage), maltotriose (glucose-glucose-glucose; α-1,4- linkages) and a mixture of dextrins containing the α-1,6- branches and averaging six glucose residues per molecule. Hydrolysis of glycogen, which also contains α-1,6- branches is similar to that of amylopectin. Amylose, a minor constituent of starch, contains only α-1,4- linkages, and its digestion products are maltose and maltotriose. Hydrolysis in the duodenum is very rapid. Measurement of enzyme concentration in samples recovered from the human intestine after a meal shows that amylase is maximally concentrated in the duodenum; the amylase contained in only 1 ml of duodenal contents can hydrolyze 5 gm of starch per hour at 37 C. Within 10 min after starch enters the duodenum it is converted to fragments containing, on the average, three hexoses, and complete hydrolysis soon follows. The products of starch digestion, unlike starch itself, are osmotically active; hydrolysis of 10 gm of starch to glucose creates 60 mOsm. Were not glucose absorbed almost as rapidly as starch enters the duodenum, 200 ml of water would be required to bring duodenal contents to isotonicity after hydrolysis of this small amount of starch.

Hydrolysis of Oligosaccharides

The activity of enzymes that hydrolyze oligosaccharides is very low in intestinal contents; the enzymes present are not secreted but come from desquamated cells. Consequently, oligosaccharides are not split in the lumen as they are liberated from starch. On the other hand, only monosaccharides appear as absorption products in the portal blood.

Major enzymes catalyzing hydrolysis of oligosaccharides are maltase (maltose and maltotriose hydrolyzed to glucose), lactase (lactose to glucose and galactose), sucrase (also called invertase; sucrose to glucose and fructose and maltose to glucose), and isomaltase (also called oligo-1,6- glucosidase or α-dextrinase; branched dextrins to glucose and maltose). Although samples of intestinal mucosa taken by biopsy from the whole length of the small intestine of normal subjects contain all these enzymes, their concentration per gram wet weight of tissue is highest in mid-jejunum and upper ileum and lowest in the duodenum and terminal ileum. The enzymes are components of the brush border of epithelial cells; and two of them, sucrase and maltase, have been identified as 60Å particles attached to the luminal surface of the microvilli. Brush borders (microvilli and terminal web) can be isolated virtually uncontaminated by other cell components, and such preparations contain all the oligosaccharidases of the cells and phosphomonoesterases as well. Enzymes of the brush border also hydrolyze cellobiose, trehalose, α-glycerophosphate, hexose diphosphate, adenosine triphosphate and a variety of polyphosphates.

Oligosaccharides are hydrolyzed on contact with the brush border, and their hydrolysis products can be absorbed or can return to the lumen. Glucose and galactose molecules liberated are picked up by an active transport system that resides in cell membranes close to the hydrolytic enzymes, and the two hexoses accumulate within epithelial cells before passing into portal blood. Fructose, not being actively transported, is absorbed more slowly. In man, the rate of hydrolysis is faster than the rate of absorption of glucose, and glucose is absorbed 3–6 times faster than fructose. The result is that some glucose and much more fructose transiently appear in luminal contents during sucrose digestion. Maltose is hydrolyzed as rapidly as sucrose, but lactose is split only half as quickly.

Data on hydrolysis and absorption of sugars by man have been obtained by sampling, at various times and levels, the contents of the small intestine by means of an indwelling

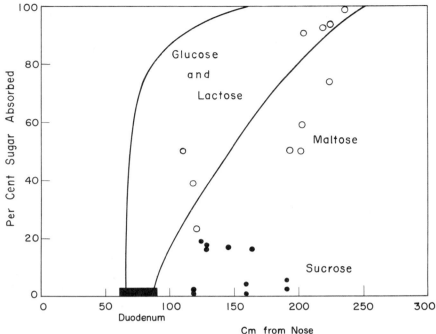

Fig 15–1.—Absorption of mono- and disaccharides by the human intestine. Area between the solid lines encloses 95% of the measurements of absorption of glucose and lactose in 16 separate experiments in which more than 100 samples were obtained. The meal given contained 75 gm of carbohydrate as glucose and lactose in 500 ml. *Open circles* show absorption of maltose from a 400-gm liquid meal containing 56.3 gm of maltose as the only sugar. *Filled circles* show absorption of sucrose from a similar meal. (Adapted from Borgstrom, B., *et al.*: J. Clin. Invest. 36:1521, 1957; and Dahlqvist, A., and Borgstrom, B.: Biochem. J. 81:411, 1961.)

tube after a 500-ml liquid meal has been placed in the stomach. When the meal contained 20 gm of lactose and 55 gm of glucose (in addition to 30 gm of fat and 25 gm of protein), up to 40% of the carbohydrate was absorbed in the duodenum (Fig 15–1). All carbohydrate was absorbed by the time the residue of the meal reached the end of the jejunum. When lactose was given as the only carbohydrate, it was absorbed at the same rate in the same place. The lactase activity of the cells of the upper small intestine is adequate for hydrolysis of all lactose of the diet. Maltose and isomaltose are not absorbed in the duodenum; 50–70% of the maltose fed in a 400-gm meal containing 56.3 gm of maltose, 21.8 gm of protein, 27.3 gm of fat and 295 gm of water was absorbed in the jejunum

and the rest in the upper ileum. Most maltose of the meal was absorbed in 4 hours. When sucrose-fat-protein meals were given to 5 human subjects, 90–100% absorption of sucrose occurred in the jejunum. Capacity to hydrolyze and absorb sucrose is lower in the ileum than in the jejunum; the human jejunum absorbs sucrose from a solution of 73 or 146 mM sucrose made isotonic with NaCl at the rate of 128 ± 26 mM per hour, whereas the ileum absorbs at the rate of 55 ± 24 mM per hour.

Induction and Deficiency of Oligosaccharidases

The concentration of oligosaccharidases depends to some extent upon the diet. In human subjects whose jejunal mucosa was repeatedly sampled by biopsy, a change from a diet containing no sucrose to a high-sucrose diet was followed in 2–5 days by a doubling of sucrase and maltase activity. Lactose, galactose and maltose do not induce changes in enzyme activity.

Some young children cannot hydrolyze and absorb lactose; all of the sugar they ingest appears in the stool either as lactose or as lactic and fatty acids, products of fermentation of lactose by colonic flora. High osmotic pressure and stimulation of peristalsis by the acids cause diarrhea with dehydration and negative electrolyte balance, and because lactose is the sugar of milk, the patients become severely undernourished as well. The immediate cause of lactose intolerance, an inborn error of metabolism, is absence of lactase from the intestinal mucosa. This enzyme is the one that catalyzes hydrolysis of the rare sugar cellobiose, and children intolerant of lactose cannot digest and absorb cellobiose either. However, they can hydrolyze any of the other disaccharides, and they can survive if their difficulty is correctly diagnosed and if sucrose is substituted for lactose in their diet. Other children who have no invertase in their intestinal mucosa cannot tolerate sucrose; if they are fed sucrose, half of it appears unhydrolyzed in the stool and the other half appears as lactic and fatty acids. Absence of isomaltase causes isomaltose intolerance. Trehalose, which occurs in mushrooms and insects, is not a major form of carbohydrate in the human diet, but deficiency of trehalase has been identified. All these intolerances are the result of the genetically determined absence of a single enzyme. Isomaltose intolerance and sucrose intolerance often occur together, although the activities of the two enzymes, isomaltase and invertase, are independent of each other. Because maltose is hydrolyzed by at least four separate enzymes, intolerance of it can occur only if all four are missing, and such intolerance has not been discovered.

In addition to genetic factors, there is a problem of development as well. Some infants with well-documented lactose intolerance lose it about the age of 6 years, and one child who had no lactase in his intestinal mucosa shortly after birth did have the enzyme at 2 years.

On the other hand, many children who tolerate lactose become deficient in lactase and intolerant of lactose between 2 and 6 years of age. As teen-agers and adults, they have flat blood sugar curves when they are given 100 gm of lactose by mouth, and they have abdominal discomfort, cramps, loose stools or diarrhea and flatulence when they drink a glass of milk. Adult lactose intolerance is genetically determined. Adult members of milk-drinking African tribes can hydrolyze lactose, but those who do not habitually use milk products cannot. For example, less than 20% of Batutsi are milk intolerant, but 100% of adult Zambians are. Because most American blacks are descendants of lactase-deficient ancestors, more than 70% of adult American blacks are likewise lactase deficient.* Other groups with very high frequencies of lactose intolerance include Ashkenazic Jews, Arabs, Greek Cypriots, Japanese, Formosans and Filipinos. Although only 5–15% of adult white Americans of North European descent are deficient in lactase, milk intolerance is frequently encountered in the mixed American population. Eighty-six of 166 unselected patients in one Veterans Administration Hospital were found to be lactose intolerant, and most of them refused to drink the milk offered on their luncheon trays. Consequently, the physician who orders milk for his patients must consider the implications of the fact that a large fraction of his patients may not tolerate it.

The intestinal mucosa of sea lion pups contains no oligosaccharidases, and they

---

*This is the incidence repeatedly given by authorities. The American blacks in the medical classes of The University of Michigan say that the frequency of lactose intolerance among them is far lower. Perhaps this is another example of discrimination by the Admissions Committee.

have diarrhea when fed lactose or sucrose; fortunately for them, the milk of their mothers contains no detectable carbohydrate.

Not all untoward consequences of milk drinking are the result of lactose intolerance. In some children and adults, signs and symptoms of milk intolerance, including flat blood sugar curves after feeding lactose or even glucose, can be induced by feedimg milk proteins

In Vitro Studies of Sugar Absorption

In vitro preparations of rat, guinea pig and hamster intestine have given knowledge of the means by which sugars are absorbed. One method of preparation is to suspend a length of intestine in a well-oxygenated salt solution whose composition is close to that of extracellular fluid. Through the lumen is perfused, with or without recirculation, another solution containing the sugar to be studied. The amount of sugar transported is deter-

mined by analyzing the fluid on the mucosal and serosal sides. Another method is to evert a short piece of intestine so that its mucosal side is outward, fill it with fluid of known composition, tie it to form a sausage-like sac and incubate it in a solution containing sugar. Again, analysis of the two solutions allows calculation of the rate of sugar movement across the intestine. These methods demonstrate that passage of some sugars through the mucosa is a passive process.

When net water absorption occurs, passively absorbed sugars may be carried by solvent drag. In the absence of water absorption, a passively absorbed sugar is not transported against a concentration gradient; its movement is only down a gradient, and its rate of movement is proportional to the magnitude of the gradient. It moves as well from serosal to mucosal sides as in the opposite direction. Anaerobiosis or metabolic poisons do not decrease these rates.

Actively transported sugars are moved

Fig 15–2.—Specificity of active transport of sugar by intestine; transport of sugars with modifications around carbon 6. Sacs of hamster intestine were everted and filled with salt solution containing the sugar at the concentration indicated, and they were incubated in 3–5 ml of the same solution. Concentrations of sugar on each side of the intestine were determined after 60 or 90 min. (From Wilson, T. H., and Landau, B. R.: Am. J. Physiol. 198:99, 1960.)

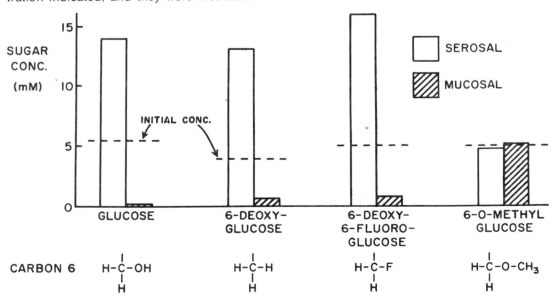

from mucosal to serosal sides against a concentration gradient (Fig 15–2). The transport mechanism is highly specific. Glucose and galactose are the chief naturally occurring hexoses actively absorbed. Glucose derivatives with some modifications about carbon 1 are transported; but when the substituted group is too large, the mucosa is incapable of transporting the compound. Gold thioglucose, in which —S-Au is substituted for —OH on carbon 1, is not transported. Any modification of the hydroxyl group at carbon 2 (for example, in 2-deoxyglucose) impairs transport. Modification of the hydrogen substituent at position 2 does not affect transport unless the substituent group is too large. In a like manner, substitution of large groups, but not of small ones, at other carbon positions prevents active transport. Steric orientation of substituents is important; reversal of orientation of hydroxyl groups at carbon 3 (allose) or carbon 4 (galactose) does not prevent transport, but reversal of both (gulose) does.

Xylose is the chief pentose actively absorbed.

Active Transport of Glucose and Galactose

Three processes occur in active absorption of glucose and galactose: diffusion through the unstirred layer, transport across the cell membrane and extrusion from the cell into the interstitial fluid.

A sugar diffuses through the unstirred layer to the cell membrane within the brush border. The rate of diffusion, J, is proportional to the diffusion constant of the sugar, D, divided by the thickness of the unstirred layer, d, and the difference between the sugar's concentration in the bulk phase, C_1, and its concentration at the interface between the unstirred layer and the cell membrane, C_2.

$$J = D/d\,(C_1 - C_2) \qquad (15.1)$$

At the cell membrane the sugar combines with a carrier, which actively transports the sugar across the cell membrane into the cell.

Although the nature of the transport machinery is incompletely known, it has well-defined characteristics. Transport is rapid but capable of saturation; one carrier transports glucose and galactose, which consequently are in competition with each other; and the transport machinery uses metabolic energy. As the concentration, C_2, of the sugar at the cell membrane rises, the quantity transported per unit time rises along a hyperbolic curve until a maximum rate of transport is reached. The relation between velocity and concentration is expressed by the equation

$$J = \frac{V'_{max}\,(C_2)}{K'_a + C_2} \qquad (15.2)$$

V'_{max} is the apparent maximal transport velocity, and K'_a is the apparent Michaelis-Menten constant for the carrier-mediated process. Equation (15.2) is similar in form to the Michaelis-Menten equation, which describes the relation between the velocity of an enzyme-catalyzed reaction and the substrate concentration.

Diffusion and carrier-mediated transport are in series, and therefore

$$J = D/d\,(C_1 - C_2) = \frac{V'_{max}\,(C_2)}{K'_a + C_2} \qquad (15.3)$$

The sugar accumulates in the cell at high concentration, and it may be extruded across the basal and lateral borders of the cell by simple diffusion. During active absorption the concentration of glucose in the interstitial fluid is high.

Glucose and galactose compete for the same absorbing mechanism; absorption of one sugar depresses the rate at which the other is absorbed. The magnitude of mutual interference can be predicted from the transport constants. Because glucose has a greater affinity for the transport mechanism, that is, a lower concentration at which its rate of transport is half maximal, its absorption should depress galactose absorption relatively more than galactose absorption depresses transport of glucose. When galactose alone is present on the mucosal side of a 10-cm length of guinea pig intestine, it is absorbed at the rate of $37\mu M$ per hour. The addition of an equal concentration of glucose reduces absorption of galactose by 80%; addition of galactose to a preparation absorbing glucose reduces the rate of glucose transport 20–30%. Other actively transported sugars also inhibit galactose transport, but fructose and other sugars that are not actively absorbed have no effect.

Although the intestine of fetal and newborn hamsters can transport glucose anaerobically, absorption by older animals requires oxygen. Glucose absorption is inhibited by 2,4-dinitrophenol, by phlorhizin and by the cardiac glycoside ouabain. Ouabain probably affects glucose absorption by virtue of its ability to inhibit Na^+-K^+-activated ATPase and thereby block sodium transport associated with glucose absorption.

Sodium and Glucose Absorption

In a normal person, presence of glucose in the lumen of the intestine is not necessary for sodium absorption. However, sodium is required for at least part of the absorption of glucose.

The concentration of sodium within the lumen is high: 100–140 mEq per kg of water. Its concentration within intestinal epithelial cells is low: 50 mEq or less per kg of cell water. This concentration gradient is maintained by continuous pumping of sodium out of the lateral borders of the cells into interstitial fluid. In addition, there is an electric gradient across the apical surface of epithelial cells, the cell interior being about 10 mV negative with respect to the luminal fluid. The combination of chemical and electric gradients comprises an electrochemical gradient tending to drive sodium from lumen to cell interior.

A barrier to free diffusion of sodium exists in the plasma membrane at the apical border of the epithelial cells, immediately beneath the microvilli. Sodium crosses this membranous barrier by attaching to a carrier at the membrane's luminal face. The carrier loaded with sodium moves to the cytoplasmic face of the barrier and there discharges the sodium ion. The carrier then returns to the luminal surface to pick up another sodium ion. Energy consumed in this transport is provided by sodium's electrochemical gradient.

The postulation that the sodium carrier has at least one other binding site to which glucose (and galactose) can attach accounts for the coupling of hexose absorption with sodium. The sugar molecule cannot attach to the carrier unless the carrier is already loaded with sodium. Once the carrier has combined with sodium and then with glucose, sodium's electrochemical gradient drives it from the luminal to the cytoplasmic face of the membrane, where it discharges sodium at a low concentration and glucose at a high one. Accumulation of glucose within cell water occurs at the expense of energy provided by the sodium gradient. Glucose, having a high concentration within the cell, diffuses into interstitial fluid and thence into portal blood. This scheme is described in Figure 15–3.

Coupling of glucose and sodium transport is useful in the management of cholera. In this disease, a very large volume of fluid pours into the jejunum and ileum. Diarrhea may occur at the rate of 600 ml per hour, and stool composition is about Na^+ 140, K^+ 10, Cl^- 110 and HCO_3^- 40 mEq per liter. Intravenous administration of fluid of the composition of that lost in the stool restores and maintains extracellular fluid volume. The same fluid taken orally is not absorbed but is added to the stool. If, however, the patient is allowed to drink a solution of Na^+ 100, K^+ 10, Cl^- 70 and HCO_3^- 40 mEq per liter to which 120 mM per liter of glucose is added, all fluid drunk is absorbed. Galactose, because it, too, is absorbed with sodium, is as effective as glucose; but fructose, which is absorbed without sodium, is not.

Malabsorption of glucose and galactose, a rare disease in man, is caused by a reduction in the number of functioning sugar carriers in

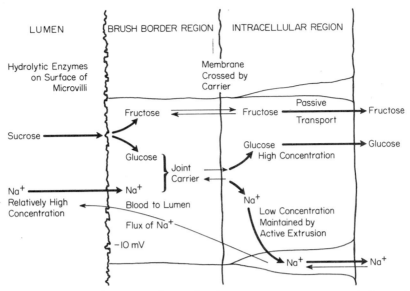

Fig 15–3.—Scheme showing coupling of glucose absorption by the intestinal epithelial cell with sodium transport along sodium's electrochemical gradient. (Adapted from Crane, R. K.: Absorption of sugars, in Code, C. F. [ed.]: *Handbook of Physiology:* Sec. 6. *Alimentary Canal,* Vol. III [Washington, D.C.: American Physiological Society, 1968], pp. 1323–1351.)

the microvillar membrane of the intestinal epithelial cells.

Glucose Absorption in Man

The concentration of glucose in the intestine during digestion of a starch-containing meal may be 75 mM, and in this circumstance a fraction of the glucose is absorbed by diffusion.

When the rate of glucose absorption by normal subjects is plotted against the luminal concentration, a curve is obtained that can be fitted by equation (15.3) (Fig 15–4). It is difficult to determine how much of the absorption depends upon coupling of glucose and sodium for the reason that the concentration of sodium on both sides of the epithelial cells cannot be manipulated, as it can in experimental animals or in vitro preparations.

Evidence for the role of sodium in absorption of glucose by the human small intestine has been obtained in experiments in which

the ileum was perfused with sodium-free solutions. The ileal mucosa, in contrast with that of the duodenum and jejunum, is sufficiently impermeable to sodium so that, when perfused with a sodium-free solution, the concentration of sodium in the luminal fluid may be less than 3mN. When the ileum is perfused with a glucose-containing solution in which sodium chloride is replaced by mannitol, the rate of glucose absorption is reduced by only 20%. This does not demonstrate that sodium is unnecessary for glucose absorption, because the concentration of sodium in the unstirred layer may be high. The microvilli of the epithelial cells are covered with a fuzzy coat, or glycocalyx, consisting of copiously sulfated polysaccharides. This fixed layer is negatively charged, and electric neutrality is attained by the presence of cations in the aqueous phase of the glycocalyx. Under normal circumstances, these cations are chiefly sodium ions. It is probable that even during perfusion of the lumen with a sodium-free solution, there is

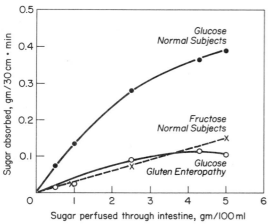

Fig 15–4. – Mean rates of absorption of glucose and fructose in the jejunum of normal subjects and of glucose in the jejunum of patients with gluten enteropathy. The perfusion fluid was delivered to a 30-cm length of jejunum at the rate of 20 ml per min. Data for glucose absorption in the normal subjects can be fitted by equation (15.1), suggesting that absorption occurs by way of a saturable transport system. (From Holdsworth, C. D., and Dawson, A. M.: Gut 6:387, 1966.)

still a high concentration of sodium within the glycocalyx at the surface of the epithelial cells. If those ions are free to participate in glucose absorption, the fact that glucose is rapidly absorbed from a solution that in the bulk phase is essentially sodium free cannot be taken as evidence that sodium is unnecessary for glucose absorption.

If the ileum is perfused with a glucose-containing solution in which sodium has been replaced by magnesium and not by mannitol, glucose absorption is reduced by 45%. Magnesium ions, which are only very slowly absorbed in the ileum, can replace sodium ions in the glycocalyx. In this circumstance, there will be fewer sodium ions at the surface of the epithelial cells to participate in glucose absorption, and that part of glucose absorption that depends upon simultaneous absorption of sodium is reduced. If this argument is correct, a substantial part of glucose absorption in the human ileum does not require sodium.

The major part of glucose absorption in man occurs in the duodenum and upper jejunum, and the sodium concentration in the lumen of these segments is always about 140 mN. Consequently, it is impossible to determine how much absorption of glucose in the human intestine does not depend upon simultaneous absorption of sodium.

When a man drinks glucose solutions ranging from isotonic 310 mOsm (5.4%) to 3,100 mOsm (54%), the more concentrated solutions are slightly diluted in the stomach. As the stomach empties concentrated glucose solutions, the contents of the duodenum and jejunum become hypertonic, and gastric emptying is delayed. By the time the solutions reach the middle or lower jejunum or the ileum, they become isotonic. Hypertonic solutions increase intestinal motility, so that they are quickly dispersed through the intestine, exposed to a large mucosal surface and brought to isotonicity by dilution and absorption. Adjustment of tonicity in the duodenum is chiefly by dilution and in the jejunum by absorption. When a drink of 100 ml of 50% glucose is taken, the rate of absorption in the duodenum is 6 – 20 gm per hour. Rarely as much as half of a drink of 25 – 100 gm of glucose taken as 5 – 50% solution reaches the jejunum, and in most instances the amount passing into the jejunum is less than 5% of that ingested.

Hydrolase-Related Transport of Hexoses

Hexose components of disaccharides are also accumulated by intestinal epithelial cells by a process that does not depend on sodium transport. The sugar to be accumulated must arrive at the brush border as a glycoside capable of being hydrolyzed by an enzyme present in the membrane of the microvilli. The glycoside binds to the enzyme and is split into its component hexoses. These then are transported into the cells; almost none of the products of hydrolysis escapes into the lumen. Thus, sucrose is hydrolyzed to glucose and fructose, both of which are immediately transported into the cells. Likewise, maltose is hydrolyzed, and its component glucose molecules are accumulated. It is likely that the hydrolytic enzymes are themselves the carriers that move the sugars. This process of hydrolase-related transport is additive with sodium-dependent transport, and in some instances its capacity to transport sugars is greater than the capacity of the sodium-dependent system transporting glucose and galactose.

Glucose Metabolism and Transport

In the course of being absorbed, glucose enters the cells of the mucosa, but there is no evidence that glucose uptake is facilitated by insulin, as it is in muscle cells. Once within the cells, some glucose enters into cellular metabolism; 10% or more of the glucose absorbed may be completely oxidized, chiefly by the phosphogluconate pathway. Mucosal cells have a high rate of aerobic glycolysis, and the lactate produced is transported to the serosal side of the intestine. An isolated piece of intestine can establish a lactate concentration gradient, the concentration on the serosal side being 4–20 times that on the mucosal side. However, conversion of glucose to lactate is not the chief means of glucose absorption. When glucose that is uniformly labeled with radioactive carbon (^{14}C) is placed in a loop of dog intestine and the total venous blood draining the loop is collected, only 7–17% of the glucose absorbed is recovered as lactate; the remainder appears as glucose. Lactate itself is absorbed in the jejunum, more rapidly at pH 2.8 than at physiological pH, because at low pH lactic acid is un-ionized and is relatively fat soluble.

Fructose Absorption

In the range of concentrations in which fructose is present in the intestine during absorption of sucrose, fructose is absorbed at a rate directly proportional to its luminal concentration (see Fig 15–4). This rate is greater than that of passively absorbed mannose and less than that of actively absorbed glucose. When, in perfusion studies, the luminal concentration of fructose is raised to 100 mM or more, the curve relating rate of absorption to luminal concentration becomes concave downward, demonstrating that the absorption process is saturable.

In intestinal epithelial cells, fructose is largely converted to glucose or to lactic acid.

Children placed on a diet in which the carbohydrate is 70% fructose exhibit a slight fructosuria and retain more nitrogen than when the dietary fructose is less than 5%; clearly, a considerable fraction of the fructose must have been absorbed as such. Some adults with diabetes mellitus have the same response to oral as to intravenous fructose, and in them, too, fructose must be absorbed as fructose.

REFERENCES

Bayless, T. M., Rothfeld, B., Massa, C., Wise, L., Paige, D., and Bedine, M. S.: Lactose and milk intolerance: Clinical implications, N. Engl. J. Med. 292:1156, 1975.

Bieberdorf, F. A., Morawaski, S., and Fordtran, J. S.: Effect of sodium, mannitol, and magnesium on glucose, galactose, 3-O-methylglucose, and fructose absorption in the human ileum, Gastroenterology 68:58, 1975.

Crane, R. K.: Absorption of sugars, in Code, C. F. (ed.): *Handbook of Physiology:* Sec. 6. *Alimentary Canal,* Vol. III (Washington, D.C.: American Physiological Society, 1968), pp. 1323–1351.

Crane, R. K.: A concept of digestive-absorptive surface of the small intestine, in Code, C. F. (ed.): *Handbook of Physiology:* Sec. 6. *Alimentary Canal,* Vol. V (Washington, D.C.: American Physiological Society, 1968), pp. 2535–2542.

Fisher, R. B., and Gardner, M. L. G.: Dependence of intestinal glucose absorption on sodium, studied with a new arterial infusion technique, J. Physiol. 241:235, 1974.

Gray, G. M.: Carbohydrate digestion and absorption, Gastroenterology 58:96–107, 1970.

Kretchmer, N.: Lactose and lactase—a historical perspective, Gastroenterology 61:805, 1971.

Lewis, L. D., and Fordtran, J. S.: Effect of perfusion rate on absorption, surface area, unstirred water layer thickness, permeability, and intraluminal pressure in the rat ileum in vivo, Gastroenterology 68:1509, 1975.

Ramaswamy, K., Malathi, P., Caspary, W. F., and Crane, R. K.: Characteristics of the disaccharidase-related transport system, Biochim. Biophys. Acta 345:39, 1974.

Ransome-Kuti, O., Kretchmer, N., Johnson, J. D., and Gribble, J. T.: A genetic study of lactose digestion in Nigerian families, Gastroenterology 68:431, 1975.

Wilson, T. H.: *Intestinal Absorption* (Philadelphia: W. B. Saunders Co., 1962).

Wiseman, G.: *Absorption from the Intestine* (New York: Academic Press, 1964).

Intestinal Digestion and Absorption of Protein

ADULTS REQUIRE approximately 0.5–0.7 gm of protein per kg of body weight per day to remain in nitrogen balance; growing children require 4 gm per kg of body weight per day at the age of 1–3 years. All but a negligible fraction of ingested protein is broken down completely and absorbed as amino acids and small polypeptides. Ten to 30 gm of protein contained in digestive juices and about 25 gm of protein derived from desquamated cells are added to ingested protein, and these, too, are digested and absorbed. Only about 10% of the daily intake escapes into the stool in the form of bacteria, desquamated cells and mucoproteins.

Gastric Digestion of Protein

Gastric digestion of protein is dispensable, for persons with achlorhydria (atrophy of the gastric mucosa resulting in failure to secrete acid and pepsin) can remain in nitrogen balance while eating protein in its usual forms. When the stomach is functioning normally, the extent of gastric digestion varies widely, depending, as it does, on factors affecting secretion, the size and division of protein foods swallowed, mixing in the body and antrum and the rate of gastric emptying. At the worst, only a small amount of protein may be attacked in the stomach; and at best, 10–15% may be broken down to amino acids. Therefore the protein of the diet delivered to the intestine is a mixture of completely undigested bundles of muscle fibers, native protein in solution and products of peptic digestion, ranging from large polypeptides to a few free amino acids. Only the last are ready for absorption; consequently, the major part of protein digestion must occur in the intestine.

Absorption of Native Protein

Mammalian fetuses do not synthesize their own antibodies, and in some species, including man, passive immunity at birth results from placental transfer of maternal antibodies. This does not occur in ruminants or rodents, and the plasma of their newborn is devoid of γ-globulin.

Whey of colostrum of ruminants contains IgG antibodies, and these are ingested by the suckling young. The globulins are protected against digestion by trypsin inhibitor in colostrum and by neonatal achlorhydria. During the first 36 hours of life, the IgG antibodies are absorbed through intestinal epithelial cells by pinocytosis.

In the rat, the process of absorption is

selective and prolonged. Ingested IgG molecules bind to receptors on the microvillar membrane of jejunal epithelial cells, and binding is apparently responsible for specificity of absorption. Rat intestine absorbs rat globulins 50 times faster than albumin; it absorbs rat globulins twice as fast as globulins from monkey or rabbit; and it does not absorb cow or fowl globulins at all. Invagination of the surface of the cells produces membrane-bound vesicles, called phagosomes, within the epithelial cells, and these migrate toward the base of the cells. During migration, there is some digestion of antibody within the phagosomes, but most of the antibody is delivered to the extracellular space by exocytosis. Absorbed globulins reach the general circulation by way of the lymph. Ability to absorb large quantities of antibodies lasts for 18–20 days after birth. In this period, administration of cortisone alters the morphology of intestinal cells so that they are indistinguishable from adult ones and partially suppresses their ability to absorb native protein. Ability to absorb large molecules is retained by the adult rat. If one is fed tritiated bovine serum albumin, 2% of the amount fed can be identified as intact albumin in the lymph draining the intestine.

Human infants do not absorb antibodies from maternal colostrum. Nevertheless, γ-globulins and other macromolecules can enter an infant's intestinal epithelial cells by nonselective pinocytosis. Only a very small fraction of these reaches the circulation, for most are digested within the cells. Native proteins absorbed in minute amounts can be immunologically important. Babies who had never received egg white were sensitized to it by intradermal injection of egg albumin. When the babies were subsequently fed the same protein, a wheal, developing at the site of sensitization, showed that absorption of native egg albumin had occurred.

Most adults can absorb immunologically detectable amounts of whole protein. Although the amounts are nutritionally negligible, they may at times be responsible for local intestinal or systemic disorders.

Intestinal Digestion of Protein

The rate at which protein is delivered to the duodenum, and therefore its rate of digestion and absorption, varies widely. Fifty percent of a finely divided lean-meat meal may leave the stomach in the first hour, and 83% by the end of the third hour. Once in the duodenum, protein is rapidly digested. Fluid aspirated from the human duodenum shortly after feeding contains 200–800 μg of trypsin and chymotrypsin per ml. This is enough to convert 50% of the protein in duodenal contents to trichloroacetic acid-soluble material in 10 min; consequently the rate of digestion of protein does not limit its absorption.

The time course and extent of protein digestion in man have been followed by aspirating samples through an indwelling tube from various levels of the intestine. When the protein content of a mixed meal is 5%, the concentration of fed protein at the end of the duodenum ranges from 1–4%, with the most frequent value being about 2%. By the time the residue reaches the end of the duodenum, 50–60% of the fed protein has been digested and absorbed. Fed protein appears in the jejunum 30 min after feeding and is detectable there for 4 hours. It does not appear in the ileum until 2 hours have passed, and its concentration in the ileum reaches a peak at 4 hours. About one-tenth escapes digestion in the small intestine and enters the colon, where its digestion is completed by microorganisms. The protein contained in the stool, equaling about 10% of the protein eaten, is not food protein but comes from bacteria and cellular debris.

Endogenous Protein

Between 10 and 30 gm of protein enters the intestinal lumen in secretions each day, and 80–90 gm of desquamated cells contributes another 10 gm. In addition, plasma proteins enter the digestive tract. Capillaries of the gut are more permeable to proteins than are capillaries of muscle, and the interstitial fluid of the gut contains a high concentration of plasma proteins. The concentration of plasma albumin in gastric interstitial fluid is about 80% of that in plasma, and the concentration of fibrinogen is about 30%. Some plasma proteins escape into the lumen and are digested. In normal persons, an average of 1.9 gm of albumin is shed into the stomach; this represents 11% of the daily degradation of albumin. In protein-losing gastroenteropathies, the rate of plasma protein loss may be enormous, and hypoalbuminemia results.

Enzymatic Hydrolysis of Protein

Proteins in the intestinal lumen are digested by pancreatic enzymes and by enzymes contained in the brush border of the epithelial cells. The digestion products are free amino acids and oligopeptides containing 2, 3 or 4 amino acid residues. Both free amino acids and oligopeptides are absorbed through the cells into the portal blood.

Pancreatic proteolytic enzymes are responsible for the major part of protein digestion within the intestinal lumen. The flow of pancreatic juice begins 10–20 min after a meal is eaten, and the concentration of pancreatic enzymes remains high throughout the jejunum and ileum over the whole period of digestion and absorption. Pancreatic enzymes are present in the stool of normal men, dogs and rats; but the fraction of the total quantity secreted by the pancreas that is lost in the stool is unknown.

In the absence of pancreatic juice, protein digestion is impaired. When the pancreatic ducts of dogs are ligated or the pancreas is completely removed, the percentage absorption of fed protein ranges between 22 and 85. The quantity absorbed varies from day to day in the same animal, for unknown reasons. Although much fed protein and unchanged muscle fibers may appear in their stool, dogs completely devoid of pancreatic juice can be made to absorb enough protein to support normal growth simply by increasing total protein intake. Children with cystic fibrosis, in whom pancreatic enzymes are almost completely absent, have low absorption of protein, but absorption can often be improved by feeding pancreatic enzymes. In two tests on a human subject whose pancreas had been completely resected for carcinoma of its head, 38% and 54% of 75 gm of fed protein were excreted in the stool. When the subject was fed pancreatic enzymes, protein loss fell to 16% and 23%.

Polypeptides liberated by peptic and tryptic digestion are hydrolyzed to free amino acids and oligopeptides by peptidases in the brush border. The concentration of peptidases is high in animals fed a high protein diet and low in animals fed a low protein diet. The process of hydrolysis at the brush border, called "membrane hydrolysis" by some, yields 30% basic and neutral amino acids and 70% small peptides. As polypeptides are digested, there is a small rise in the amino acid concentration of luminal contents.

Absorption of Amino Acids

Absorption is an active process supported by metabolic energy. Three transport systems have been identified by in vitro methods similar to those used in the study of sugar absorptions:

1. Neutral amino acids are carried by a single transport system, and they compete with each other for absorption. There is some, but not absolute, optical specificity; D-methionine is transported at about one third the rate of L-methionine, and high concentrations of the D forms inhibit absorption of the L forms. The acid transported must have a free carboxyl group; esterified acids or those reduced to an alcohol are not transported. There must be a free α-hydrogen; no substitution is allowed. The side chain must be neutral. In the aliphatic series, almost anything from H (glycine) to NH_2-CO-CH_2-

CH_2-CH_2 (citrulline) is allowed, and in the aromatic series, permitted groups range from phenyl (phenylalanine) to imidazole (tryptophan). The affinity of the amino acid (K_t) for its carrier decreases with increasing polarity of the side chain.

2. Basic amino acids are carried by a separate system at 5 – 10% of the rate of transport of neutral acids. Amino acids transported include L-arginine, L-lysine, DL-ornithine and L-cystine.

3. A third system transports L-proline, hydroxyproline, sarcosine, dimethylglycine and betaine. Proline and hydroxyproline have a much stronger affinity for this carrier system than for the neutral one; therefore they are always transported by it rather than by the neutral system, for they would be blocked from the neutral carrier by other amino acids always present in the intestine.

Absorption of free amino acids is a carrier-mediated, saturable process. An amino acid combines with a carrier in the membrane and enters the cell. Sodium can combine with the same carrier, or both an amino acid and a sodium ion can do so together. There is no preferred order of combination.

During absorption, amino acids accumulate in the mucosal cells; they enter the mucosal side faster than they leave the basal side to enter the blood. Entry into the cells is the most rapid step in absorption; when absorption of one amino acid is competitively inhibited by another, its concentration within the mucosal cells is lower than that prevailing during uninhibited absorption. Most amino acids, once inside the mucosal cells, are not extensively metabolized, the exceptions being glutamic and aspartic acids. A fraction of these two undergoes transamination with pyruvic acid so that alanine is formed and released into portal blood during their absorption. Ammonia is itself absorbed by active transport in the hamster ileum; whether it is similarly absorbed in man is not known. Glutamine is rapidly absorbed, and it is metabolized within the cells. More than half is oxidized to carbon dioxide, and the rest is converted into tissue proteins, citrulline, pro-

line and amino acids. A third of the nitrogen in the absorbed glutamine goes to citrulline, and the rest goes to alanine, proline and ammonia.

Absorption of Oligopeptides

Dipeptides and tripeptides and some larger oligopeptides are absorbed faster than are free amino acids, and this accounts for the fact that complete absorption of protein digestion products is faster than complete hydrolysis of proteins to free amino acids.

Phenylalanine, alanine, histidine, threonine, serine and tyrosine are more rapidly absorbed as components of peptides than as free amino acids. There are several peptide-absorbing systems, but none is sodium dependent. Dipeptides and tripeptides compete for absorption. Absorption of each free amino acid is depressed by the same amino acid in peptide form, and, on the other hand, some amino acids inhibit absorption of the corresponding peptides. Peptides containing D-amino acids are absorbed very slowly.

Most peptides are hydrolyzed within the epithelial cells by peptidases in the cytosol, which are different from those in the brush border. Only peptides resistant to hydrolysis escape into the portal blood. An example is thyrotropin-releasing factor (PyroGlu-His-Pro-NH_2), which cannot be attacked by aminopeptidases or by carboxypeptidases, and consequently the hormone is effective by mouth in large doses.

Hydrolysis and Absorption of Nucleoproteins

The cell surface and the intracellular milieu are the loci of hydrolysis of other compounds. Phosphatases occur on the surface of mucosal cells. Pyrimidine nucleotides in contact with the mucosa are hydrolyzed to nucleosides and inorganic phosphate, although the hydrolytic enzymes responsible cannot be found in the intestinal contents. Other compounds hydrolyzed on the surface of the intestinal cells include glucose-1-

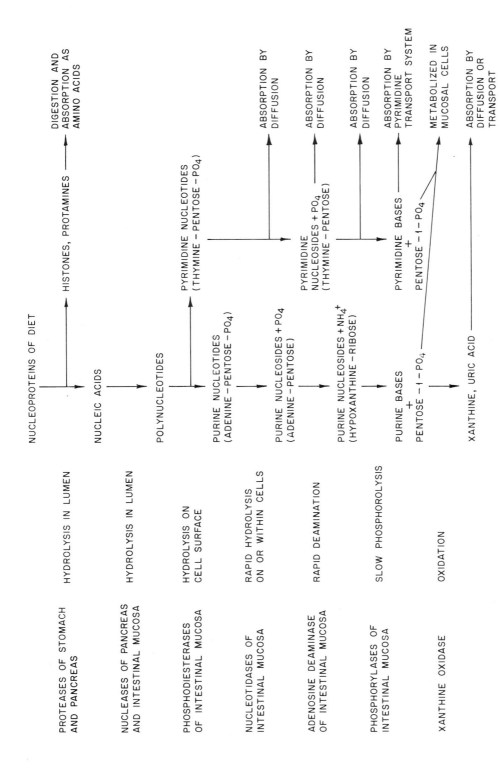

Fig 16–1.—Digestion and absorption of nucleoproteins.

phosphate, phenyl phosphate, β-glycero-phosphate, adenylic acid, hexose diphosphate, adenosine triphosphate, fructose-6-phosphate and many polyphosphates. The course of digestion and absorption of nucleoproteins is summarized in Figure 16–1.

Hypoxanthine and xanthine are both oxidized to urate within the mucosal cells, a process that can be inhibited by allopurinol. In the hamster, urate is actually secreted into the lumen; whether similar secretion in man provides for extrarenal excretion of uric acid is unknown. The methylated xanthine caffeine, because it is fat soluble, is rapidly absorbed by diffusion.

Defects in Protein Absorption

In pancreatic and biliary deficiency resulting in steatorrhea, loss of dietary protein nitrogen in the stool strictly parallels loss of dietary fat.

There are functional similarities between the small intestine and the renal tubules with respect to amino acid absorption. In each, transport of amino acids from luminal fluid to blood is inhibited by probenecid. Patients with cystinuria have impaired renal tubular absorption of cystine, lysine, ornithine and arginine; their ability to absorb the same amino acids in their intestine is also reduced. Some of the unabsorbed amino acids are converted in the colon to cadavarine and putrescine, which are absorbed and appear in the urine. The aminoaciduria of Hartnup disease involves, among many other amino acids, threonine, phenylalanine, tyrosine, tryptophan and histidine; jejunal absorption of at least tryptophan is also subnormal. Defective absorption of tryptophan results in pellagra-like symptoms, which can be prevented by feeding nicotinamide. The intestinal defect in Hartnup disease and in other forms of renal tubular malabsorption of amino acids appears to be confined to the absorption of free amino acids. The amino acids in peptides can be absorbed, and, for example, a patient with Hartnup disease can absorb phenylalanine and tryptophan in di-

peptides but not as free amino acids. Consequently, the defects are nutritionally insignificant. However, unabsorbed amino acids are degraded by bacterial metabolism to potentially toxic products, and absorption of these in the colon may account for neurologic abnormalities.

Gluten Enteropathy

In gluten enteropathy (also called celiac disease, idiopathic steatorrhea and nontropical sprue) the intestinal mucosal surface is flat, and the luxuriant array of long, thin microvilli on the surface of the epithelial cells is usually replaced by a few clubbed microvilli. There is malabsorption of all nutrients. The underlying defect appears to be the absence from the brush border of peptidases required for the complete hydrolysis of gluten of wheat, rye and oats. If gluten is completely eliminated from the diet, there is prompt and lasting remission.

Glutamine-containing polypeptides obtained from gliadin after exhaustive digestion with pepsin and trypsin evoke the signs and symptoms of gluten enteropathy when fed in small amounts to a patient in remission. Damage to the intestinal epithelial cells occurs within 8 hours, and it progresses to the basement membrane and capillaries. Disaccharidases of the brush border are depressed within 24 hours. Many patients challenged with gliadin respond by secreting an IgM gliadin antibody.

REFERENCES

Adibi, S. A., and Mercer, D. W.: Protein digestion in human intestine as reflected in luminal, mucosal, and plasma amino acid concentrations after meal, J. Clin. Invest. 52:1586, 1973.

Gardner, M. L. G.: Absorption of amino acids and peptides from a complex mixture in the isolated small intestine of the rat, J. Physiol. 253: 233, 1975.

Gray, G. M., and Cooper, H. L.: Protein digestion and absorption, Gastroenterology 61:535, 1971.

Matthews, D. M.: Intestinal absorption of peptides, Physiol. Rev. 55:537, 1975.

Milne, M. D.: Genetic disorders of intestinal

amino acid transport, in Code, C. F. (ed.): *Handbook of Physiology:* Sec. 6. *Alimentary Canal,* Vol. III (Washington, D.C.: American Physiological Society, 1968), pp. 1309–1322.

Morris, C. C.: Gamma globulin absorption in the newborn, in Code, C. F. (ed.): *Handbook of Physiology:* Sec. 6. *Alimentary Canal,* Vol. III (Washington, D.C.: American Physiological Society, 1968), pp. 1491–1512.

Walker, W. A., and Isselbacher, K. J.: Uptake and transport of macromolecules by the intestine. Possible role in clinical disorders, Gastroenterology 67:531, 1974.

Wilson, T. H.: *Intestinal Absorption* (Philadelphia: W. B. Saunders Co., 1962).

Wiseman, G.: Absorption of amino acids, in Code, C. F. (ed.): *Handbook of Physiology: Sec. 6. Alimentary Canal,* Vol. III (Washington, D.C.: American Physiological Society, 1968), pp. 1277–1308.

Wiseman, G.: *Absorption from the Intestine* (New York: Academic Press, 1964).

17

Intestinal Digestion and Absorption of Fat

MOST INGESTED FAT is delivered unchanged to the duodenum because little fat digestion occurs in the stomach. The extremes of human daily fat intake range from below 25 gm (or less than 12% of the caloric intake) among rice-eating peoples such as Japanese coal miners to 140–160 gm (or 42% of caloric intake) among Los Angeles funeral directors. The difference in the intake of fats is due to the use, by the latter group, of animal fats containing saturated fatty acids, for populations increase their fat intake by raising their use of meat and dairy fats. The intake of unsaturated fatty acids is nearly the same in the two extreme groups cited.

Because gastric emptying is slowed by fat in the duodenum, the rate at which fat enters the intestine is self-regulated. The emptying of a very fatty meal containing 50 gm of fat is spread out over 4–6 hours, so that the upper limit of the burden placed on the intestine's ability to deal with fat is between 8 and 12 gm per hour. A normal intestine absorbs all ingested fat; the fat that is always present in the feces is not unabsorbed residue but is derived from the intestinal and colonic mucosa or is synthesized by bacteria.

Emulsification and Hydrolysis of Fat

Most dietary fat of either animal or vegetable origin consists of triglycerides: glycerol combined in low-energy ester linkages with three fatty acids. The acids contain an even number of carbon atoms, the saturated acids being almost entirely palmitic (C_{16}) and stearic (C_{18}) and the unsaturated ones being oleic (C_{18}, 1 double bond) and linoleic (C_{18}, 2 double bonds). Only milk fat, which contains 3–9% of C_4 to C_{14} acids, contributes any significant quantity of shorter chain lengths. In general, the melting points of fats containing longer chain, saturated fatty acids are higher than those containing shorter chain, unsaturated fatty acids. Phospholipids in which 1 glycerol alcohol is linked to a phosphoric ester of an organic base, most frequently inositol or choline, also occur in small quantities in the diet. Natural fat and the lecithin of human bile, α-palmityl-β-oleyllecithin, which occurs in the amount of 1–5 gm per liter, mix with intestinal contents and are digested and absorbed along with exogenous fat.

Fats are insoluble in water and immiscible

with chyme, and most artificial or natural emulsions are broken in the stomach. Fats are prepared for attack by pancreatic lipase and for absorption by emulsification in the duodenum. The emulsifying agents are fatty acids, monoglycerides, cholesterol, lecithin and lysolecithin derived from it, protein and bile salts. Bile salts themselves are poor emulsifying agents, but a mixture of bile salts and a polar lipid—lecithin, lysolecithin or monoglyceride—has much greater emulsifying power. Lysolecithin and monoglycerides are products of the action of pancreatic lipase on lecithin or triglycerides; thus enzyme action tends to stabilize the emulsion. Fats are dispersed into droplets $0.5-1.0\,\mu$ in diameter, negatively charged and stable over the intestinal pH range of 6.0–8.5. Because the droplets are larger than a wavelength of light, the emulsion is cloudy. Pancreatic lipase adsorbs to the surface of the emulsified particles. Emulsions are most readily formed with liquid fat; and a fat like tristearin, which is solid at body temperature, is by itself poorly emulsified and incompletely digested and absorbed. The mixing of liquid with solid fats reduces the mixture's melting point and brings solid fats within the range of intestinal digestion.

In the course of digestion and absorption, triglycerides are partially hydrolyzed and resynthesized. If a triglyceride containing a fatty acid labeled with ^{14}C in position 1 is ingested, the ^{14}C fatty acid is found distributed among positions 1, 2 and 3 in the triglycerides recovered from lymph. Most of the ester bonds in the 1 and 1′ position are hydrolyzed. A triglyceride can be doubly labeled, with one isotope marking the glycerol moiety and another isotope identifying the fatty acid in the 2 position. When such a triglyceride is fed and the triglycerides of the lymph are analyzed, the two labels are found to be contained in many of the same molecules. This demonstrates that a large fraction, 50–75%, of the ester bonds in the 2 position of triglycerides escapes hydrolysis throughout digestion and absorption. Most hydrolysis

occurs in the lumen catalyzed by pancreatic lipase; only a small amount of hydrolysis takes place within the mucosal cells under the influence of intestinal lipases.

Pancreatic lipase is a group-specific esterase, which hydrolyzes triglycerides to fatty acids and 2-monoglycerides. It hydrolyzes 2-monoglycerides at a rate that is only 1–2% of that at which it breaks primary ester bonds. Because the energy of ester bonds is low, there is a tendency for liberated fatty acids to recombine with a glycerol alcohol group. However, the pK_a of free fatty acids in micelles within the intestinal lumen is 6.5; therefore, most of the free fatty acids are ionized. Ionized fatty acids are not capable of re-esterification, and removal of an end-product of hydrolysis by ionization favors completion of hydrolysis. There is spontaneous migration of fatty acids from one alcohol group to another. Consequently, a 2-monoglyceride may be converted into a 1-monoglyceride, which is more rapidly hydrolyzed by pancreatic lipase, the products being a free fatty acid and glycerol. Pancreatic lipase also catalyzes the hydrolysis of the primary ester bond of lecithin of bile, forming 2-lysolecithin. Pancreatic juice contains a phospholipase-A, which breaks the bond in the second position of lecithin, the product being 1-lysolecithin.

The chief products of luminal hydrolysis of triglycerides are 2-monoglycerides and free fatty acids; 1-monoglycerides and glycerol are quantitatively less important.

Fat Digestion in Man

Samples of adult human intestinal contents have been obtained by transintestinal intubation. A tube having an internal diameter of 2 mm is passed the whole length of the intestine; and through a hole in its wall, samples are withdrawn for analysis. By means of marks on the tube, the distance between the hole and the nose is determined, and from fluoroscopic observations of the relation of the hole to intestinal landmarks, the site from which samples are drawn is deduced. A fat meal of known composition, emulsified and stabilized with protein, is placed in the stomach. If the appropriate tags are added to the meal, its change in volume and composition resulting from addition by secretion, subtraction by absorption or mutation by digestion can be calculated. When a 500-ml liquid meal containing 30 gm of corn oil is placed in the stomach, emptying occurs over 4 hours.

The concentration of fat and fatty acids, which was 6% in the original meal, is between 1% and 3% in the upper duodenum as a result of dilution; and although the total volume of intestinal contents falls sharply in the duodenum, fat concentration in the upper jejunum is below 1%. By the time the ileum is reached, 4 hours after feeding, the fat in the small residual volume is 0.3–0.7%. Pancreatic secretion begins 10–20 min after feeding, and lipase concentration in intestinal contents is highest in the first hour. The gallbladder empties within 30 min, and bile concentration is also high (Fig 17–1). During the course of digestion, 4–5 gm of bile salts enters the intestine; since this quantity is larger than the body's pool of bile salts, the

salts recirculate during the digestion of a single meal. Phospholipids and cholesterol contained in bile are added to intestinal contents. Most of the meal is absorbed from a milieu containing a high concentration of lipase, bile salts and lysolecithin derived from action of pancreatic lecithinase-A on the phospholipids. A mixture of digestive products, the exact composition depending on the nature of fat fed, the locus of sampling and the elapsed time, accumulates in the lumen.

The concentration of fat digestion products falls sharply as chyme moves through the duodenum into the jejunum, and by mid-jejunum almost all fat of the diet is absorbed.

In one experiment, normal men were fed pure triolein as the only glyceride, and in it was dissolved a trace of free fatty acid labeled with isotopic carbon (^{13}C). Samples collected near the duodenal-jejunal junction contained, on the average, 42% of their total fat as free fatty acids, 8% as 1-monoglycerides, 10% as 2-monoglycerides, 16 as diglycerides and 24% as triglycerides. The concentration of labeled free fatty acid in the mixture had diminished, showing that free fatty acids had been absorbed more rapidly

Fig 17–1.—Left, concentration of lipase and trypsin in contents of human duodenum and proximal jejunum following a 500-ml liquid meal containing fat, protein and carbohydrate.

Right, concentration of bile salts in the same samples. (Adapted from Borgstrom, B., et al.: J. Clin. Invest. 36:1521, 1957.)

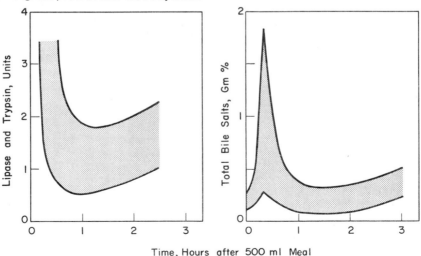

Time, Hours after 500 ml Meal

than the other components; but the accumulation of free fatty acids demonstrated that hydrolysis was occurring faster than the free fatty acids were absorbed. In addition, ^{13}C-labeled fatty acid was found esterified in the mono-, di- and triglycerides. The quantity of labeled fatty acid exchanged with unlabeled fatty acids in ester bonds was 30% of the maximal value to be expected if complete randomization of labeled among unlabeled acids had taken place. On the average, one fatty acid had exchanged in each triglyceride.

Physical and Chemical State of Fat during Absorption

Fat undergoing digestion and absorption is distributed among the emulsified fat droplets, micelles that are small, hydrated polymolecular aggregates and molecular solution.

If intestinal content is heated to destroy lipase and centrifuged at high speed ($100,000 \times g$ for 12 hours at 37 C), an oily top phase and a completely clear bottom phase are obtained. The oil is derived from the large, relatively unstable fat droplets; it contains all the triglycerides of intestinal contents, most of the diglycerides, some free fatty acids and very little monoglyceride. Although the lower aqueous phase shows no opalescence or light scattering, it contains as much fat as does the oil: most of the monoglycerides and fatty acids, a large fraction of the cholesterol, only a small amount of diglycerides and almost no triglycerides. The fat is dispersed in highly stable micelles, which, being $4-6$ mμ in diameter, have one millionth the volume of fat droplets of the oily phase. Their diameter is a little less than twice the length of the paraffin chains of the fatty acids; each probably consists of about 20 fat molecules whose hydrocarbon chains interdigitate within a fluid interior and whose polar groups form a negatively charged spherical shell surrounded by cations in aqueous solution.

When the concentration of a bile acid is above a certain value, its *critical micellar concentration*, the bile acid spontaneously aggregates with monoglycerides to form micelles. Figure 17–2 shows the relation between the amount of monoglyceride carried in micellar solution and the concentration of two bile acids. The critical micellar concentration of the conjugated bile acid, glycodesoxycholate, is only about 0.26 mM; above that concentration it associates with a large amount of monoglyceride. Under normal circumstances, the concentration of conjugated bile acids in intestinal contents is always above the critical micellar concentration. Consequently, monoglycerides, as soon as they are liberated from triglycerides, form micelles with bile acids. Once formed, micelles dissolve free fatty acids, and the amount dissolved increases, up to a limit, with the amount of monoglyceride contained in the micelles. Micelles also dissolve other lipids: lysolecithin, cholesterol and fat-soluble vitamins. Micelles obtained from human intestinal contents during digestion of a meal were found to contain 1.4 M of fatty acids, 0.15 M of lysolecithin and 0.06 M of cholesterol for each mole of bile acid. On a molar basis, their content of monoglyceride was less than that of free fatty acids.

Unconjugated bile acids have a much higher critical micellar concentration than do conjugated bile acids. This explains why fat absorption is reduced when bile acids are deconjugated in the small intestine: monoglycerides and free fatty acids cannot be held in micellar solution.

Each bile acid has a temperature, its Krafft point, below which it cannot form micelles. The Krafft point of all naturally occurring bile acids, except lithocholic acid, is well above body temperature.

The concentration of free fatty acids and 2-monoglycerides in molecular solution in intestinal contents is not precisely known; it is probably of the order of 0.01 mM. There is very rapid exchange of fatty acids and monoglycerides between micelles and molecular solution; the mean residence time of a particular molecule in a micelle is about 10 msec.

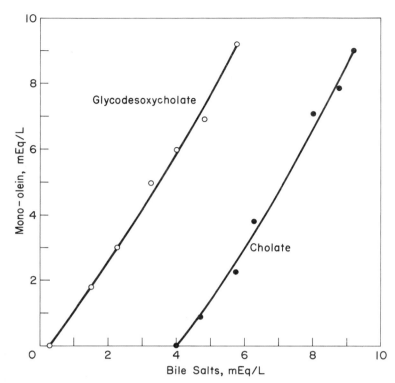

Fig 17–2. — Relation between bile salt concentration and quantity of monoglyceride brought into micellar solution at pH 6.3, 37 C and [Na+] of 150 mN. The critical concentration for glycodesoxycholate is 0.25 mM and for cholate 4 mM. (Adapted from Hofmann, A. F., and Borgstrom, B.: Fed. Proc. 21:43, 1962.)

Exchange between micelles and solution keeps the solution saturated.

Absorption of Fat Digestion Products

Fat digestion products accumulate in the bulk phase of intestinal contents, and the bulk phase is well mixed by segmentation and peristalsis. The contents of the lumen are progressively less and less well mixed in the radial direction from the center of the lumen to the surface of the intestinal epithelial cells. A thin layer, 0.05–2 mm thick, on the surface of the epithelial cells is poorly mixed or mixed not at all. This is the unstirred layer.

Lipid molecules can diffuse through the lipoprotein membranes of the epithelial cells, because they are fat soluble. Once inside the cells, free fatty acids and monoglycerides are

picked up by the metabolic machinery in the endoplasmic reticulum and are rapidly resynthesized to triglycerides and phospholipids. Consequently, the concentration of fat digestion products at the membranous surface of the epithelial cells is nearly zero. There is a diffusion gradient from the well-stirred bulk phase through the unstirred layer to the surface of the epithelial cells. Free fatty acids and 2-monoglycerides diffuse down the gradient through the unstirred layer to be absorbed by the epithelial cells. The flux of molecules through the layer is the product of the diffusion gradient and the diffusion constant of the molecules in solution. The concentration of free fatty acids and monoglycerides in solution is so low that their flux through the unstirred layer is not great enough to achieve the rate of absorption that actually occurs.

TABLE 17–1.—DIFFUSION OF FATTY ACIDS AS MOLECULES IN
SOLUTION, AS FREE FATTY ACIDS AND AS CONSTITUENTS OF
2-MONOGLYCERIDES THROUGH AN UNSTIRRED LAYER

FORM	DIFFUSION GRADIENT (μM CM^{-3})	DIFFUSION COEFFICIENT (CM^2 SEC^{-1} 10^{-6})	FLUX OF FATTY ACIDS (μM CM^{-1} SEC^{-1} 10^{-6})	RELATIVE FLUX
Molecules	0.01	7	0.07	1
Micelles	10.00	1	10.00	142

Most of the free fatty acids and monoglycerides diffuse through the unstirred layer aggregated in micelles. Diffusion coefficients decrease as the molecular weight increases. The weight of micelles is about 300–400 times as great as that of an individual fatty acid, and the diffusion coefficient of a micelle is approximately one-seventh that of a fatty acid. However, the concentration of fatty acids and 2-monoglycerides in micelles is 1,000 times as great as their concentration in solution. Therefore, the flux of fatty acids in micelles through the unstirred layer is more than 100 times as great as the flux of fatty acids in solution (Table 17–1, Fig

17–3). Formation of micelles from the products of triglyceride hydrolysis prevents segregation of hydrolysis products into poorly soluble oils, keeps the luminal solution saturated and provides the vehicle that carries most of the free fatty acids and 2-monoglycerides through the unstirred layer to the absorbing surface of the intestinal epithelial cells.

The constituents of micelles are absorbed at different rates at different sites; micelles themselves are not absorbed. Monoglycerides and free fatty acids are absorbed chiefly in the duodenum and upper jejunum, and most of the conjugated bile acids are ab-

Fig 17–3.—The role of micelles in fat absorption: keeping the solution saturated with free fatty acids *(FFA)* and 2-monoglycerides and carrying fat digestion products through the unstirred layer.

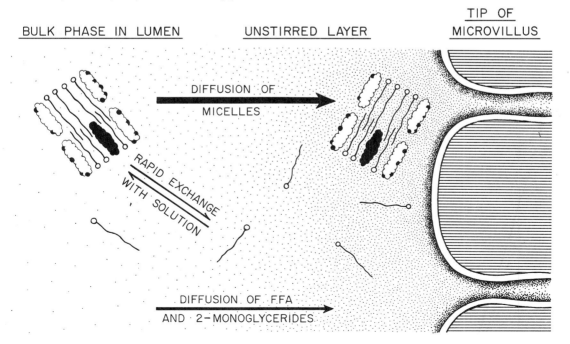

BULK PHASE IN LUMEN   UNSTIRRED LAYER   TIP OF MICROVILLUS

DIFFUSION OF MICELLES

RAPID EXCHANGE WITH SOLUTION

DIFFUSION OF FFA AND 2-MONOGLYCERIDES

sorbed in the terminal ileum. Because cholesterol is absorbed more slowly than are free fatty acids or monoglycerides, the concentration of cholesterol rises in micelles as chyme moves down the intestine.

Lipids that are relatively soluble in water need not be carried in micelles. Short-chain and medium-chain fatty acids and the triglycerides of those acids are sufficiently soluble in solution so that diffusion through the unstirred layer is not the rate-limiting step in their absorption.

Lysolecithin apparently enters the mucosal cells as such. Within the cells its base is removed by phosphodiesterase and its fatty acid by lysophosphotidase. The final cleavage to phosphate and glycerol is catalyzed by nonspecific phosphatases.

Intracellular Fat Metabolism

Most monoglycerides absorbed into the mucosal cells are re-esterified to triglycerides or phospholipids without further hydrolysis. The lipase contained in mucosal cells hydrolyzes ester bonds of short-chain acids, e.g., trioctanoin, but not those of long-chain acids, such as tripalmitin (C_{16}); therefore it plays little role in the intracellular metabolism of most dietary fat. Human subjects with cannulated thoracic lymph ducts have been fed 2-monoglycerides labeled with ^{3}H in the glycerol and with ^{14}C in the fatty acid; total recoveries of the label in lymph ranged from 35 – 55%. In the first 6 hours after feeding, trigylcerides in lymph had the same ratio of labels as that of the monoglyceride fed, and the labeled fatty acid was in the 2-position. After 6 hours, the proportion of labeled glycerol to labeled fatty acid decreased, and there was some change in position of the fatty acid in the triglyceride; some hydrolysis and isomerization had occurred. The small quantity of 1-monoglycerides produced during digestion is absorbed and re-esterified to triglycerides without intermediate hydrolysis.

Long-chain free fatty acids, but not short-chain ones, are resynthesized into triglycerides and phospholipids before they leave the mucosal cells to enter the lymph. In one experiment in which free palmitic acid (C_{16}) labeled with ^{14}C was fed to rats, either alone or mixed with triglycerides, 92% of the labeled acid was recovered in the lymph. Of this, 80 – 90% was incorporated into triglycerides and the remainder into phospholipids. None of the labeled acid occurred as free fatty acid. The same results have been obtained when labeled long-chain acids esterified as triglycerides were fed; despite very extensive intraluminal hydrolysis, all labeled acids recovered in lymph were in ester form. Experiments using human subjects from whom intestinal lymph can be recovered give qualitatively similar results.

When mucosal cells of the duodenum or jejunum are fractionated, the enzyme system resynthesizing triglycerides is found in the microsomal fraction. Microsomes are the broken remnants of the endoplasmic reticulum, and in the intact cells, resynthesized triglycerides accumulate within the reticulum. Synthesis of triglyceride ester bonds is not simple reversal of lipase hydrolysis. The ester bonds of 2,2-dimethylstearin are not split by pancreatic lipase, and when 2,2-dimethylstearic acid is mixed with triglycerides and pancreatic lipase, either in vitro or in the lumen of the intestine, the acid is not incorporated into triglycerides. Nevertheless 2,2-dimethylstearic acid is absorbed, and it appears in lymph in triglycerides and phospholipids. Catalyzed by a kinase requiring Mg^{++}, fatty acids react with ATP and coenzyme A to form fatty acyl~CoA; this product in turn reacts with monoglycerides to give diglycerides or with diglycerides to give triglycerides. One fatty acyl~CoA can also react with L-α-glycerophosphate to form lysophosphatidic acid. This product can react with a cytidine intermediate to become a phospholipid, or it can lose its phosphate and become a diglyceride. Mucosal cells can also synthesize long-chain fatty acids, chiefly stearic

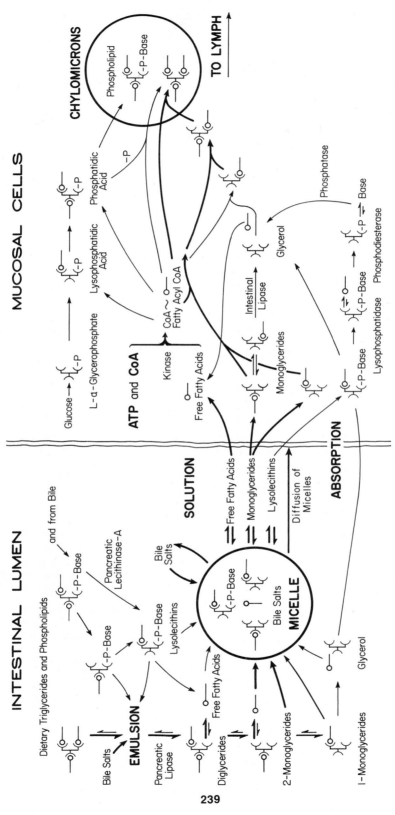

Fig 17–4.—Scheme of fat digestion, absorption and resynthesis. The *heavy arrows* indicate the more important pathways.

239

(C_{18}), from acetic acid, and they can lengthen palmitic acid by two carbons to form stearic acid.

Part of the glycerol liberated by hydrolysis of triglycerides in the intestinal lumen is oxidized by the mucosal cells to carbon dioxide, and some finds its way to the liver, where it is transformed to glycogen. The remainder is used for resynthesis of triglycerides. When 1.67 gm of glycerol-1-^{14}C was fed along with free fatty acids to a patient with chyluria, over 93% of the ^{14}C was recovered in the glycerol of lipids of the urine in the next 24 hours. In two experiments, 14% and 19% of all the triglycerides recovered in 24 hours were labeled, and as much as 60% of the lipids of a single sample contained glycerol-1-^{14}C. Glycerol used for resynthesis is also derived from glucose, the immediate precursor being L-α-glycerophosphate.

Chylomicrons delivered to the lymph are enclosed in a layer of phospholipid, part of which is derived from dietary phospholipid and the rest of which is synthesized within mucosal cells. Lecithin of the diet or bile is hydrolyzed to lysolecithin and absorbed as such. Within the cells lysolecithin is reacylated to lecithin before being incorporated into chylomicrons. A second component of the layer covering the chylomicrons is β-lipoprotein synthesized within the mucosal cells. Chylomicrons form a very stable emulsion, but the emulsion can readily be cleared if the phospholipid envelope is attacked by lecithinase.

During fat absorption, the phospholipids of the intestinal mucosa turn over rapidly, although their concentration remains constant. When phosphate labeled with ^{32}P is fed along with triglyceride, the specific activity of the phospholipids of the mucosa rises rapidly. Peak concentration of ^{32}P in phospholipids is reached 4 hours after feeding, but the various fatty acids are incorporated into phospholipids at different rates. Stearic acid enters more rapidly than do palmitic and myristic acids. Intracellular phospholipid metabolism may be part of the process by which fat droplets are provided with an envelope. Phospholipid turnover is greatly accelerated during transport of particles through the endoplasmic reticulum, and phospholipids may be a major component of the membranes surrounding the particles.

Fat hydrolysis and resynthesis are summarized in Figure 17–4.

Intracellular Fat Transport

In animals just beginning to absorb fat, the apical cells at the tips of the villi contain more fat than do those at the sides, and the fat is restricted to the supranuclear part of the cell. Small discrete fat particles first appear in the endoplasmic reticulum just beneath the terminal web. Microsomes derived from the reticulum contain the enzymes that resynthesize triglycerides, and fat droplets deposited within the reticulum are probably newly synthesized by the adjacent membrane. The shape of the vesicles shows that they are actually contained in a labyrinthine system consisting of small cavities connected by slender trabecular bridges. This endoplasmic reticulum is similar to that of the pancreatic acinar cell. In addition, irregular masses of fat not included in a membranous envelope lie in the cytoplasm. There is, however, no evidence that the cytoplasmic matrix contains any increased concentration of fatty acids or dissolved fat; resynthesis apparently keeps pace with absorption. As absorption continues, fat droplets fill the entire reticulum and move to the supranuclear part of the cell, along the way acquiring an envelope of phospholipid. Fat droplets are discharged from the sides of the cells at or below the level of the nucleus, and they collect at the base of the epithelium above the basement membrane.

In animals rapidly absorbing fat, droplets occur scattered through the lamina propria and in the connective tissue spaces of the villi. Fat droplets rarely pass into the capillaries; but, entering the central lacteal in streams, they are carried into the major lymph channels.

Absorption of Fat into Lymph and Portal Blood

Fat delivered to lymph is aggregated into droplets or chylomicrons $0.1-3.5$ mμ in diameter. The droplets are covered with a membrane that is a mosaic of a small amount of protein, free cholesterol and saturated triglycerides in a monolayer of phospholipid. There is only enough protein to cover 10% of the particle's surface. The droplets contain $89-93\%$ triglyceride by weight, $5-9\%$ phospholipids, $0.7-1.5\%$ cholesterol and $1-7\%$ free fatty acids. Fat in chylomicrons only remotely reflects the composition of dietary fat, for most short- and medium-chain fatty acids and some glycerol are shunted to the portal blood, and saturated, long-chain fatty acids are added by mucosal synthesis.

Diffusion probably carries water-soluble components of fat into capillaries. Capillaries of the villi are fenestrated, but the fenestrae are stopped by an uninterrupted basement membrane that envelops the endothelium. These fenestrae may be the ports of entry of water-soluble compounds into capillary blood. Particles the size of chylomicrons cannot pass through the basement membrane. On the other hand, chylomicrons can enter lacteals because there are open channels between the interstitial spaces and the lymphatic lumen. The endothelial wall of the lacteal is relatively thick, but an enveloping basement membrane is absent. There are endothelial separations or intercellular spaces through which droplets may pass. The cells forming the walls of the lacteal contain many vesicles, which may be pinocytotic ones capable of carrying fat droplets across them. Once within the terminal lacteals, the fat droplets are carried centrally by the tidal flow of lymph caused by contraction of smooth-muscle fibers of the villi or by more gross movements of the mucosa. Once delivered into the blood, chylomicrons do not recirculate in the lymph.

Lymph flow increases during fat absorption, and absorbed fat may appear in the lymph as long as 13 hours after fat has entered the intestine. The fat content of the lymph may be as high as 6%, and only a few hundred milliliters of lymph is required to carry the absorbed fat. When the thoracic duct is ligated, absorption of fat is temporarily reduced; after a few days, absorption returns to normal, probably because other lymph-to-vein channels open. However, fat absorption is grossly impaired when intestinal lymphatic channels are blocked by intestinal lipodystrophy (Whipple's disease), in which condition the lymphatics are filled by accumulation of a mucopolysaccharide.

Not all long-chain fatty acids are carried in the lymph. Some, perhaps as much as 15% of absorbed oleic acid, for example, escapes re-esterification and is transported to the liver in portal blood.

Medium-Chain Triglycerides

Short-chain and medium-chain fatty acids are not abundant in the diet. Depot fat contains few fatty acids shorter than 10 carbons, and less than 10% of the fatty acids in milk are shorter than 14 carbons. Synthetic triglycerides containing medium-chain (C_6 to C_{12}) fatty acids are fed as a source of energy to patients who are unable to digest and absorb triglycerides made of long-chain fatty acids.

Intraluminal hydrolysis of medium-chain triglycerides is faster than that of long-chain ones; and, because rapid isomerization between the 2- and 1-positions of glycerol occurs, hydrolysis is more complete. However, prior hydrolysis and micelle formation are not required for absorption. Medium-chain triglycerides, absorbed intact, are hydrolyzed within the mucosal cells by intestinal lipase, and most of their constituent glycerol and fatty acids pass into the portal blood. Very few medium-chain fatty acids are re-esterified. When the rare patient with chyluria whose abdominal lymph vessels drain into his renal pelvis is fed triglycerides of mixed long-chain and medium-chain fatty acids, no fatty acids of 8 or 10 carbons appear in the triglycerides of his urine, and the proportion of lauric (C_{12}) and myristic (C_{14}) fatty acids to the longer fatty acids is less than that in the ingested triglycerides. Only an occasional medium-chain fatty acid may be incorporated into a triglyceride containing two long-chain fatty acids and find its way into the chylomicrons of lymph.

Malabsorption: Pancreatic Deficiency

A classification of the physiological disturbances leading of malabsorption of fat is given in Table 17–2.

Normal men are capable of absorbing up to 150–200 gm of fat a day. No more than

TABLE 17–2.—ERRORS IN FAT DIGESTION AND ABSORPTION LEADING TO STEATORRHEA*

LOCATION	STEP	PHYSIOLOGICAL DISTURBANCE	DISEASE STATE
Intraluminal	Emulsification of triglycerides	Impaired emulsification	Deficiency of conjugated bile salts: biliary fistula or obstruction, iliectomy, bacterial deconjugation
	Hydrolysis to fatty acids and monoglycerides	Pancreatic lipase deficiency	Pancreatic disease
		Absolute or relative bicarbonate deficiency	Pancreatic disease or gastric hypersecretion
	Formation of micellar phase	Conjugated bile salt deficiency	Biliary fistula or obstruction; bacterial deconjugation
		Altered partition of fatty acids between oil and micellar phases from absolute or relative deficiency of bicarbonate	Pancreatic disease or gastric hypersecretion
Intraluminal to cellular	Absorption of fatty acids and monoglycerides by mucosal cells	Decreased cell uptake: reduction in number, activity or surface area of cells	Intestinal resection or bypass; tropical sprue; gluten enteropathy
		Cells already saturated with fatty acids and monoglycerides	Alteration in cellular steps of triglyceride synthesis, chylomicron formation or transport from cell
		Decreased time of contact of micellar phase with mucosal cells	Increased transit time
	Resynthesis of fatty acids and monoglycerides to triglycerides	Deficiency of resynthesizing enzymes	Not described
Cellular	Chylomicron formation	Deficiency in synthesis of chylomicron protein	A-β-lipoproteinemia
Cellular to extracellular	Transport of chylomicrons from cell via lymph channels to blood	Lymphatic obstruction or lymphangiectasia	Lymphosarcoma; intestinal lipodystrophy; protein-losing enteropathy

*Adapted from Hofmann, A. F.: Gastroenterology 50:56, 1966.

2–4 gm of the 3–6 gm of fat excreted in the stool each day is of dietary origin; the remainder comes from intestinal secretions and desquamation and from the colonic flora. About 40% of this fecal fat is in the bodies of bacteria; the remainder is in nonbacterial solids. In the complete absence of bile or pancreatic juice, fat escapes into the stool roughly in proportion to the amount in the diet (Fig 17–5). In biliary deficiency, half the dietary fat is not absorbed, and in pancreatic deficiency two-thirds is not. Although fat in the stool, or steatorrhea, demonstrates absorptive defects, the really striking fact is that fat can actually be absorbed at all when there is no pancreatic lipase or bile acids.

There is a large pancreatic reserve, for four fifths to nine tenths of the gland may be removed without interfering with fat digestion or absorption. Only 15% of patients with disturbances of pancreatic secretion show malabsorption of fat, and those are the most advanced cases. In steatorrhea, the composition of fecal fat differs from that of dietary fat. Short-chain fatty acids are absent, and abnormal fats produced by bacterial modification of unabsorbed fats are present. One such product is a hydroxy fatty acid structurally similar to ricinoleic acid, the fatty acid of castor oil; it may cause diarrhea.

In pancreatic deficiency, a large fraction of fat is hydrolyzed and absorbed. The lipase concentration of jejunal contents of one patient with 75% pancreatectomy was found to be 70% of normal value; this patient did not have steatorrhea on a high-fat diet. Seven patients, each having greater than 95% complete pancreatectomy, had no demonstrable functioning pancreatic tissue. They were fed 83–130 gm of triglycerides a day, and they absorbed 61–91% of the fat. In them, the lipase concentration of jejunal contents was one-tenth that of control subjects. In two of these patients, one with moderate and the other with minimal steatorrhea, jejunal contents were aspirated after a fatty meal. The aspirates contained, on a molar basis, 8% of the fatty acids in triglycerides, 14% and 25% in diglycerides and 22% and 39% in monoglycerides; 39% and 45% of the fatty acids were free. The extent of hydrolysis was similar to that of normal men. Feeding 1 gm of pancreatin with meals usually decreases fecal fat in patients with pancreatic deficiency who eat a low or moderately fatty diet; it is ineffective when patients eat a high-fat

Fig 17–5.—Relation between dietary fat and total fecal fat in man. **Left,** in complete absence of bile. Slope of the regression line is 0.46 ± 0.05. **Right,** in complete absence of pancreatic juice. Slope of the regression line is 0.47 ± 0.09. (Adapted from Annegers, J. H.: Quart. Bull. Northwestern Univ. M. School 23:198, 1949.)

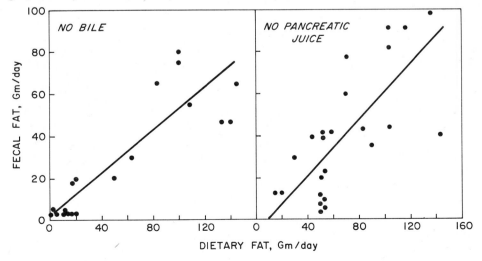

diet. Although the pH of intestinal contents of patients with pancreatic deficiency may be within the normal range after a meal, fat absorption is sometimes improved by feeding sodium bicarbonate.

Malabsorption: Bile Deficiency

When bile acids are totally excluded from the intestine, fat absorption is reduced, and voluminous fatty stools are passed containing, in man, approximately half the dietary fat. The defect is not in fat hydrolysis but in absorption. Hydrolysis is adequate, for between 80% and 100% of fat in the stool is in the form of free fatty acids when only triglycerides are ingested. Fed fatty acids are as poorly absorbed as are triglycerides. In a cholecystonephrostomized dog, all bile is diverted from the intestine to be excreted in the urine. Four hours after such a dog is fed 25 gm of corn oil, the fat recovered from the intestine is found to be greater in quantity than in a normal dog, but hydrolyzed to the same degree. Emulsification of fat sufficient for hydrolysis must, therefore, be accomplished by other agents, chiefly lysolecithin. Most lecithin of the diet and the bile is converted to lysolecithin by lecithinase-A of pancreatic juice. Bile acids are not totally indispensable for micelle formation, for lysolecithin in high concentrations also forms micelles in which monoglycerides and fatty acids are carried. Consequently, some hydrolyzed fat can be absorbed in the absence of bile, the defect being relative rather than absolute.

When ileal disease or ileal resection interrupts the enterohepatic circulation, deficiency of bile acids is periodic. After the first meal of the day, duodenal bile acid concentration is about half normal. After subsequent meals, the concentration is very low, for the first meal has flushed out the pool of acids accumulating overnight. During bacterial overgrowth, the concentration of conjugated bile acids is reduced. In both states, bile acid concentration is below the critical micellar concentration; the percentage of luminal fat in the micellar phase may be only 2%, compared with the normal value of 50–60%. Consequently, fat absorption is seriously impaired.

Other Absorptive Defects

Malabsorption of fat occurs widely in the tropics, and the disease tropical sprue has been attributed to deficiency of pteroylglutamic acid, vitamin B_{12} and adequate protein. In prolonged deficiency, irreversible atrophy of the small intestinal mucosa occurs. Nontropical sprue, or idiopathic steatorrhea, occurs in temperate climates and is unrelated to dietary deficiency. There are thinning and irregular dilatation of the small intestinal mucosa, with atrophy of both mucosa and muscularis. Microvilli on the epithelial cells become short and clubbed. Degeneration of myenteric and submucous plexuses may also occur. The total fat content of the stool amounts to 40–50% on a dry-weight basis, or about 50 gm per day. This equals 450 calories lost. Both the absorptive and the morphologic defects can be reversed by elimination of gluten from the diet. In persons who have been relieved of steatorrhea in this way, another attack can be precipitated by feeding purified gluten or fragments of it produced by peptic digestion. For this reason, the name idiopathic steatorrhea is being dropped in favor of gluten enteropathy.

Patients with steatorrhea may have osteomalacia as the result of reduced calcium absorption and a consequent negative calcium balance. Loss of calcium in fatty stools has been attributed to vitamin D deficiency and to the carriage of calcium into the stool as calcium soaps. However, the patients often do not respond to intramuscularly administered vitamin D, and there is no consistent relation between the degree of steatorrhea and loss of calcium.

Cholesterol Absorption

Cholesterol, which is present in all diets, chiefly as the free alcohol, is absorbed by adult human beings at the rate of about 10 mg per kg per day. When 10–20 eggs are fed, cholesterol absorption is doubled. The upper limit of absorption when a large amount is fed is about 2 gm per day. Peak delivery of cholesterol into the circulation by way of the lymph occurs, in man, 9 hours after feeding, compared with the peak of fat absorption 3 hours earlier. When cholesterol absorption is studied by means of trace amounts of cholesterol-^{14}C, the compound is found to be rapidly absorbed into the intestinal mucosa, reaching its highest concentration in 3 hours. Then it is slowly released into the lymph.

Cholesterol in the intestinal lumen is derived from two sources in addition to the diet: from the bile, which in the adult man carries 1–2 gm per day, and from the secretions or desquamated intestinal mucosal cells. The latter source has been found, in three adults with complete obstruction of biliary and pancreatic ducts, to be 0.25–0.4 gm per day. When a patient with chyluria was fed a low cholesterol diet containing 0.04 gm per day, she excreted 0.34 gm into her urine. Her shunt from intestinal lymphatic channels to renal pelvis was estimated to divert 40% of her intestinal lymph; therefore she absorbed and synthesized in her mucosa about 0.85 gm per day, or 21 times as much as she ate. Cholesterol not absorbed enters the colon, where it is reduced by colonic flora to coprostanol. No reduction occurs in the small intestine, for coprostanol is not present in fluid collected from an ileostomy. The fecal excretion of total sterols by a normal 70-kg man on a mixed diet is 0.5 gm per day, with a range of 0.12–0.90 gm.

Cholesterol in the intestine is mixed with bile acids, fats and cholesterolesterase present in pancreatic juice, and most esterified cholesterol is hydrolyzed before being incorporated into micelles. Cholesterolesterase requires bile acids to be active. In one sample of intestinal contents recovered from the duodenum and upper jejunum of a human subject and centrifuged at $100,000 \times g$ for 24 hours, the supernatant oil contained 0.19 mg

of cholesterol per ml, and the subnatant micellar solution contained 0.15 mg per ml, of which only 0.04 mg was esterified. Differences in the capacity of sterols to form micelles probably account for selectivity in their absorption. Ergosterol, which is poorly absorbed, does not form micelles as readily as cholesterol; the critical concentration of bile acids required to carry ergosterol into micelles is more than 3 times that needed for cholesterol, and ergosterol's concentration in micelles is about half that of cholesterol. In the absence of bile, no cholesterol is absorbed. Cholesterol is not absorbed at the same rate as other constituents of micelles. When a 50-cm length of upper jejunum of a human subject was perfused with a micellar solution made of bile acids, 1-monoglycerides and cholesterol, none of the bile acids, all of the monoglycerides and 73% of the cholesterol were absorbed. A quarter of the cholesterol in the solution recovered at the lower end of the perfused segment sedimented during high-speed centrifugation, showing that this fraction had been liberated from micelles and was no longer soluble. Concurrent absorption of exogenous fat is not required for cholesterol absorption. Cholesterol fed in a fat-free diet can be absorbed, probably because endogenous fat sufficient for micelle formation is always present in intestinal contents. Simultaneous absorption of unsaturated fats stimulates cholesterol absorption to a greater degree than does absorption of an equivalent quantity of saturated fats. Most or all esterified cholesterol is hydrolyzed by pancreatic cholesterolesterase before it is absorbed. When a cholesterol ester of a labeled fatty acid is fed, less than 2% of the label is recovered in the cholesterol esters of lymph. The more rapidly a fed cholesterol ester is hydrolyzed by the esterase, the more rapidly it is absorbed.

Once absorbed into the mucosal cells of the duodenum and upper jejunum, cholesterol mixes with the large pool of cholesterol within them. All epithelial cells from stomach to colon synthesize cholesterol from acetate, those of the ileum and distal colon being most active. This synthesis of cholesterol is controlled by chenodeoxycholic acid; the rate of synthesis increases 4 times in a man whose bile acids are diverted from his intestine, and expansion of the bile acids pool by ingestion of chenodeoxycholic acid decreases cholesterol turnover. A tracer dose of cholesterol-^{14}C remains detectable within the mucosal pool for 48–72 hours. In the fasting human subject, intramucosal turnover results from continuous transfer of intracellular cholesterol to lymph and its continuous replacement by new synthesis and by absorption of cholesterol secreted in bile. A total of 1–3 gm of cholesterol from all sources is delivered from the intestinal tract to the lymph each day. Synthesis of cholesterol in the liver adds to the circulating pool; when absorption of cholesterol is low, hepatic synthesis increases. Fed cholesterol is therefore diluted by cholesterol already circulating and synthesized by intestine and liver.

Although by far the largest fraction of cholesterol within the mucosa is the free alcohol, cholesterol is esterified in the cells, and two thirds or more of it in lymph is esterified with fatty acids drawn both from fed fat and from the intracellular fatty acid pool. Seventy percent of the cholesterol of lymph is in the chylomicrons; the rest is in the aqueous phase of lymph. Chylomicrons of human lymph contain 0.7–3.5% cholesterol ester and about 1% free cholesterol. The most abundant fatty acids in the esters are the unsaturated linoleic and linolenic acids.

Absorption of Other Sterols and Vitamins

Other ingested sterols are absorbed to varying extents. The rate of absorption is roughly related to the sterol's structure: the more lipoidal the greater the rate; the more polar and water soluble, the lower. Phytosterols and ergosterol with additional double bonds in the B ring are very poorly absorbed. When phytosterols uniformly labeled with ^{14}C are fed to a rat, they enter the intestinal mucosa, where they occur as free alcohols. Only 2% of the label on these phytosterols appears in the lymph in 24 hours, and about one third of the phytosterols is metabolically degraded in the intestinal mucosa until it is no longer identifiable as sterol. When the human small intestine is perfused with a micellar solution of bile acids, monoglycerides and cholesterol, phytosterols appear in the perfusate; they are apparently excreted by the mucosal cells. Administration of these and other related sterols blocks absorption of cholesterol, so that fecal excretion of cholesterol rises and its lymph concentration falls.

Fat-soluble carotene and vitamins D, E and K are dissolved in mixed micelles. Vitamin K is absorbed in the proximal small intestine by an energy-requiring process, and the vitamin can be absorbed only when bile is present in the small intestine. Absorption of vitamin A and of its precursors, the carotenoids, is grossly impaired in the absence of bile. A large fraction, 70% or more, of the absorbed fat-soluble vitamins is transported in chylomicrons of lymph to the blood. Absorption of the vegetable steroid vitamin D is promoted by fat, for the healing of rickets in rats is greatly accelerated by the addition of 5% fat along with vitamin D to a low-fat rachitogenic diet.

The sterol derivatives—testosterone, cortisol, cortisone acetate and corticosterone—are not transported in lymph after absorption; they enter the circulation by way of the portal blood.

REFERENCES

Brunner, H., Northfield, T. C., Hofmann, A. F., Go, V. L. W., and Summerskill, W. H. J.: Gastric emptying and secretion of bile acids, cholesterol, and pancreatic enzymes during digestion, Mayo Clin. Proc. 49:851, 1974.

Dietschy, J. M.: The role of the intestine in the control of cholesterol metabolism, Gastroenterology 57:461, 1969.

Dietschy, J. M.: Mechanisms of bile acid and fatty acid absorption across the unstirred layer and brush border of the intestine, Helv. Med. Acta 37:89, 1972.

Hofmann, A. F.: Functions of bile in the alimentary canal, in Code, C. F. (ed.): *Handbook of*

Physiology: Sec. 6. *Alimentary Canal,* Vol. V (Washington, D.C.: American Physiological Society, 1968), pp. 2507–2533.

Johnson, J. M.: Mechanism of fat absorption, in Code, C. F. (ed.): *Handbook of Physiology:* Sec. 6. *Alimentary Canal,* Vol. III (Washington, D.C.: American Physiological Society, 1968), pp. 1353–1376.

Kayden, H. J., Senior, J. R., and Mattson, F. H.: The monoglyceride pathway of fat absorption in man, J. Clin. Invest. 46:1695, 1967.

Treadwell, C. R., and Vahouny, G. V.: Cholesterol absorption, in Code, C. F. (ed.): *Handbook of Physiology,* Sec. 6. *Alimentary Canal,* Vol. III (Washington, D.C.: American Physiological Society, 1968), pp. 1407–1438.

Weiner, I. M., and Lack, L.: Bile salt absorption; enterohepatic circulation, in Code, C. F. (ed.): *Handbook of Physiology:* Sec. 6. *Alimentary Canal,* Vol. III (Washington, D.C.: American Physiological Society, 1968), pp. 1439–1456.

Wilson, F. A., and Dietschy, J. M.: Characterization of bile acid absorption across the unstirred water layer and brush border of the rat jejunum, J. Clin. Invest. 51:3015, 1972.

Wilson, T. H.: *Intestinal Absorption* (Philadelphia: W. B. Saunders Co., 1962).

Wiseman, G.: *Absorption from the Intestine* (New York: Academic Press, 1964).

18

Absorption and Excretion by the Colon

THE COLON receives more than 500 ml of chyme a day from the ileum; from the chyme the colon absorbs sodium and water, and to the residuum the colon adds potassium and bicarbonate. About 100 ml of water, containing sodium and potassium in amounts less than the dietary intake, is lost in the stool each day; but when diarrhea occurs, losses of water, sodium and potassium may be catastrophic. The abundant microflora inhabiting the colon of omnivorous animals, such as man, partially digests the residues of plant cells and contributes trace nutrients. Bacterial fermentation adds hydrogen and, in 3 of 10 persons, methane to the gas arriving from the small intestine, and about 1,000 cc per day of gas is passed as flatus. Frequency of bowel movements is extremely variable, but at some low frequency the symptoms of constipation appear. These are largely the result of chronic distention of the rectum. Under normal circumstances, any potentially toxic compound absorbed from the colon is prevented by the liver from entering the general circulation.

Electrolyte Exchanges in the Human Colon

Electrolyte exchanges in the colon are measured by transintestinal intubation. A multilumen tube is passed by mouth until its most distal luminal opening enters the cecum. Fluid of known composition containing an unabsorbable marker, such as polyethylene glycol, is infused at a constant rate into the cecum; and contamination of cecal contents is prevented by aspiration of ileal fluid through tubes with more proximal openings. Unidirectional fluxes are measured by including appropriate isotopic tracers in the perfusion fluid. Outflow is collected at the anus; and after a steady state has been attained, absorption and secretion are determined by comparing the volume and composition of the fluid collected with those of the fluid infused.

The major part of exchanges occurs in the ascending and transverse colon. Sodium is absorbed against a concentration gradient from fluid containing as little as 25 mEq per liter. Absorption of sodium is accomplished by two pumps, one of which is electrogenic and capable of generating a potential difference of about 10 mV, the lumen being negative with respect to interstitial fluid. The other pump is electrically neutral, for sodium reabsorption is coupled with hydrogen ion secretion. Chloride is likewise absorbed against a concentration gradient, and to some extent chloride absorption is coupled with secretion of bicarbonate into the lumen. Secretion of bicarbonate leaves acid behind in the interstitial fluid, and persistent colonic reabsorption of chloride and secretion of bicarbonate accounts for the hyperchloremic acidosis that occurs when the ureter is transplanted into the colon. The net movement of potassium is into the lumen as the ion diffuses along its electrochemical gradient. For this reason, severe and even fatal potassium depletion can be caused by repeated flushing of the colon with tap-water enemas. Water movement is likewise passive along the osmotic gradient, and net absorption of water follows upon net absorption of sodium and chloride. The normal human colon has the capacity to absorb 460 mEq of sodium and more than 2,000 ml of water a day.

The concentrations of electrolytes in terminal ileal fluid sampled in normal human subjects by transintestinal intubation are given in Table 18–1. In the same subjects, outflow of fluid from ileum to cecum was measured over 24 hours. Ileal outflow was least overnight, averaging 0.1–1.7 ml per min, and it rose to a peak of 6–8 ml per min 1–2 hours after a meal. Comparison of the quantities of water and electrolytes delivered to the colon with those in the stool demonstrates the importance of the colon in the conservation of water, sodium and chloride.

A patient with an ileostomy is denied the conservative function of the colon. Immediately after an ileostomy is created, fluid loss is high. Eventually, fluid loss falls; the average in 21 patients with well-established ileos-tomies was found to be 690 ml per day. Some clinicians believe that patients learn to reduce their fluid intake and therefore their fluid loss; others have found no relation between fluid intake and ileostomy loss. The mass of sodium lost in ileostomy fluid averages 85 mEq a day, with a range of 44–197. Because this loss is equal to or greater than the usual sodium intake, a patient with an ileostomy is in danger of dehydration. The mass of potassium lost is far below the normal intake, and there is little possibility that negative potassium balance will occur as the result of an ileostomy.

Stool weight is strongly influenced by the amount of undigestible fiber in the diet. The weight of individual stools of persons on an average American diet ranges from 30–230 gm, with a mean of 123.6 ± S.D. 40.2 gm. Because the average interval between stools is 27.6 ± S.D. 9.5 hours (range 9–57 hours), the mean stool weight per day is 119 gm. Stool water content is 70% or greater, and therefore roughly 20–160 ml of water is lost in the stool each day. Increased absorption of water in persons with constipation is the result of long transit time. The colon secretes a few mEq of calcium into the stool and absorbs a few mEq of magnesium. Although the net movement of bicarbonate is into the lumen, its concentration in stool water is low because it reacts with organic acids formed by bacterial fermentation. The concentration of organic anions is 133–238 mEq per liter. Stool water also contains ammonium (mean 14, range 2–34 mEq per

TABLE 18–1.—FLUID AND ELECTROLYTE LOAD AND ABSORPTION IN THE COLON OF 5 NORMAL HUMAN SUBJECTS*

	CONCENTRATION IN TERMINAL ILEAL FLUID (mEq/L)	QUANTITY (PER 24 HR)		
		TERMINAL ILEUM	STOOL	ABSORBED
Water Mean	—	1,524 ml	39 ml	1,485 ml
Range	—	1,255–1,751	22–43	1,192–1,718
Na⁺ Mean	127	196 mEq	1 mEq	195 mEq
Range	101–139	139–243	0–2	137–242
K⁺ Mean	6	9 mEq	5 mEq	5 mEq
Range	5–7	7–11	1–8	1–7
Cl⁻ Mean	67	103 mEq	1 mEq	103 mEq
Range	50–89	63–123	0–1	62–123

*Adapted from Phillips, S. F., and Geller, J.: J. Lab. Clin. Med. 81:733, 1973.

liter), calcium (mean 38, range 6–72 mEq per liter) and magnesium (mean 49, range 13–98 mEq per liter). The stool is hyperosmotic, with a mean osmolality of 376 (range 336–423) mOsm, because formation of osmotically active organic compounds outstrips osmotic equilibration.

Although aldosterone has only trivial effects upon sodium exchange in the small intestine, it regulates sodium absorption in the colon. The injection of 1 mg of d-aldosterone increased the capacity of the colon of each of 12 normal men to absorb sodium by 60%; water and chloride absorption concomitantly increased, but there was little effect upon potassium exchange. In conditions in which adrenal secretion of aldosterone rises, fecal excretion of sodium falls; and, conversely, when aldosterone secretion falls, fecal excretion of sodium rises.

Diarrhea

Diarrhea occurs when an abnormally large volume of fluid entering the colon from the small intestine overwhelms the colon's ability to absorb water and electrolytes, or it occurs when the absorptive capacity of the colon is reduced. This latter state may be secondary to some malfunction of the small intestine.

If ileal fluid is delivered to the colon at a steady rate, the colon of a normal man can absorb more than 2 liters per day. One patient with pancreatic cholera passed 2,390 ml. of water in his feces, but the colon had already absorbed 6,394 ml of water delivered to it by the ileum. If the rate of delivery is greater or is grossly irregular, the colon cannot absorb the fluid it receives. When a man forces himself to drink an isotonic solution (Na^+ 140, K^+ 10, Cl^- 105, HCO_3^- 35 mEq per liter) at the rate of 2–6 liters per hour, his fluid output per rectum ranges from 1–5 liters an hour. The toxin of *Vibrio cholerae* causes the mucosa of the jejunum and ileum to pour an enormous volume of an isotonic solution of sodium, potassium, chloride and bicarbonate into the lumen of the small intestine. A cholera victim may lose as much as 60 liters of water containing more than 8,000 mEq of sodium and 1,200 mEq of potassium in a 5-day period. Diarrhea of intestinal origin also occurs when ileal contents delivered to the colon, in volume within the normal range, contain something that alters the function of the colon. A simple example is the watery stool that follows ingesting of magnesium sulfate, a poorly absorbable osmotically active solute. Those persons who, because of intestinal deficiency of oligosaccharidases, fail to digest and absorb one or another carbohydrate, have diarrheal stools containing acid fermentation products.

Because conjugated bile acids are actively reabsorbed in the terminal ileum, only a small quantity normally enters the colon. When, because of ileal resection or disease, bile acids are not completely reabsorbed, they are lost in the stool. This causes diarrhea by two means: depression of sodium and water reabsorption by the colon and maldigestion of long-chain triglycerides.

When bile acids are lost, increased hepatic synthesis of new bile acids tends to maintain the bile acid pool. If hepatic synthesis is capable of keeping the bile acid pool at a nearly normal level, fat digestion and absorption are only moderatey impaired. Loss of fat in the stool occurs at less than 16 gm a day. In this case, the only effect on the colon is that produced by dihydroxy bile acids, which reduce the rate of reabsorption of sodium and water and thereby cause diarrhea. This kind of diarrhea can be alleviated by feeding cholestyramine, a resin that adsorbs bile acids. If, however, hepatic synthesis is not capable of maintaining the bile acid pool, maldigestion and malabsorption of long-chain triglycerides result in steatorrhea of more than 20 gm of fat a day. This causes diarrhea. Bacterial action on unabsorbed long-chain fatty acids produces hydroxy acids; these may stimulate intestinal motility. The diarrhea of gross steatorrhea cannot be controlled by adsorption of bile acids alone; but it can be controlled by a combination of adsorption of bile salts on cholestyramine and substitution of medium-chain triglycerides (which do not require bile acids for digestion and absorption) for long-chain ones in the diet.

Intractable watery diarrhea occurs in patients who have tumors or hyperplasia of nonbeta cells of the pancreas. These cells secrete large amounts of vasoactive intestinal peptide (VIP), which stimulates the small intestine to secrete large amounts of an isotonic fluid. The tumors may secrete other hormones as well.

Solids of the Stool

Organic materials entering the colon are mucus, desquamated cells and enzyme secretions of the upper digestive tract together with undigested food residues. Most of the digestible carbohydrate, fat and protein has already been attacked and absorbed. When fed protein is tagged with radioactive iodine, little or none of the radioactive isotope reaches the stool. The small fraction of dietary protein that enters the colon is attacked by bacteria; and the major part of stool protein is contributed by bacteria themselves, which comprise about 10% of the dry weight of the stool. There are few bacteria within the stomach or small intestine. Most ingested bacteria are killed by acid in the stomach. The bacterial population increases in the ileum and reaches its maximum in the colon. When intestinal stasis occurs, the bacterial population of the static fluid in the lumen rises quickly. It is likely that the relatively slow movement of colonic contents, rather than any particular environmental factor, accounts for its abundant flora.

The microorganisms of the colon (bacteria, yeasts, fungi) also digest organic compounds not attacked by digestive enzymes. Most of the food of herbivorous animals, such as the rabbit, is confined within cellulose walls, and microbial digestion in the large sacculated cecum is an important digestive function. In strictly carnivorous animals in which the cecum is small or absent, such digestion is negligible. If a man is fed 50 gm of citrus pectin a day together with a mixed diet, only 10% of the pectin is recovered in the stool. Since 75–95% of the fed pectin can be recovered from an ileostomy, most of its degradation and absorption must occur in the colon. Most cellulose fibers fed to men having bowel movements once a day are recovered in the feces, but only a small percentage is recovered from constipated persons. The same is true of agar-agar, which passes unaffected through the normal human colon. However, a person with a "greedy colon," in which there are abnormally slow motility and high absorption, may excrete only 0.8 gm of a 12-gm dose of agar-agar. Digestion occurs within the central part of the fecal mass, and acid products may be absorbed.

For man, the colonic contribution to total energy balance is small, and the value of nutritional enemas containing glucose or amino acids is doubtful. The colon does not actively absorb these compounds. Trace nutrients synthesized by colonic flora are absorbed. These include riboflavin, nicotinic acid, biotin and folic acid. In both man and animals, the administration of sparingly soluble bacteriostatic agents such as sulfaguanidine reduces the colonic flora and, when vitamin B intake is low, may precipitate signs of deficiency. The prothrombin time of human infants frequently lengthens in the first few days after birth, when normal intestinal flora, which will later contribute some vitamin K, have not yet become established.

The nature of the diet influences the size and composition of intestinal flora; but, on the other hand, intestinal flora alter the structure and function of the digestive tract. Germ-free animals have shallow crypt glands, taller and more delicate villi and a thin-walled cecum.

Intestinal Gas

The normal human digestive tract contains about 150 cc (ATPS) of gas. Of this, 50 cc is in the stomach, little in the small intestine and 100 cc in the colon. The sources of intestinal gas are swallowing, neutralization, fermentation and diffusion.

Gas in the stomach accumulates from that accompanying food or drink in each swallow, from frothy saliva and from gas trapped in food. The amount swallowed varies greatly with habit, and it may be as much as 500 cc with a meal. Several liters may be quickly swallowed during a period of anxiety. Most of the gas is either eructed or passed on, but on rare occasions enough may accumulate so that the gas bubble in the stomach fills the entire upper left abdominal quadrant. Sensations of bloating and discomfort are not related to the total volume of gas, for a person with a huge bubble may have no untoward symptoms, whereas one who complains of gas on the stomach may contain less than the usual volume. A person who complains that everything he eats turns to gas has no more gas in the bowels than does a normal person; he has instead disordered motility, which does not move the gas rapidly, and abnormal sensitivity to distention.

When the gas bubble in the stomach is large enough to extend below the entrance of the esophagus, gas may escape into the esophagus when the lower esophageal sphincter relaxes. It usually does not pass the hypopharyngeal sphincter. Distention of the esophagus stimulates secondary peristalsis, and the gas is swept back into the stomach. This sequence of filling and emptying may be repeated many times before eructation occurs. Then, when the esophagus is filled with gas, the jaw is thrust forward, intrathoracic pressure rises and gas is expelled through the partially opened hypopharyngeal sphincter.

Gas not eructed passes into the small bowel, most easily if the subject is supine, and its to-and-fro movement in the antrum or its passage through the pylorus causes gushing and explosive sounds. These and other intestinal gas noises are called borborygmi. Very little gas is normally present in the small intestine, probably because it is rapidly absorbed or passed on to the colon. That

present causes soft crepitations or slow rumbles, which occur at the frequency of intestinal movements, 7–12 times a minute. Their absence in the "silent abdomen" is a sign of general cessation of intestinal motility.

When 1 mEq of acid reacts with 1 mEq of bicarbonate at body temperature, 25 cc of carbon dioxide is liberated. During emptying of a meal from the stomach, 50 mEq of acid may react with an equivalent amount of bicarbonate, and 1,250 cc of carbon dioxide is formed. The partial pressure of carbon dioxide may rise as high as 700 mm Hg, and most of the carbon dioxide diffuses back into the blood. Carbon dioxide resulting from the reaction of organic acids in the colon with bicarbonate secreted by the colonic mucosa is more slowly absorbed.

Under normal circumstances, very little gas is formed by fermentation in the small intestine. When obstruction occurs, fermentation may produce as much as 3,500 cc of gas in the small intestine in 24 hours. Normally, the chief site of fermentation is the colon. Colonic flora of all persons produce hydrogen gas at the rate of about 0.6 cc per min. Fourteen percent is absorbed and excreted through the lungs; the rest escapes in the flatus. About one third of the population produces a large amount of methane in the colon; the other two-thirds does not. Some methane is absorbed, and its measurement in the breath is a means by which its production can be followed. Only a very small amount of hydrogen sulfide is produced, and it is far less than 0.01% of the flatus. Because the smell of hydrogen sulfide is easily recognized, eggs are blamed for flatulence.

The amount of hydrogen and methane produced depends upon the nature of unabsorbed residues. The hulls of beans contain undigestible oligosaccharides, chiefly raffinose and stachyose, and when beans are eaten to the extent of 25% of the caloric intake, the rate of passage of flatus may be as high as 200 cc an hour. Deficiency of lactase leaves lactose to be fermented, and when milk is drunk by a person with lactase deficiency, he makes as much as 4 cc of gas a minute. All varieties of fermentation produce organic acids, which, when neutralized by bicarbonate, liberate carbon dioxide.

Addition of other gases to swallowed nitrogen reduces the partial pressure of nitrogen in the gut, and consequently nitrogen diffuses from blood into the gas phase at the rate of 1–2 cc a minute. Any oxygen diffusing from blood is used by colonic flora.

The human colon usually contains about 100 cc (range 30–200 cc) of gas. The volume increases when ambient pressure is rapidly reduced by quick ascent to high altitude in an unpressurized airplane or by failure of pressurization at high altitude. At a barometric pressure of 230 mm Hg, which prevails at 30,000 ft, the volume becomes 500 cc; at 140 mm Hg, corresponding to 40,000 ft, it is 1,000 cc, and this volume causes abdominal pain.

There have been no mass surveys upon which a reliable estimate of the volume and composition of the average American flatus can be based. A low figure for volume was provided by 5 male medical students untroubled by flatulence who passed from 380–650 cc in 24 hours, and a high figure has been given by a physician who collected 1,500–1,600 cc per day from himself for a week. Individual samples have been analyzed with great precision by gas chromatography or mass spectrometry, but the variation from sample to sample is enormous. Oxygen is usually nearly absent; nitrogen ranges from 12–60%; and carbon dioxide may be as high as 40%. Because hydrogen and methane may each comprise as much as 20%, flatus containing them will burn with a hard gem-like flame.

It is not the fat content but the entrained gas bubbles that cause feces to float. Feces may be made to sink or rise like a Cartesian diver by increasing or decreasing the ambient pressure.

Controlled expulsion of gas when there are feces in the rectum is accomplished by contraction of abdominal muscles; this raises pressure in the rectum. At the same time, the external anal sphincter is voluntarily contracted. When pressure in the rectum excedes pressure in the sphincter, gas escapes through a slit that is too narrow for solid feces to pass. The relation between pressure and velocity of gas in the anal canal obeys Bernoulli's principle, and consequently the lips of the anus vibrate like the double reed of a bassoon, sounding a low-pitched note.

Rate of Passage Through the Intestine

High-residue diets pass more rapidly through the intestine than do low-residue diets. Individual particles of food residues go back and forth under the influence of intestinal movements, and several particles ingested together may be evacuated at widely different times. The residue of one meal catches up and mixes with that from previous meals, and a particular stool is a blend of residues of food eaten over several previous days.

Two thirds of normal subjects eliminate all but traces of a small amount of barium sulfate taken with a meal within 3 or 4 days, but the rest require 5 or more days to do so. Since chromate is not absorbed from the intestine, rate of passage may be measured by giving a small amount of sodium chromate labeled with ^{51}Cr by mouth; individual stools are subsequently collected, and their radioactivity is measured. In one such study on 10 normal subjects, the time before the initial appearance of ^{51}Cr in the stool ranged from 4–10 hours; and the final appearance of ^{51}Cr was between 68 and 165 hours. There were distinct differences among subjects; some excreted ^{51}Cr relatively early; others late.

Constipation

"The abnormal action of the bowels in constipation may manifest itself in three different ways: defecation may occur with insufficient frequency, the stools may be insufficient in quantity, or they may be abnormally hard and dry."* This definition of constipation implies that there is some normal standard of frequency, size and quality of bowel movements, but it is difficult to specify the exact point at which normal variation shades into pathologic function. Some healthy persons open their bowels after every meal; others usually do so once a day; but many who defecate only once a week experience no difficulty attributable to costiveness. The popular notion that one must defecate once a day or be overwhelmed by the disaster of irregularity is false. Nevertheless, at some degree of infrequency the symptoms of constipation appear; these are mental depression, restlessness, dull headache, loss of appetite that is sometimes accompanied by nausea, a foul breath and coated tongue, and abdominal discomfort with heaviness and swelling.

Straining to defecate in the constipated state may cause hernia, hemorrhoids, prolapsed rectum or anal ulceration, and elderly persons may faint or have a fatal cardiovascular accident. Constipation may be colic, resulting from delay in the passage of feces through the intestine to the pelvic colon; or it may be dyschezic, resulting from difficulty in emptying the rectum. *Colic constipation* occurs in a person in whom food residues take as long as 72 hours to reach the descending colon; or it occurs in the aged, whose colons may be atrophic and dilated. Gastrocolic and gastroileal reflexes are not aroused during anorexia, and an entirely bland diet with no indigestible residue leaves little to stimulate colonic motility. Movement of the intestine is reduced by pain, particularly that arising from intestinal disease. Chronic spasm of the transverse and descending colons, narrowing of the lumen by strictures, malignant growths, cicatrization of ulcerations or obstruction by masses of

hard, dry feces delays passage of the intestinal contents to the rectum. In *dyschezia,* or the inability to defecate completely, the rectum is always filled with feces, even immediately after defecation; and in the extreme form of dyschezia, defecation without mechanical assistance is impossible. The condition results either from insufficient power of the defecation mechanism or from an obstacle to defecation. The latter may be a fecalith or even a foreign object. Foreign objects recovered from the rectum or sigmoid colon range from a 40-watt frosted electric light bulb to a tool case complete with tools. Inefficient defecation often occurs when, for one reason or another (haste, lack of privacy or unpleasant or uncomfortable surroundings), the urge to defecate is ignored; in this circumstance, constipation can more readily be cured by installing another bathroom than by consulting a gastroenterologist. Fear of painful defecation, prolonged illness in bed, spasm of the anus induced by anal ulcer or inflamed hemorrhoids may also depress the defecation reflex and lead to constipation.

Autointoxication

The miserable state of constipation is exacerbated by fear of its consequences. These fears are assiduously cultivated by vendors of laxatives and by quacks who profit from many bizarre forms of treatment, with the result that many persons believe that serious and permanent injury to health will follow the missing of a bowel movement. To most adults, feces are repulsive. (They are not so to infants, who have not yet been acculturated.) It is easy to believe that retention of feces is deleterious, and on this belief is based the theory of autointoxication, which is that toxins absorbed through the colon from retained feces are responsible for all disease, ranging from alopecia to zoanthropy.

In a person with a normal liver, autointoxication is an imaginary condition. Some products of digestion in the small intestine and

---

*Hurst, A. F.: *Constipation and Allied Intestinal Disorders* (2d ed.; New York: Oxford University Press, 1919). The definition and terminology of *colic constipation* and *dyschezia* (difficult easing) follow Hurst.

colon are potential toxins. Among these are histamine, tryptamine and cadaverine. Ammonia is formed in the colon by oxidative deamination of amino acids and by hydrolysis of urea; rate of production is about 1.8 mg per hour. Ammonia is passively absorbed into portal blood by non-ionic diffusion, and consequently less is absorbed when the contents of the colon are acid than when they are neutral or alkaline. In normal persons, ammonia arriving in the portal blood is converted to urea in the liver, and the concentration of ammonia in arterial blood is vanishingly small. In a person with cirrhosis or a portacaval shunt, ammonia escapes into the arterial blood, causing encephalopathy, which may be fatal. In such a person, reduction of ammonia production and absorption may be achieved by cleansing the colon or by treatment with neomycin.

Many experimental observations show that it is not the nature of the colonic contents but their presence in the descending colon and rectum that is responsible for the symptoms of constipation. When defecation is voluntarily restrained for several days, all the familiar effects occur. Immediately following defecation these effects disappear too quickly to be accounted for by the elimination of toxins. The same symptoms of constipation can be produced by stuffing the rectum with cotton or distending it with a balloon. When the mass is removed, the symptoms promptly disappear. Further evidence is provided by the numerous men who have retained enormous masses of feces for years without suffering any bad effects other than the burden of carrying a colon containing' from 60–100 lb of fecaliths.*

Fear of constipation encourages the use of laxative drugs, and anthropological surveys have shown that among primitive tribes, such as the inhabitants of Nebraska farm communities, it may be the custom for every person to take a dose of laxative every day. When self-medication is omitted, the colon that had been violently and artificially emptied subsides; no bowel movements occur for a day or two. This confirms the subject's suspicion that he is the victim of constipation, and he resorts to still more heroic measures. The result is that the abused colon is not given a chance to function at its normal pace. Patients may complain of "constipation," but because they take laxatives regularly they never have formed stools.

The volume of stool and the frequency of defecation can be regulated in a more normal way by adding cereal bran and vegetables to the diet. Whereas British naval ratings and their wives have an average daily stool weight of 104 gm (range 39–223 gm), British vegetarians have stool weights and bowel patterns similar to the vegetarian Ugandans. Ugandans as a group hold the Olympic record for stool weight, an average weight of 470 gm a day, with a range of 178–980 gm. Vegetables contain cellulose, hemicellulose and lignin, which are not attacked by mammalian enzymes and are thought to stimulate intestinal motility by adding bulk. Cellulose and hemicellulose are partially digested by bacteria in the colon with the formation of volatile fatty acids; these act as laxatives. Agar, which also contains hemicellulose, increases fecal output because it imbibes water and swells.

Some physicians believe that absence of crude fiber from the diet and consequent alteration of colonic function are the causes of appendicitis, diverticulosis and benign and malignant tumors of the colon.

Many foods, including milk, potatoes and eggs, which contain little or no indigestible residue, are alleged to increase fecal bulk and frequency of defecation; however, no evidence outside the realm of folklore supports their reputation. Prune juice, which is completely devoid of indigestible residue, produces catharsis, probably on account of its content of magnesium salts.*

---

*A notable example of failure to defecate for more than a year was described by Geib, D., and Jones, J. D.: J.A.M.A. 38:1304, 1902.

*Contrary to occasionally expressed opinion, the active principle of prune juice is probably not an isatin derivative. See Hubacher, M. H., and Doernberg, S.: J. Pharm. Sci. 53:1067, 1964.

REFERENCES

Berk, J. E. (ed.): Gastrointestinal gas, Ann. New York Acad. Sc. 150:1, 1968.

Burkitt, D. P., Walker, A. R. P., and Painter, N. S.: Effect of dietary fibre on stools and transit-time, and its role in the causation of disease, Lancet 2:1408, 1972.

Calloway, D. H.: Gas in the alimentary canal, in Code, C. F. (ed.): *Handbook of Physiology: Sec. 6. Alimentary Canal,* Vol. V (Washington, D. C.: American Physiological Society, 1968), pp. 2839–2860.

Cummings, J. H.: Laxative abuse, Gut 15:758, 1974.

Daniel, E. E., Bennett, A., Misiewicz, J. J., Edmonds, C. J., Hill, M. J., and Cummings, J. H.: Symposium on colon function, Gut 16:298, 1975.

Devroede, G. J., and Phillips, S. F.: Conservation of sodium, chloride, and water by the human colon, Gastroenterology 56:101, 1969.

Fordtran, J. S.: Speculations on the pathogenesis of diarrhea, Fed. Proc. 26:1405, 1967.

Ihre, T.: Studies on anal function in continent and incontinent patients, Scand. J. Gastroenterol. [Suppl.] 9:1, 1974.

Lasser, R. B., Bond, J. H., and Levitt, M. D.: The role of intestinal gas in functional abdominal pain, N. Engl. J. Med. 293:524, 1975.

Levitt, M. D.: Production and excretion of hydrogen gas in man, N. Engl. J. Med. 281:122, 1969.

Levitt, M. D., and Bond, J. H., Jr.: Volume, composition, and source of intestinal gas, Gastroenterology 59:921, 1970.

Levitt, M. D., and Duane, W. C.: Floating stools—flatus versus fat, N. Engl. J. Med. 286:973, 1972.

Phillips, S. F., and Geller, J.: The contribution of the colon to electrolyte and water conservation in man, J. Lab. Clin Med. 81:733, 1973.

Newman, A.: Breath-analysis tests in gastroenterology, Gut 15:308, 1974.

Wolpert, E., Phillips, S. F., and Summerskill, W. H. J.: Ammonia production in the human colon, N. Engl. J. Med. 283:159, 1970.

Index